D0419552

THE NEUROBIOLOGY OF BEHAVIOR: An Introduction

THE NEUROBIOLOGY OF BEHAVIOR: An Introduction

GORDON J. MOGENSON

THE UNIVERSITY OF WESTERN ONTARIO

LAWRENCE ERLBAUM ASSOCIATES, PUBLISHERS
1977 Hillsdale, New Jersey

DISTRIBUTED BY THE HALSTED PRESS DIVISION OF

JOHN WILEY & SONS
New York Toronto London Sydney

Lawrence Erlbaum Associates, Inc., Publishers
62 Maria Drive
Hillsdale, New Jersey 07642

Distributed solely by Halsted Press Division
John Wiley & Sons, Inc., New York

Library of Congress Cataloging in Publication Data

Mogenson, Gordon J 1931-
The neurobiology of behavior.

Bibliography: p.
Includes indexes.
1. Neuropsychology. 2. Human behavior.
3. Animals, Habits and behavior of. 4. Neurobiology.
I. Title.
QP360.M6 599'.05 77-18283
ISBN 0-470-99341-3

Printed in the United States of America

Contents

Preface **ix**

1. INTRODUCTION **1**

The Nature and Determinants of Motivated Behaviors 3
Factors that Initiate and Influence Motivated Behavior 7
Differential Contributions of Factors and Determinants of Motivated Behaviors 13
Ontogeny of Motivated Behaviors 14
The Neural Substrates of Behavior 15
Summary 23

2. BIOLOGICAL FOUNDATIONS, EXPERIMENTAL STRATEGIES, AND TECHNIQUES FOR THE STUDY OF MOTIVATED BEHAVIORS **25**

Biological Foundations 26
Research Strategies and Experimental Techniques for the Study of the Neural Substrates of Motivated Behaviors 41
Summary 57

3. THERMOREGULATORY BEHAVIOR **58**

Body-Temperature Regulation 59

Characteristics of Thermoregulatory Behavior 66
Factors that Contribute to Thermoregulatory Behavior 67
Ontogeny of Thermoregulatory Behaviors 72
Neural Integrative Systems Subserving Physiological and Behavioral Thermoregulatory Responses 74
Summary 84

4. FEEDING BEHAVIOR 86

Characteristics of Feeding Behavior 86
Food Intake and Body-Energy Homeostasis 87
Factors that Initiate and Contribute to Feeding Behavior 93
Ontogeny of Feeding Behavior 99
Neural Substrates of Feeding Behavior 100
Summary 118

5. DRINKING BEHAVIOR 121

Characteristics of Drinking Behavior 120
Water Intake and Water Balance 122
Factors that Initiate and Contribute to Drinking Behavior 127
Ontogeny of Drinking Behavior 135
Neural Substrates of Drinking Behavior 137
Summary 149

6. SEXUAL BEHAVIOR
by Boris Gorzalka and Gordon Mogenson **151**

Characteristics of Sexual Behavior 151
Factors that Influence and Determine Sexual Behavior 152
Ontogeny of Sexual Behavior 158
Hormonal Determinants of Sexual Behavior 163
Neural Determinants of Sexual Behavior 169
Future Directions of Research 180
Summary 184

7. SLEEP AND WAKING 187

Characteristics of Sleep 188
Ontogeny of Sleep 194

Factors that Influence Sleep and Waking 197
The Neural Substrates of Sleep and Waking 202
Summary 210

8. BRAIN SELF-STIMULATION BEHAVIOR
by E. T. Rolls and G. J. Mogenson **212**

Characteristics of Brain Self-Stimulation 214
Factors that Influence Brain Self-Stimulation 217
The Neural Basis of Brain-Stimulation Reward 221
Summary 235

9. EMOTIONAL BEHAVIORS **237**

Characteristics of Emotional Behavior 237
Factors that Contribute to Emotional Behavior 241
The Neural Substrates of Emotional Behaviors 249
Neural Substrates of Abnormal Behavior and Psychiatric Disorders 258
Summary 264
Appendix A 265

10. PROBLEMS AND STRATEGIES FOR FUTURE RESEARCH **269**

Problems of Measurement and of Experimental Techniques 270
The Problem of Nonspecificity of Experimental Manipulations of the CNS 271
The Virtues and Limitations of Theoretical Models 274
The Hypothalamus and Motivated Behavior: Current Status 276
Current and Future Developments 279

References **282**

Glossary **312**

Author Index **317**

Subject Index **327**

Preface

This book has been written for senior undergraduate students with some background in biology and psychology and, ideally, in physiology and neuroscience. The gestation period coincided with the time that I offered a course on The Physiological Psychology of Motivated Behavior in the Department of Psychology and a course on Regulatory and Behavioral Physiology in the Department of Physiology at the University of Western Ontario. It was conceived at the time Dr. Franco Calaresu and I were planning and editing a monograph in memory of our colleague, Professor James A. F. Stevenson, entitled *Neural Integration of Physiological Regulations and Behavior,* and reflects the philosophy of that earlier volume. The writing of *The Neurobiology of Behavior: An Introduction* was accomplished during intervals of "spare" time when I was preoccupied with research and the directing of graduate students so that certain sections may also be of interest to research workers in the field.

My interests are primarily the physiology of the nervous system—in discovering the "secrets of brain function"—but in the course of writing this book, I became more acutely aware of the fact that one of the prerequisites for this endeavour is the understanding of behavior. Our limited knowledge of the neural substrates of behavior is due in part to the complexity of the nervous system and the relative crudity of our experimental techniques. Another important reason is that neurobiologists have tended to have only a secondary interest in behavior and have frequently lacked sophistication in behavioral analysis. If someday there is to be an adequate neurobiology of behavior, the importance of contributions in ethology and in comparative and experimental psychology must be recognized. This is reflected by the progressive approach of this book: I begin with the characteristics and determinants of each of the behaviors and consider them in detail before dealing with what is known about the neural substrates.

Following an introductory chapter in which this unique approach is formulated, and a chapter concerned with biological foundations and with experimental techniques for investigating brain function and behavior, there are chapters dealing with thermoregulatory, feeding, drinking, sexual, and emotional behaviors and with sleep-waking and brain stimulation reward. The final chapter deals with difficulties in brain-behavior research in relation to experimental strategies and with crucial problems for future investigation.

The objective of the book is to examine the mechanisms by which the multiple factors or determinants—homeostatic deficits, hormonal influences, circadian rhythms, experiential and cognitive factors—become translated by the central nervous system into thermoregulatory, feeding, sexual, aggressive, and other behaviors. A conceptual framework has been used that reflects relevant contributions from biology, regulatory physiology, physiological psychology, and other neuroscience disciplines.

I wish to express sincere gratitude to the many people who helped to make this book possible. Chapter 6 was written in collaboration with Dr. Boris Gorzalka, the University of British Columbia, and Chapter 8 in collaboration with Dr. Edmund Rolls of Oxford University, who, in each case, did most of the work while graciously permitting me to make the necessary adjustments of the presentation so that the organization of these chapters is similar to that of the rest of the book. I thank Charles Brimley, Franco Calaresu, Jan Cioe, Graham Goddard, John Kucharczyk, Surendra Manchanda, Bill McClelland, Peter Milner, Francisco Mora, Paul Morris, Tony Phillips, Don Richardson, Ann Robertson, Morikuni Takigawa, and Dick Weick, who read various sections of the book and provided helpful criticisms and suggestions. Special acknowledgment should be made to Ron Kramis, who offered constructive and detailed critical feedback on several of the chapters, and to Philip Zeigler, who, as editorial consultant, pointed out deficiencies in my presentation as well as in my understanding and logic, in the process teaching me a good deal and making the book more organized and readable. Finally, very special thanks go to Blanche Box, who assisted with all aspects of the book and prepared the glossary and index; to Rebecca Woodside, who prepared the illustrations; to Vince Nicol, who photographed the illustrations; and to Audrey Gibson, Valerie Maday, and Marilyn Allen, who typed the chapters of this book—several times in fact—with skill and good humor, tolerant of the "emotional aberrations" that are sometimes associated with strongly "motivated" behavior such as completing a book.

GORDON J. MOGENSON

THE NEUROBIOLOGY OF BEHAVIOR: An Introduction

1
Introduction

This book is concerned with brain function and behavior and specifically with the neural substrates of thermoregulatory, feeding, sexual and aggressive behaviors. These and other behaviors shown in Figure 1.1 are usually designated as motivated behaviors, because they are goal-directed and persistent: a tiger hunts for food; a zebra searches for water; a chimpanzee builds a nest and goes to sleep; a gull drives other birds from its home territory; a child plays with a toy; a mathematician solves a complex problem. Motivation and motivated behaviors are not very satisfactory as scientific terms, however, because they have not been defined precisely, and there have been vigorous controversies about their meaning.[1] These terms are used in this book with some reservation and are used only in a descriptive sense and as a shorthand to facilitate communication. When the neural substrates of these behaviors are better understood, these terms and such inferred concepts as drive will be unnecessary. In the meantime motivation and related terms must be used with caution, the main justification being their role as a reminder to the neurobiologist of the nature and dimension of the task of providing an experimental and theoretical analysis of behavior.

The term motivation was introduced 35 or 40 years ago to recognize that certain behaviors are goal-directed and persistent but more importantly to recognize that the so-called motivated behaviors could not be accounted for in terms of external stimuli—as reflex responses or sequential chains of reflexes. Rather, the initiation of feeding, drinking, sexual, and other behaviors seemed to depend on the current physiological state of the animal, on hormones, and past experience, as well as on events or stimuli in the external environment. When a serious

[1]For discussion of the theoretical issues associated with motivation see Bolles (1967) and Cofer and Appley (1963). The controversies and debates have long since become fruitless and sterile, and the term motivation is used here to designate a field of research; it identifies a problem area rather than providing a solution or scientific explanation.

FIG. 1.1 These pictures illustrate the variety of behaviors that have been designated as motivated behavior. *Top left*–Canadian wolves stalk their prey, the buffalo. (From Young & Goldman, 1944.) *Top middle*–Thirsty dog drinking water from a stream. *Top right*–Hostile aggressive behavior of a dog toward another dog. (From Darwin, 1872/1965. Reprinted by permission of the University of Chicago Press. © 1965 by The University of Chicago.) *Bottom left*–A male baboon exposes his large canine teeth to intimidate spectators. (Photo courtesy of Brian Soper). *Bottom middle*–Monkey engaged in attempting to solve a mechanical puzzle. (Photo courtesy of Dr. H. F. Harlow.) *Bottom right*–A man endeavouring to scale a mountain peak.

interest in investigating the neural substrates of these behaviors began in the 1940s and 1950s, dramatic experimental observations and the theoretical models to account for them emphasized certain factors (e.g., biological deficits), ignored others (e.g., experiential influences), and directed attention to the hypothalamus as a critical neural integrative site. These empirical and theoretical advances were strongly influenced, as indicated later in this chapter, by important events in biology and in regulatory physiology—feeding, thermoregulatory, and other behaviors were shown to have biological significance in that they contributed to the homeostasis of the internal environment as well as in adapting the animal to the external environment. Accordingly, the theoretical models that emphasized homeostatic factors and hypothalamic integrative mechanisms had a scientific credibility that made them the center of attention for a number of years, and the more nebulous experiential and cognitive determinants of behavior, which are associated with more complex and sophisticated regions of the forebrain about which little is known, were ignored.

Events of the last few years have indicated the shortcomings of this approach and led to a reassessment of classical concepts and, in some instances, to a reinterpretation of earlier experimental findings. A more balanced approach is needed that takes into account the role of various factors and determinants of behavior. The distinctive feature of this book is the recognition that several factors contribute to thermoregulatory, feeding, aggressive, and other behaviors and that they contribute differentially depending on the species, the sex of the animal, the behavior, the internal physiological state, the circumstances of the environment, past experience, and the stage of ontogenetic development. The major objective is to consider the neural substrates of certain behaviors that have been investigated in the laboratory, but in each case, this is done from the perspective of the characteristics and determinants of the behavior.

THE NATURE AND DETERMINANTS OF MOTIVATED BEHAVIORS

Characteristics of Motivated Behaviors

The behaviors considered in this book, such as seeking water or food, building a shelter from the cold, escaping from a predator, or defending a ''home territory,'' are typically goal-directed. It was Charles Darwin (1859), in the middle of the nineteenth century, who suggested that such behaviors are ''purposive'' because of their biological significance in that they are adaptive to the habitat and ''life style'' of the animal and directed toward goals: feeding behavior has a preservative function; sleep a restorative function; aggression a protective function; and sexual behavior contributes to species perpetuation. Darwin suggested that the evolution of these species-specific adaptive behaviors is as important to the survival of the animal as the evolution of morphological characteristics. The biological significance of behavioral responses was further demonstrated more than 75 years later in a series of experiments by Curt Richter (1943), who showed that a number of behaviors contribute to homeostasis (Cannon, 1932)—they are ''self-regulatory''.[2]

Another prominent characteristic of many motivated behaviors is their strength and persistence (Figure 1.1): A wolf will pursue its prey for hours; a zebra will go for miles to reach water; a mother rat will cross an electrified grid tolerating very painful shock to reach her pups; a mountain climber struggles for days to reach the summit; a young man studies for several years for a medical degree in order to practice medicine. When behavior becomes intense, erratic, and difficult to

[2]These conceptual developments from Darwin and Bernard to Cannon and Richter provided the biological foundations for the concepts of drive and central motive state that gained acceptance in the 1930s and 1940s—homeostatic imbalance or deficits being considered the source of drives. The drive-reduction theory of behavioral reinforcement formulated during this period by Hull (1943), and which had a strong influence on the experimental study of behavior for a number of years, was derived in part from these developments.

inhibit, such as aggressive behavior, the designation *emotional behavior* is frequently used. It has been suggested that behaviors of this sort are associated with higher levels of arousal or activation of the central nervous system (Lindsley, 1951; Hebb, 1955).

A number of motivated behaviors occur in a periodic fashion: A rat feeds every few hours; a cow drinks water two or three times a day; a female dog mates twice a year; a bear hibernates in the late fall. Such periodicity may be the result of the disappearance or removal of the internal or external stimuli that initiated the behavior and the occurrence of other stimuli, or it may be due to hormonal changes, or to the operation of a ''biological clock.'' Under certain conditions, two or more motivated behaviors may compete with one another; ''time-sharing'' is a prominent feature of motivated behaviors (McFarland, 1974). For example, feeding may be postponed in the presence of strong ''hunger signals'' because of the higher priority of another activity, such as escape by the buffalo from a predator or sexual behavior in the receptive female rat. The Great Tit *Papus Major,* which eats two or three meals per hour during the day, does not feed at night and is presumably very hungry in the morning. Yet feeding is postponed while it engages in territorial behavior (McFarland, 1975). It appears from these examples that ''motivational time-sharing'' may operate on a competitive basis. There is, however, an overall organization of motivated behaviors so that they contribute to ''biological fitness'' and survival (McFarland, 1974).

In summary, motivation—a rather unsatisfactory term from the scientific point of view—refers to a heightened level of consciousness or arousal resulting from alterations in CNS activity and usually accompanied by responses of the whole animal in its external environment (behavioral responses). These behavioral responses are frequently characterized by their *purposiveness, persistence,* and *periodicity*. Because several factors that initiate various motivated behaviors may be present at the same time, the particular behavior observed may reflect *priorities* and compromises in accordance with changes in the internal and external environments; this is ''motivational time-sharing.''

Many investigators of the neural substrates of feeding, drinking, copulation, and aggression have only a secondary interest in behavior. They are concerned primarily with trying to understand how the brain works. Physiologists in particular, but also a number of physiological psychologists, interested in how the brain controls feeding, drinking, sleep, and waking, have not been very sophisticated and thorough in the behavioral aspects of their investigations and in the techniques of behavioral analysis (for further discussion, see Chapter 10). It is not uncommon to designate an inferred central state (e.g., the animal is attentive or fearful) rather than to objectively describe a particular behavioral pattern. Even Charles Darwin sometimes departed from providing a description of the actual behavior of animals. A number of examples may be found in chapter 3 of Darwin's *The Descent of Man* (1874); for example, ''In the Zoological Gardens I saw a baboon who always got into a furious rage when his keeper took out a letter or book and read it aloud to him [p.87].''

Therefore, before going on to deal with the determinants of motivated behaviors, it seems appropriate to consider briefly what is actually observed or measured when behavior is being investigated.

The Observation and Description of Motivated Behaviors

According to Hinde (1970), there are two ways to describe behavior: in terms of the ''strength, degree, and patterning of muscular contractions [p. 11]'' or in terms of the ''consequences'' of the sequence of muscular contractions.

This distinction is illustrated by the classical studies by Tinbergen (1951) of courtship behavior in the stickleback (Figure 1.2). The left part of the figure illustrates the sequential response components, and the lower right portion of the figure illustrates the final consequence. Whether the behavior is dealt with in terms of ''spatio–temporal patterns of muscle contractions'' or ''description by consequence'' may sometimes depend on whether one is concerned with the

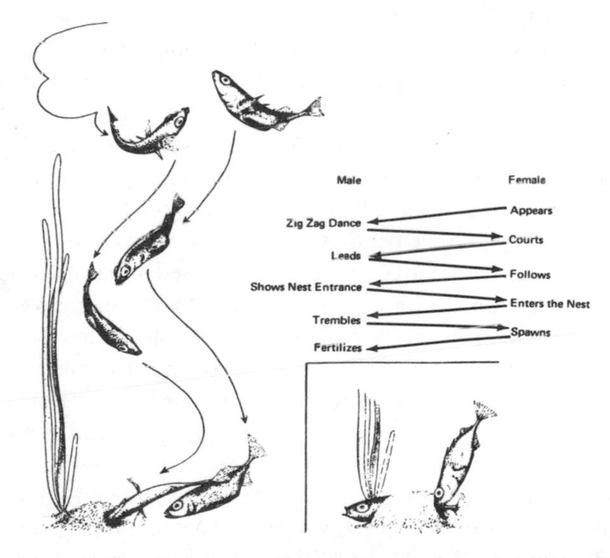

FIG. 1.2 Courtship behavior of the three-spined stickleback illustrating the mutually releasing actions of the male and female partners. Ethologists refer to the highly stereotyped and stable sequence of motor responses as fixed action patterns, and one of their major tasks has been to try to identify the specific stimuli (sign stimuli or releasers) that trigger the fixed action patterns. Releasing stimuli, like the fixed action patterns they initiate, are assumed to be species-specific and include the odor or some visual stimulus provided by another behaving animal, and in particular such stimuli provided by a member of the same species as shown here. (From Tinbergen, 1951, as adapted by Eibl-Eibesfeldt, 1970. By permission of the Oxford University Press.)

appetitive phase or the consummatory phase of motivated behavior.[3] The consummatory phase is usually more stereotyped—for example, rats lap at a water spout at a stable rate of seven laps per second—and an account in terms of spatio–temporal patterns of skeletomotor responses is possible. The appetitive phase is typically more complex and variable, and, because it is difficult to provide a description in terms of patterns of muscle contractions, it is, according to Hinde, more feasible and appropriate to use description by consequence.

Behavior is frequently "measured" by the cumulative record of lever presses to turn on a heat lamp, number of grams of food eaten, number of laps at a water spout, or for sexual behavior, number of intromissions and ejaculations. For some of these "measures of behavior," it is not even necessary to observe the animal directly; an account or description of "spatio–temporal patterns of muscle contractions" is not being used, rather a quantitative measure of the consequences of the behavior. However, relying on "description by consequence" can result in the erroneous interpretation of experimental observations, and, as indicated in later chapters, it may be necessary to use other behavioral measures including analysis of sequential response components.

To assist in scientific investigation, the strategy of "simplification of phenomenon" is frequently used. The experimental study of behavior has been no exception, and because behavior is exceedingly complex, this approach is useful and often essential. An effective and widely used procedure for simplifying an animal's behavior is to simplify its environment—a strategy of research used to great advantage by Skinner (1938) and his followers (see Teitelbaum, 1976). The animal is provided an opportunity to make an arbitrary response to obtain food or other appropriate reward, usually by placing a manipulandum in the experimental chamber, such as a lever to be pressed or a disk to be pecked (see Figures 1.3 and 3.8). The rate of making the operant response is used as a criterion and quantitative index of motivation. Although this research strategy enables the investigator to perform a well-controlled study of behavior and to obtain quantitative measures that are highly reliable, it has the shortcoming that the behavior is often rather artificial and that certain factors or determinants that normally initiate or influence the behaviors are operative only to a minimal extent or are even absent.

[3]The distinction between the "appetitive phase" and the "consummatory phase" of motivated behaviors was first made by American ethologist Craig (1918) in an article concerned with the nature of what he called instinctive behaviors. By appetitive phase Craig meant the "readiness to act," which he suggested was associated with the physiological state of the animal and appropriate stimuli in the environment. The consummatory phase consists of the species-specific behavioral responses to achieve the goal. One of Craig's examples is nest building in the dove. The appetitive phase is characterized by the nest-call, a distinctive posture, and selecting a nest site, and the consummatory phase is characterized by the "well-ordered movements" to construct the nest. Unfortunately, this distinction between appetitive phase and consummatory phase has frequently not been recognized, and as a consequence there have been difficulties in the interpretation of the results of experiments in which lesion and stimulation techniques have been used to investigate the neural substrates of feeding, drinking, and other behaviors. For details, see Chapters 4 and 10.

FIG. 1.3 A porpoise jumps out of the water to receive a food reward from the keeper's hand.

Some ethologists have been particularly critical of these limitations of laboratory studies of behavior, suggesting that they are sometimes analogous to the study of behavior in "solitary confinement" (see Collier et al., 1976, p. 31). They point out that an understanding of behavior must take into account an animal's evolutionary history and its habitat and ecological niche, as well as its physiological state and present environmental circumstances. This is one of the major reasons why the neural substrates of behavior are considered, in subsequent chapters of this book, from the perspective of the various factors or determinants of the behaviors.

FACTORS THAT INITIATE AND INFLUENCE MOTIVATED BEHAVIOR

It was noted in the introduction to this chapter that a number of factors initiate and contribute to thermoregulatory, feeding, drinking, sexual, and other behaviors. They include *internal deficit signals, hormones, circadian rhythms, external stimuli,* and *experiential and cognitive factors*. In this section the role of these factors in motivated behaviors is considered briefly, and it is seen that they contribute differentially depending on the behavior. For example, thermoregulatory behavior may be readily elicited by exposure to cold or by a drop in body temperature and feeding behavior by energy deficits following a period of food deprivation. On the other hand, sexual behavior is not initiated by homeostatic deficits of this sort but depends very much on hormones and experiential and cognitive factors. The role of these factors also depends on age and experience and on the sex and species of animal. For example, sexual behavior depends less on hormones in the male animal than in the female, especially in primates and higher mammals with previous sexual experience.

Internal (or Homeostatic) Deficit Signals

There is a popular saying that "you can lead a horse to water but you can't make it drink." A reliable way to get a horse, a laboratory rat, or other animal to drink is to deprive it before making water available. It is well known from laboratory studies of animal learning that a rat water-deprived or food-deprived for 24 hours will traverse a complex maze or press a lever for long periods of time to obtain water or food. Clearly, water, energy, or other homeostatic deficit signals can be strong initiators of motivated behaviors with water or food as effective rewards (Figure 1.3).

Because laboratory studies were done for a number of years using animals that were food- or water-deprived, there was a strong emphasis on internal deficit signals and on the so-called primary drives (e.g., hunger, thirst). Theoretical considerations of motivated behaviors in terms of their contributions to homeostasis also stressed the importance of internal deficit signals (Richter, 1943). According to the drive-reduction theory (Hull, 1943), popular for a number of years, animals are motivated by drives, and learning depends on reinforcement resulting from drive reduction.

Although homeostatic deficit signals are potent initiators of certain motivated behaviors, their role and significance has been overemphasized, as indicated earlier, because of the rather artificial nature of the experimental preparations frequently used and because of theoretical considerations. As indicated in the following, other factors also contribute. It is important in our thinking about the various motivated behaviors and the neural substrates that subserve them to consider under what conditions homeostatic deficit signals are utilized, particularly when an animal is in its natural habitat.

Hormones

Hormones contribute in different ways and in differing degrees to motivated behaviors. In certain cases hormones may be directly involved in their initiation; for example, estrogen is essential for copulatory behavior in the female rat, and angiotensin is the initiator of copious drinking in several animal species. In other cases hormones do not directly initiate behavioral responses but exert regulatory influences that complement the contributions of the behaviors to homeostasis. For example, hormones are clearly involved in the control of energy metabolism and heat production, and the classical experiments of Richter (1943) demonstrated the importance of behavioral thermoregulatory responses when these hormonal mechanisms were disrupted by removing one of the endocrine glands (see Chapter 3 for further details). Because hormones control the metabolic activity of body tissues, they, in conjunction with mechanisms that control feeding, make an essential contribution to the regulation of body-energy content (body-weight regulation).

When the effects of hormones on the various behaviors are considered, it will be important to identify those effects that are specific to a particular behavior (such as the effects of estrogen on female copulation) and those that are nonspecific effects on basic bodily functions, indirectly influencing various motivated behaviors (such as thyroid hormones or certain pituitary hormones).

Circadian and Other Biological Rhythms

One of the prominent characteristics of a number of motivated behaviors, as indicated earlier, is their periodicity. In some cases the periodicity of behavior is very striking: The female rat comes into estrous every fourth or fifth day, and copulation begins between 12:00 p.m. and 2:00 a.m.; estrous and copulatory behaviors occur every 6 or 7 months in the female dog and once a year in the ewe; rats feed and drink mainly at night, and 80–90% of the food intake occurs during the 12-hour dark period; bears go into hibernation in the late fall. Perhaps the most significant biological rhythm from the behavioral point of view is the circadian rhythm of sleep–waking, because a number of motivated behaviors depend on a certain level of consciousness and tend to occur during the waking period. The sleep–waking cycle is usually superimposed on a circadian rhythm of rest–activity (Figure 1.4).

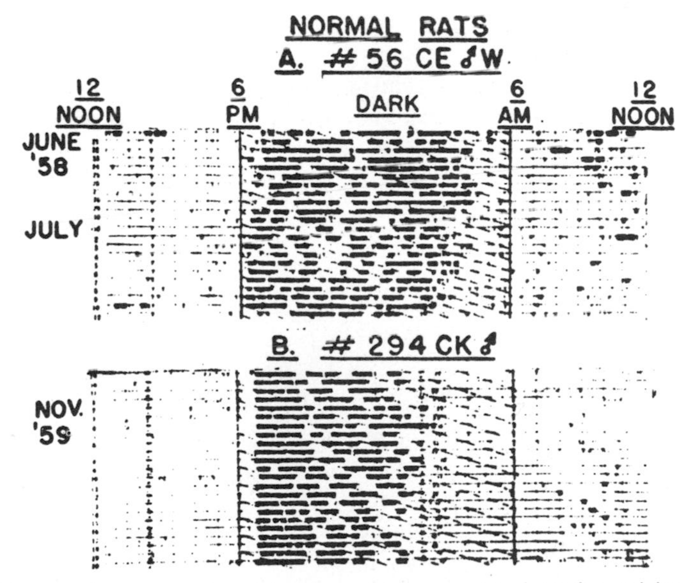

FIG. 1.4 Samples from an event recorder of spontaneous nocturnal running activity when the laboratory was in darkness. (From Richter, C.P., *Biological Clocks in Medicine and Psychiatry*, 1965. Courtesy of Charles C Thomas, Publisher, Springfield, Illinois.)

Circadian and other biological rhythms have been somewhat neglected in earlier analyses of motivated behaviors, particularly in the formulation of theoretical models and in the consideration of neural substrates. This is unfortunate, because such rhythms are obviously of great importance in seeking an understanding of animal behavior. In recent years a number of biological and behavioral scientists have become interested in this field, and the significance of biological rhythms for understanding the neurobiology of motivated behaviors is beginning to be appreciated (e.g., Rusak & Zucker, 1975).

Circadian and other biological rhythms contribute to behavioral and biological adaptations and to the survival of the individual animal and of the species. For example, Rusak and Zucker (1975) have suggested that circadian rest–activity cycles ''prevent wasteful and dangerous expenditure of energy at nonoptimal times of day'' and thereby ''ensure that each species is active only when its activity has the greatest potential survival value [p.159].'' Another example is seasonal breeding, which ensures that the offspring are born at the time of year favorable for survival.

Oatley (1973) has published a scholarly treatment of the biological significance of circadian rhythms, and we can do no better than to quote a relevant passage from his article:

> Animals typically exhibit circadian rhythms of physiological and behavioral processes, and these are part of a genetically determined strategy fitting the animals to their particular ethological niche. Thus an animal which depends heavily on photopic vision will operate well in daylight and be disadvantaged at night. Its motivational mechanisms have to suit this fact, and the alternating sequence of sleeping and waking exhibits just such a fit. However, it is no good having a mechanism which simply detects light in order to initiate and maintain the active phase. An animal sleeping in a dark hole would never know that day had come. Instead the animal needs a model of the rotation of the earth, or perhaps more prosaically, a working model of the alternation of light and darkness which itself exhibits daily oscillations and will keep in time with the rotation of the earth to predict day and night. This is precisely the role of mechanisms giving rise to circadian rhythms [p. 214].

External Stimuli

The circadian rhythms of sleep–waking, feeding, etc. are closely related to the onset and offset of light; circadian rhythms are said to be ''entrained'' to light. Visual and other external stimuli also influence motivated behaviors in ways other than to entrain rhythms. For example, the sight of water or the smell of food can serve as cues to guide the behavior of the animal that has been without water or food for several hours and in which water- and energy-deficit signals are strong. Do such external stimuli merely serve to guide the behavior of the animal once ''motivation'' has been induced by homeostatic or deficit signals, or do they also contribute to the initiation of such behaviors? This has been an important theoretical issue for a number of years. It has become clear that certain stimuli, such as the sweet taste of a saccharin solution or the sight of a highly palatable food, may be strong initiators of motivated behaviors (Pfaffmann, 1960; Young, 1959). These stimuli are said to have incentive properties (Bolles, 1970). Visual and other stimuli can produce curiosity and exploratory behavior especially if they are novel or occur in unusual circumstances (Figure 1.5).

External stimuli are often more effective in initiating behavior if the animal is already "motivated" by hunger, thirst, or some other drive state. How do external and internal stimuli interact in determining motivated behaviors? Can external stimuli arising from homeostatic disturbances or other internal signals initiate motivated behaviors in the absence of a drive state? Is motivated behavior always a combination of "push" and "pull" mechanisms in the sense that the animal is "pulled" by stimuli in the external environment on the one hand and "pushed" by the stimuli of the internal environment on the other hand (Bindra, 1968). Or can external stimuli initiate motivated behaviors on their own even in the absence of a drive state? In later chapters we see that both internal and external stimuli contribute to initiating motivated behaviors, and that for some behaviors, particularly in man and higher mammals, external stimuli are very important.

FIG. 1.5 "Curiosity" motivates this cat to explore a novel object in its environment. In the 1950s motivated behaviors in response to extrinsic stimuli were ignored by popular theories that emphasized internal deficit signals. (From Berlyne, 1960; photograph by Roc Walley.)

Experiential and Cognitive Factors

Motivated behaviors are not only initiated or influenced by internal deficit signals and stimuli in the external environment; memory, learning, perception, and other cognitive processes also contribute. This is especially evident in man and higher animals in which the central nervous system has become more complex, with greater capacity to process and store sensory information. Furthermore, cognitive or higher mental processes not only contribute to guiding the behavior in the external environment, to establishing behavioral priorities, and to the selection of a particular goal object (e.g., not just food but a hamburger or a strawberry sundae); behavioral responses, in turn, provide sensory input (or sensory support) to the neural systems that subserve these cognitive processes (Hebb, 1955). In higher animals, particularly man and primates, behavior is more complex and variable, less stimulus-bound, related less directly to homeostatic factors, and often characterized by activities that provide novel sensory experience and that challenge cognitive capacities (Figure 1.6). Exploratory, manipulatory, and problem-seeking behaviors occur more frequently in man and higher animals; man may be concerned more with chess or golf, for example, than with water balance or temperature regulation. This assumes, of course, that man has been able to utilize his cognitive and intellectual abilities to organize his environment and his behavioral activities so that his basic biological needs are readily looked after.

FIG. 1.6 A chess player concentrating on the next move. Man has devised many games and sports to "stimulate" his own brain. (Photo by B. Box.)

DIFFERENTIAL CONTRIBUTIONS OF FACTORS AND DETERMINANTS OF MOTIVATED BEHAVIORS

This section is included here in order to provide further examples that illustrate multideterminants of motivated behavior. It is intended for students who are newcomers to this field. They may find it helpful to reread this section after completing Chapters 3 to 9. The more advanced reader should merely skim or bypass this section.

A reliable way to elicit a behavior, as indicated earlier, is to deprive it of food or water or to expose it to a low ambient temperature. This is why professional animal trainers employ food rewards in food-deprived animals trained to perform a circus act. Clearly homeostatic or deficit signals can be powerful initiators of motivated behaviors. However, there are behaviors, such as copulatory and maternal responses and brain self-stimulation, that are not associated with homeostatic, deficit signals. Aggressive, fearful, and other emotional behaviors are also not initiated by homeostatic signals, although such signals may make an animal hyperirritable and thus more responsive to emotion-provoking stimuli in the external environment. Furthermore, the occurrence of feeding and drinking are not always associated with the presence of energy- and water-deficit signals. On the one hand, feeding, drinking, and thermoregulatory behaviors are sometimes postponed in the presence of energy and/or water deficits because of ''higher priorities'' and motivational time-sharing. For example, a female in ''heat'' or an animal being pursued by a predator may not feed or drink and thus lose weight and become dehydrated. On the other hand, feeding and drinking may occur in the absence of such deficit signals, especially if the food or fluid are highly palatable.

Hormones are important contributors to many motivated behaviors but have an especially direct and essential role in sexual behavior both in differentiation of the brain (''male'' brain versus ''female'' brain) and in the initiation of the appetitive phase of the mating pattern in many species. For example, the sexual receptivity and mating behavior that occurs every 4 or 5 days in the female rat is determined by the action of estrogen and progesterone on the preoptic region of the brain. Angiotensin II, a hormone produced by the kidney, mediates drinking behavior in response to intracellular hypovolemia by acting on receptors in the brain. Hormones are less important and only indirectly involved in the initiation of feeding, thermoregulatory, aggressive, and emotional behaviors. For example, by regulating metabolic processes, hormones control the rate of energy utilization and expenditure and thereby determine indirectly the frequency and duration of feeding behavior necessary to maintain energy balance. Certain hormones such as adrenalin or thyroxin may influence emotional behaviors by altering reactivity of the central nervous system to emotion-provoking stimuli.

The contribution of hormones to sexual behavior depends upon several variables. Estrogens and androgens, for instance, are more or less effective depending on the sex of the animal, its age and sexual experience, and the complexity of its central nervous system. An example, considered in Chapter 6, is the effect of removal of the gonads. In female rats and dogs, reduction of estrogen levels by removal of the ovaries eliminates sexual receptivity, but a similar reduction in adult primates produces little effect on sexual behavior. In most mammals, male sexual behavior is less disrupted by gonad removal than is female sexual behavior, especially if the castrated male was sexually experienced. For example, the castrated male dog in contrast with the ovariectomized female dog continues to copulate, especially if copulation had occurred prior to castration. Additional experiential factors (e.g., level in social hierarchy and, in humans, even symbols and gestures) contribute to sexual arousal in species with more complex brains, and, correspondingly, the effectiveness of hormones in these species is reduced.

Levels of sexual hormones in many species fluctuate according to biological rhythms, and so consequently does the occurrence of sexual behavior. Another and perhaps the most prominent biological rhythm from the behavioral point of view is the circadian rhythm of sleep–waking. Sleep may be treated as a motivated behavior in its own right, representing the consummatory phase that follows the drowsiness or appetitive phase (see Chapter 7), but the sleep–waking rhythm is of additional importance because it provides a constraint on a variety of behaviors. For example, feeding and drinking occur during wakefulness. In the rat and other nocturnal species, wakefulness occurs primarily during the night, and hence feeding, drinking, and many other behaviors occur in these species at night. Diurnal species such as humans, on the other hand, are normally awake and active during the daytime. Most motivated behaviors in these species occur in the light, and visual stimuli are frequently found to be important to the performance of these behaviors. Again, in species with more complex brains, behaviors entrained to the 24-hour light–dark cycle can be ''decoupled,'' but even in humans this usually occurs only in unusual situations and with artificial control of the 24 hour light–dark cycle.

Visual and other stimuli from the external environment not only guide motivated and emotional behaviors but also contribute to their initiation. The effect of a particular external stimulus upon a particular behavior depends upon many variables including the behavior, the species, its internal physiological state, and its past experience. Thus, whether or not a stimulus elicits feeding and the potency of its effect may depend on whether the animal has been deprived so that a ''biological need'' is present. For example, a food-deprived rat will rapidly traverse a complex maze to obtain a food pellet but will ignore the food when satiated. Some stimuli are inherently ''pleasurable'' or rewarding, such as the sweet taste of saccharin or other taste and olfactory stimuli. They elicit ingestive behaviors even in the absence of a biological need or drive and facilitate the intake of foods and fluids (Pfaffmann, 1960). It is suggested in Chapter 8 that brain-stimulation reward is the result of activating neural structures or pathways that subserve such properties of conventional rewards.

An initially neutral stimulus may acquire ''motivational significance'' when paired with an effective stimulus. By pairing a light with food, the light will come to elicit salivation and other components of the appetitive phase of feeding behavior, or a tone associated with painful electric shock will elicit fear-like behavior. According to reinforcement theory, stimuli associated with primary drive stimuli initiate secondary drives. External stimuli, especially when they are novel and complex, may initiate investigatory or exploratory behaviors (see Figure 1.5). Harlow (1953) and Hebb (1955) stressed the importance of this sort of ''extrinsic motivation'' in the 1950s to counterbalance the emphasis at that time on internal, homeostatic signals as initiators of motivated behaviors. Hebb proposed that motivated behaviors could be thought of as an attempt by the animal to maintain an optimal level of arousal and that emotional behaviors were typically associated with the extremes of arousal (fear or rage with high arousal, boredom and lethargy with low arousal).

The effectiveness of external stimuli in the initiation of behaviors depends on experiential factors. Rats and other species including man will eat a familiar food and avoid a strange food. A lamb fed by bottle and kept away from other sheep will not stay with the flock when it becomes an adult, because it has been imprinted to man (also see Figure 9.7). The neighbour's dog will greet you ''affectionately'' or run away in ''fear'' depending on how you treated him previously. Experiential and cognitive factors are relatively more important and homeostatic signals less important in man and other higher mammals in which forebrain structures are more complex and highly developed.

ONTOGENY OF MOTIVATED BEHAVIORS

The neural and other physiological mechanisms that subserve behaviors develop at different rates and begin to operate at different stages of ontogenetic development. A number of mammalian species are relatively helpless at birth, and thermoregulatory, sexual, and a number of other behaviors are absent. Newborn mice and rats are capable of consummatory suckling responses and feed if the mother is in the nest. However, the capacity to locate food or even the mother, which require sensory systems, motor responses and sensory–motor integrations, appears only much later when these mechanisms have matured sufficiently. Sexual behaviors not only require maturation of sensory and motor systems as a prerequisite but also the presence of gonadal hormones for sexual arousal and receptivity. In some species the importance of these hormones decreases with age as the animal acquires sexual experience (see Chapter 6).

Memory, learning, and other cognitive processes that contribute to motivated behaviors in higher mammals usually develop even more slowly and later than sensory and motor functions. These processes become more sophisticated and make possible greater versatility of behavioral adaptation as the brain becomes larger and more highly organized. Because of the long period of maturation of these processes, experiential and cognitive factors become more important, as compared to internal stimuli and other factors, as determinants of motivated behaviors in the adult than they were in the younger organism.

Clearly, the relative contribution of the factors that initiate behaviors depends on the age of the animal and the stage of ontogenetic development. It is appropriate in subsequent chapters, therefore, to consider the ontogeny of feeding, drinking, thermoregulatory, and other behaviors when the various factors or determinants are discussed.

THE NEURAL SUBSTRATES OF BEHAVIOR

The major objective of this book is to consider what is known about the neural substrates of thermoregulatory, feeding, drinking, sexual, and aggressive behaviors. This is done after considering the characteristics and determinants of each particular behavior. In this section we deal with brain function and behavior in general terms, beginning with a brief discussion of the study of the neural substrates of behavior in historical perspective.

Historical Background

A question that has intrigued scholars—philosophers, poets, naturalists—for centuries is what motivates man and animals to do what they do. Philosophers tried to account for motivated behaviors, in terms of man seeking pleasure and avoiding pain, as the manifestation of rational processes or as due to divine guidance. Students of animal behavior, including Charles Darwin (1859), attributed various species-specific behavioral patterns in lower animals to instinctive mechanisms. These diverse views constituted the conceptual framework inherited by those attempting an experimental analysis of behavior in the latter part of the nineteenth century.

At the beginning of the twentieth century, John Watson (1913) advocated the use of a strictly empirical approach to the study of animal behavior. According to Watson, and the Behaviorist School he founded, the behavior of animals and man should be dealt with in terms of stimuli that elicit responses; the reflex was considered the unit of behavior, and motivational and other inferred concepts were to be avoided. However, it soon became apparent that it was not always possible to account for variations in behavior in terms of external stimuli. The nature and probability of a behavioral response to changes in the environment frequently depended on a central state related to such factors as bodily deficits and past experience. Drive (Woodworth, 1918), central motive state (Lashley, 1938), and other concepts seemed to be useful, and, according to some investigators, even essential; feeding behavior was attributed to a hunger drive, sexual behavior to a sex drive, etc.[4] By the 1950s, the term *motivation* was widely used to designate an important new field of behavioral investigation.

During this period there were other important developments that emphasized the biological utility of feeding, thermoregulatory, and other motivated behaviors. Walter Cannon's classic book, *The Wisdom of the Body,* published in 1932, popularized the term *homeostasis*. Cannon used this term to designate the constancy of the internal environment—the classical concept formulated by Claude Bernard (1878) almost 50 years earlier—and presented the results of experimental investigations of homeostatic mechanisms for regulating body pH, blood sugar, the exchange of O_2 and CO_2 and other conditions of homeostasis. His work dealt with the internal bodily responses or physiological regulations that contribute to maintaining the internal environment. A few years later, Richter (1943) proposed that homeostasis also depends on behavioral or what he called *total organism* responses. In a series of simple but highly significant experiments, he demonstrated that when a physiological homeostatic mechanism is disrupted, the occurrence of appropriate behavioral responses increases. For example, rats with the adrenal gland removed and thus not able to produce hormones for the retention of sodium by the kidney, drank large amounts of water containing sodium. Sodium is essential for bodily function, and sodium homeostasis depends on behavioral as well as physiological regulations. Richter's experiments (also see Chapter 3, pp. 65–66) clearly established an important association between motivated behaviors, physiological regulations, and homeostasis.

The study of the neural substrates of feeding and other behaviors began about 1940, following the introduction of the stereotaxic procedure, which made it possible to produce discrete lesions in and to stimulate deep structures of the brain (Figure 1.7). Lesions of the ventromedial nucleus of the hypothalamus

[4]The term *central motive state* came from Lashley (1938) and Morgan (1943).

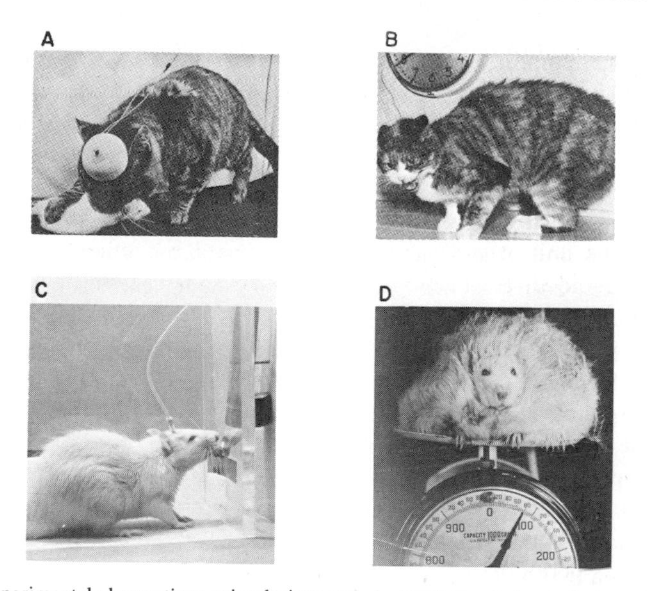

FIG. 1.7 Experimental observations using lesions and electrical stimulation techniques have implicated the hypothalamus in ingestive, aggressive, and other behaviors. A. Quiet biting attack of a rat elicited in a cat by means of electrical stimulation of the hypothalamus. (After Flynn, 1967.) B. Defense reaction elicited by hypothalamic stimulation. This picture was taken just before a vigorous flight response (jumping from table and escaping) occurred. (After Brown, Hunsperger, & Rosvold, 1969.) C. Drinking elicited by electrical stimulation of the hypothalamus of the rat. (From Mogenson & Stevenson, 1966.) D. An obese rat that has become hyperphagic following lesions of the ventromedial hypothalamic safety region. (From Stevenson, 1969.)

resulted in overeating and obesity (Brobeck et al., 1943) and lesions of other regions of the hypothalamus were subsequently observed to disrupt feeding and drinking (Anand & Brobeck, 1951) and sexual behavior (Brookhart et al., 1941; Sawyer & Robison, 1956). It should not be surprising that the interpretation of these experimental findings was strongly influenced by the contemporary motivational concepts—drive and central motive state—and by the contributions of Cannon and Richter, which suggested the biological utility of such behaviors. Accordingly, the hypothalamus, already implicated in temperature and other homeostatic regulations, was considered the locus of neural mechanisms subserving the hunger drive, thirst drive, sex drive, etc.[5] The results of these lesion

[5]The prevalent view was expressed by Teitelbaum (1966) as follows:

The hypothalamus integrates the series of instinctive acts from appetitive to consummatory behavior. It may do this by producing the appropriate drive state. In other words, the hypothalamus is necessary for the existence of a drive state which welds together appetitive and consummatory behavior into integrated instinctive activities. Therefore, when damage to the hypothalamus impairs instinctive activities such as eating, drinking, or mating, this may not occur as a result of primary damage to the nervous structures responsible for the individual components of these acts, such as chewing, swallowing, mounting, etc., but rather to the

experiments suggested that the hypothalamus contained central excitatory mechanisms and central inhibitory mechanisms for the initiation and termination of feeding, drinking, and other motivated behaviors in the way that inspiratory and expiratory systems were thought to be represented in the medulla (see Figure 2.2).[6] By the mid-1950s, this view had become designated as the *dual-mechanism model* (Figure 1.8A).

Another reason for the focus of attention on the hypothalamus was that this small structure was also shown to exert control over the endocrine system, via the pituitary gland, and over the peripheral autonomic nervous system (see Figure 1.9); it appeared that the hypothalamus had an important and even vital role in homeostatic regulations. Because feeding, thermoregulatory, and other behaviors were attributed to drive mechanisms resulting from biological needs or homeostatic deficits, the hypothalamus was considered an important integrative structure for behavioral as well as physiological regulations.

During the next 10 or 15 years, the dual-mechanism model stimulated a good deal of interest in the study of the neural substrates of feeding and other motivated behaviors. The model suggested experiments to be undertaken and also served as a conceptual framework for interpreting the results obtained; the empirical observations and the theoretical model were mutually reinforcing. The hypothalamus was the focus of attention, and when associated regions, such as limbic forebrain structures, were implicated, the model was expanded by the proposal that the limbic system exerts modulatory influences on the primary neural systems in the hypothalamus (Figure 1.8B). Although some investigators (Harlow, 1953; Hebb, 1955) pointed out that all motivated behaviors are not homeostatic and emphasized the importance of experiential, cognitive, and ontogenetic factors, the emphasis on homeostatic factors and on hypothalamic mechanisms continued, and the dual-mechanism model was widely accepted. By 1970, however, attention was beginning to shift away from the primacy of the hypothalamus, and there were increasing reservations about the adequacy of an exclusively homeostatic model of motivation; doubts were being expressed about the classical interpretation of lesion and electrical stimulation studies as supporting the view that a feeding system was in the lateral hypothalamus and a satiety system in the ventromedial hypothalamus.

> motivational processes which must energize them sufficiently for them to be elicited in response to the stimuli that normally release them. Therefore, increasing the level of motivation, either by using supernormal stimuli or by changing the internal environment, may be sufficient to elicit the behavior once again, though under normal circumstances it would appear to be permanently abolished [p. 600].

[6]Sherrington (1906), from his classic studies of the physiology of the spinal cord, had introduced the concepts of central excitatory state and central inhibitory state in his model for spinal reflexes. These concepts were found useful for dealing with supraspinal functions such as central control of respiration. It was Lashley (1938) who first used the concept of central excitatory state or mechanisms in dealing with behavior, and subsequently the term *central motive state* (Morgan, 1943) came into general use.

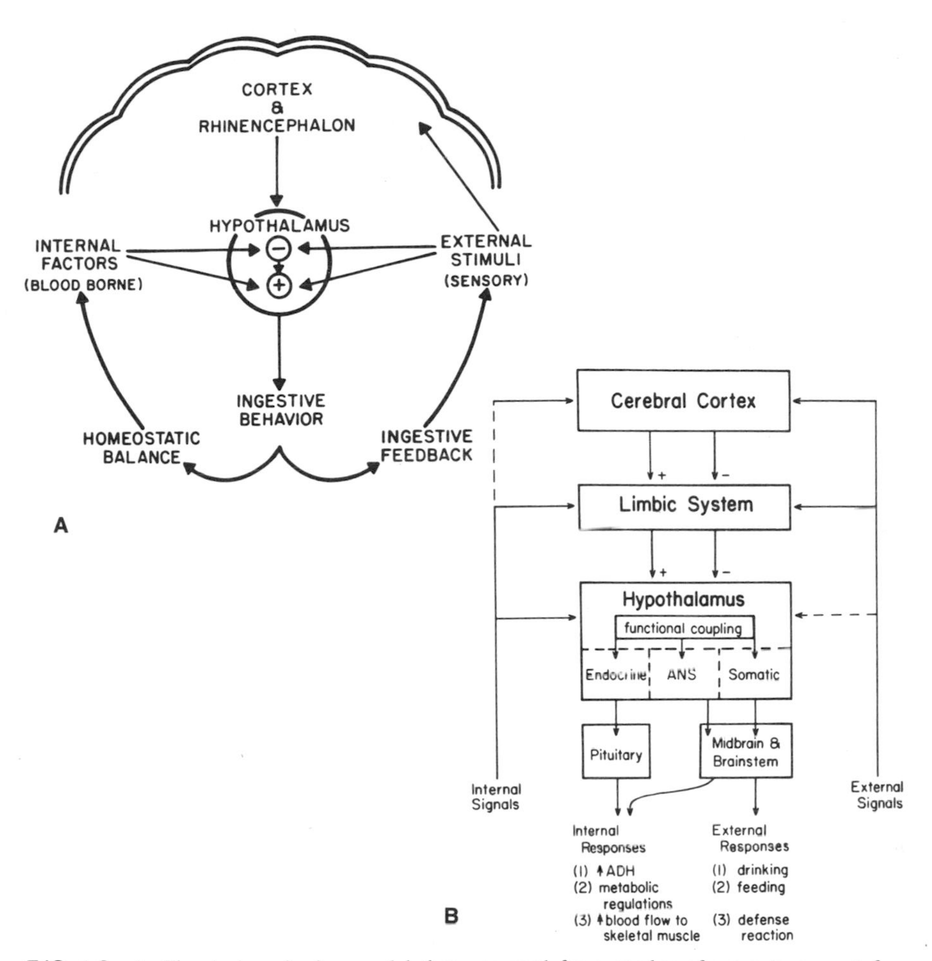

FIG. 1.8 A. The dual-mechanism model that was used for a number of years to account for hypothalamic mechanisms that control feeding and other motivated behaviors. It derived from two great traditions of Bernard–Cannon and of Sherrington,

> According to Cannon, energy balance was an important condition of homeostasis resulting from both behavioral regulation (feeding) and physiological regulation (energy utilization). Feeding was initiated and terminated by signals integrated by the central nervous system. From studies in which the hypothalamus was lesioned or stimulated, it was concluded, in accordance with Sherrington's concepts of central excitatory state and central inhibitory state, that the lateral hypothalamus was an appetite center and the ventromedial hypothalamus a satiety center (Mogenson, 1976, p. 473).

(Modified from Stellar, 1954.) B. Lesion and stimulation studies in the late 1950s and 1960s implicated limbic forebrain structures in feeding, drinking, and other motivated behaviors. The dual-mechanism was modified by assuming that limbic structures exerted modulatory effects on the drive mechanisms represented in the hypothalamus. The CNS is organized in a hierarchical manner for the control of feeding, aggressive, and other behaviors. The reflex components for such behaviors are believed to be organized in the brainstem. Superimposed are facilitatory and inhibitory mechanisms in the hypothalamus, limbic forebrain, and cerebral cortex. The hypothalamus and associated structures have a special role in the overall integration and "functional coupling" of the endocrine, autonomic, and somatic components that subserve those behaviors. (From Mogenson & Huang, 1973.)

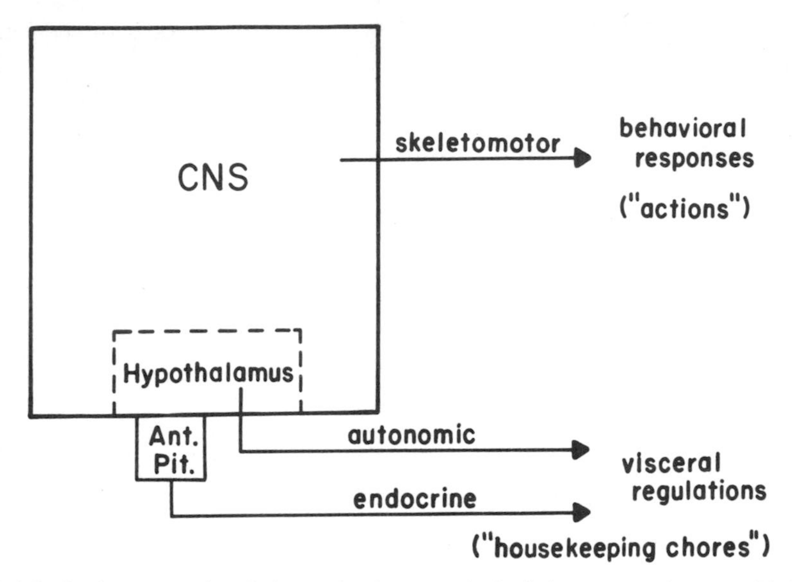

FIG. 1.9 In the course of evolutionary development, the brain has come to have two kinds of functions, as illustrated in this schematic diagram. The brain controls behavioral responses (''actions'') in the external environment, and it controls homeostatic regulations of the internal environment (''housekeeping chores'' of the body). (CNS—central nervous system; Ant. Pit.—anterior pituitary gland).

The Brain, Physiological Regulations, and Behavior

As animals have evolved into higher forms, there has been a gradual specialization of physiological systems to meet the various needs of living cells. There are digestive, respiratory, circulatory, and renal systems, for example, which transport and make available oxygen, carbon dioxide, and nutrients for energy and excrete wastes resulting from the metabolic activities of cells. At the same time, there has been development of the neural and endocrine systems for the overall integration and coordination of other systems of the body. As discussed more fully in Chapter 2, these coordinating functions of the nervous and endocrine systems enable the animal to maintain the homeostasis of the internal environment. As a consequence the higher animals' existence, in contrast to that of the amoeba and other lower forms, does not depend on being in an external environment conducive to providing directly the moment-to-moment needs for energy, for optimal temperature and pH; one of the important kinds of functions to which the brain contributes is the ''housekeeping chores'' of the body (Figure 1.9).

The second kind of function of the brain is to initiate and control the external behavioral responses or ''actions'' of the animal in relation to stimuli in the external environment. Some behavioral responses, as illustrated by Richter's experiments referred to earlier, contribute to the homeostasis of the internal environment; for example, nest building may occur in response to a fall in body temperature and the drinking of water to a body-water deficit. Behavioral responses initiated and controlled by the central nervous system also occur in the absence of homeostatic deficit signals as the animal adapts to a continuously changing external environment—escaping from a predator or investigating the source of an unusual sound. According to the terminology used by Konorski (1967) in his book, *Integrative Activity of the Brain,* for classifying behavior in terms of its biological role and significance, there are preservative responses and protective responses.

As the brain becomes more complex in higher animals in response to evolutionary pressures of the environment, the behavioral responses are more varied and complicated. This is the result of increased sophistication of the motor system and of increased capacities for processing, storing, and utilizing sensory stimuli from the environment (Figure 1.10; also see Figures 2.9 to 2.12).[7] In monkeys and man, in particular, interactions with the external environment depend increasingly on learning, memory, perception, language, and other cognitive processes, and these interactions become more complex and sophisticated in such social species. Behavioral responses utilizing cognitive processes contribute to the adaptation of the animal to the external environment as well as to the homeostasis of the internal environment; internal regulatory responses and external behavioral responses are complementary. In addition, however, the cognitive processes assume a functional autonomy and primacy enabling higher mammals, and especially man, to engage in activities that may also be goal-directed or "purposive" (see p. 22) but may not be directly associated with homeostasis and biological adaptation and survival. As a consequence, what initiates or motivates behavior, as well as the nature and expression of the behavior, becomes more complex and variable (Figure 1.1). For example, a

[7]The neural systems concerned with analyzing patterns of stimuli from the external environment and storing and utilizing this information, are designated as the "cognitive brain" by Konorski (1967). Anatomically and functionally distinct from it is the "emotive or motivational brain," which monitors the internal environment.

> The cognitive system operates through what is generally called the specific division of the brain—the division of more or less strict localization and where the pathways leading from the receptors to the centers and from the centers to the effectors have an orderly topographical arrangement. Here are specific thalamic nuclei, both projective and associative, and the neocortex, both projective and associative. We shall refer to all these parts of the brain as the *cognitive brain*. On the other hand, the emotive system operates through what is generally referred to as the unspecific division of the brain; here the principle of topographical arrangements is less strictly binding and sometimes seems to be lacking. It is represented by the reticular system, the hypothalamus, the intralaminar thalamic nuclei, and the rhinencephalon (limbic system). We shall refer to all these parts of the brain as the *emotive brain* [p.42).

Konorski reviews experimental evidence showing that lesions of the "cognitive brain" produce quite different behavioral deficits in animals than lesions of the "emotive brain" and that electrical stimulation of these two systems elicits quite different effects.

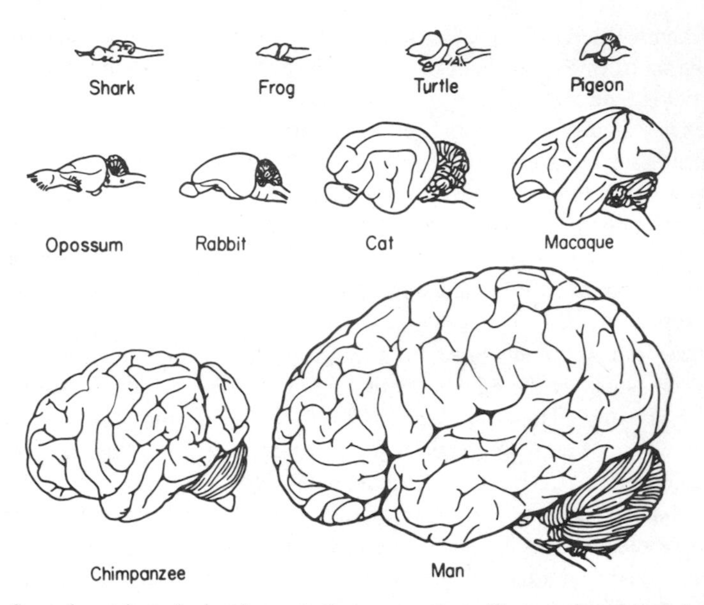

FIG. 1.10 Several vertebrate brains drawn to the same scale to illustrate the relative size of the neocortex. Increased capacity for memory, learning, and other cognitive processes is associated with increased development of the cerebral cortex. (From Eccles, 1973.)

monkey will spend a good deal of time and effort taking apart a mechanical puzzle, a college student will work late at night learning physiology for an examination the next morning, and a college professor spends his vacation attempting to reach the top of a high mountain. Thus, in man and primates, behavioral responses are not only *the result* of brain function but also *in the service of* the increased capacity for processing and storing sensory information and of cognitive processes (Hebb, 1955). In man, intellectual and conceptual growth and self-actualization are important aspects of the subject of motivation (Maslow, 1954). Language, a unique type of cognitive activity that becomes highly developed in man, provides not only an advantage in meeting biological needs and in adapting to the environment but also the capacity for intellectual, social, and aesthetic pursuits that add an important new dimension to the subject of motivation.[8]

[8]Isaacson (1974) comments on the special significance of language as follows:

With language, man can override the demands both of his internal environment and of the outer world. He can reject the feelings of thirst and hunger in favor of abstract goals, sometimes ones of his own creation. Thus we have martyrs who will starve to death to further their own "salvation" or the "salvation" of others. Man can reject the directions provided by the world and the people around him. He can, in short, be free from internal and external factors which shape the destinies of those animals without language.

To exercise this freedom, man must override and suppress the well-learned and genetically determined behaviors which are presumed to be properties of the protoreptilian core brain.

Although experiential and cognitive processes are important determinants of behavior in man and higher mammals, very little is known about the neural mechanisms that subserve these processes. This is in part because they involve complex and highly organized forebrain structures about which relatively little is known. Another reason, as indicated earlier, is that for many years homeostatic factors and hypothalamic mechanisms were emphasized, and the role of experiential and cognitive determinants of motivated behaviors was neglected. The emphasis given to experiential and cognitive factors in subsequent chapters is not justified by the experimental evidence available, but by drawing attention to these neglected mechanisms, we are more likely to have a more adequate understanding of the neural substrates of motivated behaviors sometime in the future.

SUMMARY

This book deals with the general field of brain function and behavior and specifically with the neural substrates of motivated behaviors—with thermoregulatory behavior, feeding, drinking, sexual behavior, sleep–waking, aggressive, and other emotional behaviors. These behaviors have certain common characteristics that are related to their biological significance. Motivated behaviors are goal-directed or ''*purposive*.'' They are frequently quite intense or *persistent* and occur in a periodic manner. Feeding, for example, occurs as discrete meals in many animals, and drinking, mating, and other motivated behaviors also show *periodicity* in their occurrence. Feeding is *periodic* partly because the signals for its onset and termination wax and wane with the availability of energy in the body and partly because of events occurring in the external environment. As a consequence of environmental variations, feeding ''competes'' with other behaviors, and this ''time-sharing'' is a prominent feature of motivated behaviors in man and animals. ''Motivational time-sharing'' may appear to operate on a competitive basis as suggested by Adolph's classic concept of ''*priorities and compromises,*'' but, as proposed more recently by McFarland, there is an overall organization for the initiation of motivated behaviors so that they contribute to ''biological fitness'' and survival.

Therefore, the neocortical mechanisms must overcome the habits and memories of the past.

The protoreptilian brain looks to the past. It learns and remembers but is poor at forgetting. The neocortical brain looks to the future, either to tomorrow, next week, next year, or to heavenly rewards. The neocortex is the brain of anticipation. It prepares for, anticipates, and predicts the future.

By the use of language, we can store the memories of the past. We build libraries and repositories of information which can extend our knowledge of the past for thousands of years, far beyond the life span of the protoreptilian brain. The neocortical contribution of language enables us to forecast the future and to anticipate conditions hours, weeks, years, and centuries from the present. It is this extension of time, both forward and backward, which represents the singular advance produced by the neocortex [pp. 242–243].

Previous attempts to define motivation have proved difficult and resulted in disagreement and controversy. A major reason for the lack of consensus in defining this psychological construct is that authors have recognized certain factors that contribute to motivated behaviors and ignored others, some authors stressing one factor (e.g., homeostatic signals) and other authors stressing another. The distinctive feature of this book is the recognition that several factors contribute to motivated behaviors and that they contribute differentially. Thermoregulatory behavior and feeding may be initiated by homeostatic deficit signals whereas sexual behavior is not. Hormones make a direct and essential contribution to the occurrence of mating behavior in the female rat but not to thermoregulatory behavior or to feeding behavior. Hormones do have an indirect effect on thermoregulatory behavior through their influence on metabolism and thus on body-temperature regulation. Because behavioral responses and internal physiological regulations are typically complementary in their contributions to homeostasis, certain factors that influence internal homeostatic regulations may have an indirect effect on motivated behaviors that contribute to homeostasis. Hormones also play a less important role in some animals that have had previous sexual experience, particularly in man and primates, in which forebrain structures are more complex and in which sexual arousal depends on social, experiential, and cognitive factors as well as on hormones and sensory stimulation. Feeding, as indicated previously, depends on a number of factors. The relative contributions of the various factors to the initiation of motivated behaviors depend on the species, the sex of the animal, the behavior, the internal physiological state, the circumstances of the external environment, past experience, and the state of ontogenetic development.

The approach and chapter organization used for dealing with each of the behaviors selected for consideration are similar. The various factors and their relative contribution to the particular behavior are first considered. It is from this perspective that current evidence concerning the neural substrates of the behavior is presented. This approach is particularly appropriate in view of the current reassessment of the contributions of the hypothalamus and limbic forebrain structures to motivated behaviors. Results of experiments utilizing newer techniques are presented. These challenge theoretical models widely accepted a few years ago and indicate that we know less about the neural substrates of motivated behaviors than was previously assumed. Although the long-term goal is to understand motivated behaviors in terms of the contemporary concepts of the neurosciences, the approach taken in this book follows the principle that it is the behavior of the animal that indicates what kind of neural mechanisms to look for.

The topics selected for detailed consideration have been arranged so that the behaviors dealt with in the earlier chapters are closely associated with homeostasis and may be readily elicited by deficit signals. The behaviors dealt with in later chapters depend to a greater degree on experiential and cognitive factors and are not directly associated with homeostasis.

2

Biological Foundations, Experimental Strategies, and Techniques for the Study of Motivated Behaviors

In Chapter 1 reference was made to the specialization of the various physiological systems for meeting the needs of living cells. We begin this chapter by expanding on this view with examples that illustrate the functioning of these specialized systems and with a consideration of the concomitant development of two integrative systems—the nervous and endocrine systems. These integrative systems coordinate the functioning of the other specialized systems of the body for the homeostatic regulation of the internal environment. They also contribute to the behavioral adaptations of the animal to its external environment. The nervous system—with its capacity to rapidly transmit a variety of internal and external signals; to process, store, and utilize information provided by these signals; and to send command signals to muscles and other effectors—is especially important in intiating and controlling behavioral responses.

The nervous system is highly organized and very complicated both structurally and functionally (Figure 2.1). It is not possible in this chapter to deal with all of the structures and subdivisions of the nervous system, and the student is referred to other sources at the end of the chapter. I deal only with those structures and systems that contribute to the initiation of the behaviors considered in subsequent chapters. From the accumulated experimental evidence over the last several decades, the hypothalamus and limbic forebrain structures have been implicated in motivated behaviors as well as in physiological regulations. There has been a great deal of interest recently in the studying of the monoaminergic neural pathways, which project from the midbrain and brainstem to forebrain structures, and their role in motivated behaviors. Space considerations do not permit a detailed treatment of sensory and motor systems, although both are important for the initiation and control of behavior. They are dealt with only briefly in relation to the factors that contribute to motivated behaviors. Again the student is referred to other sources for further information. The treatment of experimental techniques used for studying the neural substrates of motivated behaviors is also selective.

THE NERVOUS SYSTEM

A

CENTRAL NERVOUS SYSTEM (CNS)			
Brain			Spinal Cord
Forebrain	Telencephalon	Neocortex	
		Basal Ganglia	
		Limbic System	
	Diencephalon	Thalamus	
		Hypothalamus	
Midbrain			
Hindbrain	Cerebellum		
	Pons		
	Medulla Oblongata		

B

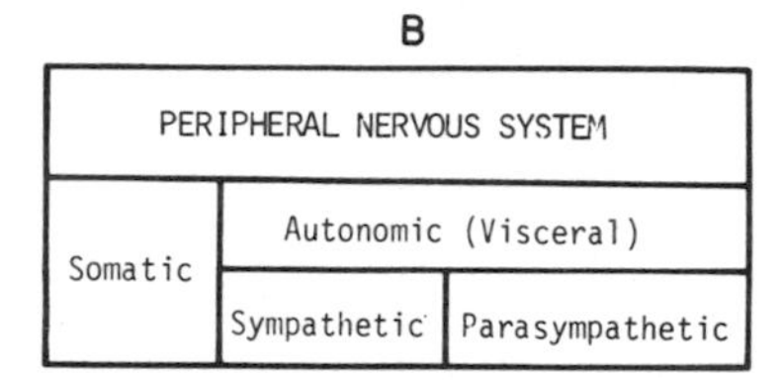

FIG. 2.1 Diagram of the major subdivisions of the nervous system. The first division is into the central nervous system (CNS) consisting of brain and spinal cord (A), and the peripheral nervous system (B) which is composed of all neural tissue outside the CNS. The basic divisions of the brain are hindbrain, midbrain and forebrain. The medulla of the hindbrain is attached to the upper spinal cord. The major components of the brain are designated. The peripheral nervous system has two major divisions—somatic and autonomic.

Rather than trying to duplicate a number of other excellent presentations, some of which are listed at the end of this chapter, we deal with the most widely used techniques in relation to the research problems and strategies in this field.

BIOLOGICAL FOUNDATIONS

Specialized Systems of the Body and Their Control by the Nervous and Endocrine Systems

The major specialized systems of the body are listed in Table 2.1, with a brief description of their physiological significance and the nature of the control over them.

TABLE 2.1
Specialized Systems of the Body
Controlled by the Nervous and Endocrine Systems

System	Function	Control
Cardiovascular	Transport of O_2, CO_2, nutrients, hormones, and water.	Primarily by NS[a] and secondarily by ES.
Respiratory	Exchange of gases (O_2 and CO_2)	NS
Alimentary and digestive	Processing of foods so nutrients can be utilized by body cells	ES and NS
Renal	Excretion of water, electrolytes, and wastes	ES and NS
Skeletomotor	Movement of body parts and locomotion of body in external environment	NS

[a] NS—nervous system; ES—endocrine system

The body requires oxygen (O_2) for the metabolism of glucose, free fatty acids, and other nutrients to provide energy for the heart, skeletal muscles, and other tissues, and carbon dioxide (CO_2), a metabolic product, must be eliminated. The respiratory system is specialized for the exchange of these gases between the body and external environment. In conjunction with the cardiovascular system, it ensures that O_2 from the air is made available in suitable quantities to body cells and that CO_2 reaches and is expired from the lungs. The levels of O_2 and CO_2 in the blood are monitored by receptors, designated chemoreceptors, located along the aorta and carotid arteries and possibly at other peripheral sites, as well as in the medulla, located in the lower brainstem. When the plasma level of CO_2 is elevated or when the level of O_2 is reduced, the rate of respiration is increased. The reflexes for the contraction of the muscles, which rhythmically expand and compress the thoracic cavity so that air enters and leaves the lungs as they expand and contract, are integrated by neural systems in the medulla (Figure 2.2).

O_2 and CO_2 are transported between the lungs and various tissues of the body by the blood, principally by combining with the hemoglobin of the red blood cells. The number of red blood cells may change over time, and thereby the capacity to transport O_2 and CO_2 changes (e.g., at high altitude when they increase), but the short-term adjustment of the capacity to transport these gases depends on how rapidly the blood circulates through the body. This in turn depends on heart rate and the resistance of the vessels to the flow of blood. Heart rate and the size of the blood vessels—especially of the smaller arterioles—are controlled by the CNS. The blood pressure is continuously monitored by

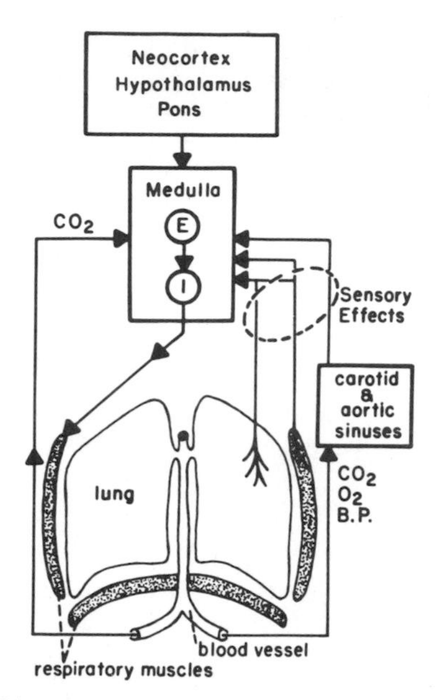

FIG. 2.2 Schematic diagram of the control of respiration. The activity of the respiratory cells of the medulla is influenced by the neocortex, hypothalamus, and pons. Two bilaterally paired areas in the medulla are concerned with respiration; the expiratory area (E) is thought to have an inhibitory effect on the inspiratory area (I) so that chest muscles relax and air is forced out of the lungs as the diaphragm rises and the rib cage collapses. Respiratory systems in the medulla are influenced by feedback systems that depend on the concentration of oxygen (O_2) and carbon dioxide (CO_2) in the blood as well as blood pressure (BP). The functional relation between E and I can be influenced directly by factors that stimulate receptors in the aortic and carotid sinuses. These receptors, along with others in the lungs themselves, send nerve impulses to the medulla via the vagus and glossopharyngeal nerves. In addition, the motor act of respiration stimulates still other receptors that feed back to the medulla and modulate neural activity. The control devices for breathing, which is a relatively simple behavior, are very elaborate and complex and even yet, not completely understood.

baroreceptors in the carotid artery; signals are initiated and reflexes integrated by the medulla, and higher brainstem structures maintain blood pressure.

Water is a major constituent of blood and other tissues of the body, and its content is kept relatively constant by regulating the exchange of water with the environment (for details, see Chapter 5). The kidney, specialized to excrete water, sodium, and wastes from metabolic activity, is under neuroendocrine control (see p. 36 for details). Water is also taken into the body at regular intervals to maintain water balance; the drinking behavior consists of sequences of skeletomotor and oral motor responses controlled by complex neural systems.

Another example of the integrative activities of the nervous and endocrine systems is the processing and utilization of energy for various bodily functions and in regulating energy stores as reflected by the relative constancy of body weight (for details, see Chapter 4). The source of energy is food, and feeding behavior depends on complex skeletomotor and oral responses controlled by motor systems of the brain. Digestion of food so that it can be utilized by body tissues requires gastrointestinal hormones and contractions of the stomach and intestine, controlled by the autonomic nervous system. The nutrients, when absorbed into the blood stream, are transported throughout the body, and the functioning of the cardiovascular system, which makes this possible, depends on neural and endocrine integrative activities (see Table 2.1). The rate at which nutrients are metabolized depends on thyroid and other hormones, and metabolism requires O_2 made available by the respiratory and cardiovascular systems.

Similarly, thermoregulatory responses, controlled by the nervous and endocrine systems, depend on the cardiovascular system, the respiratory system, the digestive system, and the skeletomotor system. When an animal is exposed to the cold, the blood vessels near the surface of the body constrict (designated "peripheral vasoconstriction"), resulting from signals along sympathetic nerves of the autonomic nervous system to reduce heat loss from the body. The animal may become more active. There is also an increase in metabolic heat production due to sympathetic nervous system discharge as well as increased release of hormones (e.g., adrenaline and noradrenaline) from the adrenal gland and increased release of thyroid hormones. The digestion of food providing nutrients for metabolic heat production also depends, as we have seen, on neural and endocrine controls; the peristaltic movements of the intestine, which move the food along, are influenced by the nervous system, and digestion requires a number of gastrointestinal hormones.

These examples are concerned with neural and endocrine integrations of physiological and behavioral responses, which maintain the internal environment. In Chapter 1 we designated these the "housekeeping" chores of the body and indicated that the homeostatic regulation of the internal environment is one of the important functions of the brain (see Figure 1.9). Behavioral responses of the animal that contribute to the homeostasis of the internal environment and to adapting to changes in the external environment also depend on complex integrative activities of the CNS concerned with sensory—motor integration and motor control—for which cardiovascular, respiratory, endocrine, digestive, and other specialized systems play a supportive role. For example, an animal that escapes from a predator coordinates its skeletomotor responses by taking into account visual, auditory, and other relevant sensory information, and at the same time, the cardiovascular, respiratory, and endocrine systems must respond to the sudden increase in activity, thus ensuring that sufficient O_2 and energy are available.

Integrative Systems and the Regulation of the Internal Environment ("Housekeeping Chores")

The internal physiological regulations referred to in the previous section are controlled by the autonomic nervous system and hormonal regulations and are coordinated by the hypothalamus and associated neural systems. The neural regulations are typically associated with relatively rapid, short-term adjustments (e.g., increasing heart rate and blood pressure). The endocrine system coordinates responses and processes with a long time-course such as the control of metabolism, the regulation of the chemical composition of body fluids, cyclical processes such as sexual receptivity, and growth and development (Figure 2.3).

The hypothalamus. The hypothalamus is a small structure at the base of the brain beneath the thalamus and just above the pituitary gland (Figure 2.4). In a rostral–caudal direction, the hypothalamus extends from the preoptic region near the optic chiasma to the mammillary bodies. The optic chiasma and mammillary bodies are clearly seen on the basal surface of the brain just rostral and caudal to the pituitary stalk.

Although the hypothalamus weighs only about 4 grams in the human brain, which has a total mass of 1200–1400 grams, it has a number of nuclei and is associated with a number of important functions (Table 2.3). The reason that such a small structure contributes to so many functions is that it is strategically located to influence the autonomic nervous system, the pituitary gland, and the endocrine system. The hypothalamus is in a position to integrate information about visceral functions and the internal environment together with the information from other CNS structures concerning the external environment and past experience, so that internal regulations, controlled by the autonomic and endocrine systems, are appropriate to the circumstances of the behaving animal. In other words, the hypothalamus (in association with limbic and other CNS structures) contributes to the higher-order integration of autonomic, endocrine, and

FIG. 2.3 A "giant" whose excess growth may be due to increased output of growth hormone from the pituitary gland. (From Lewin, 1972, p. 29.)

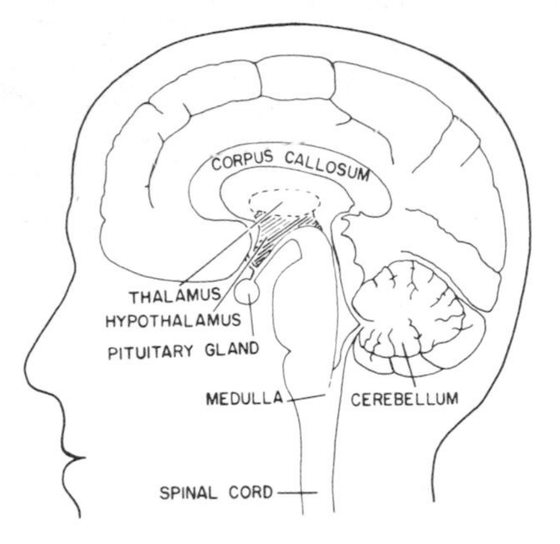

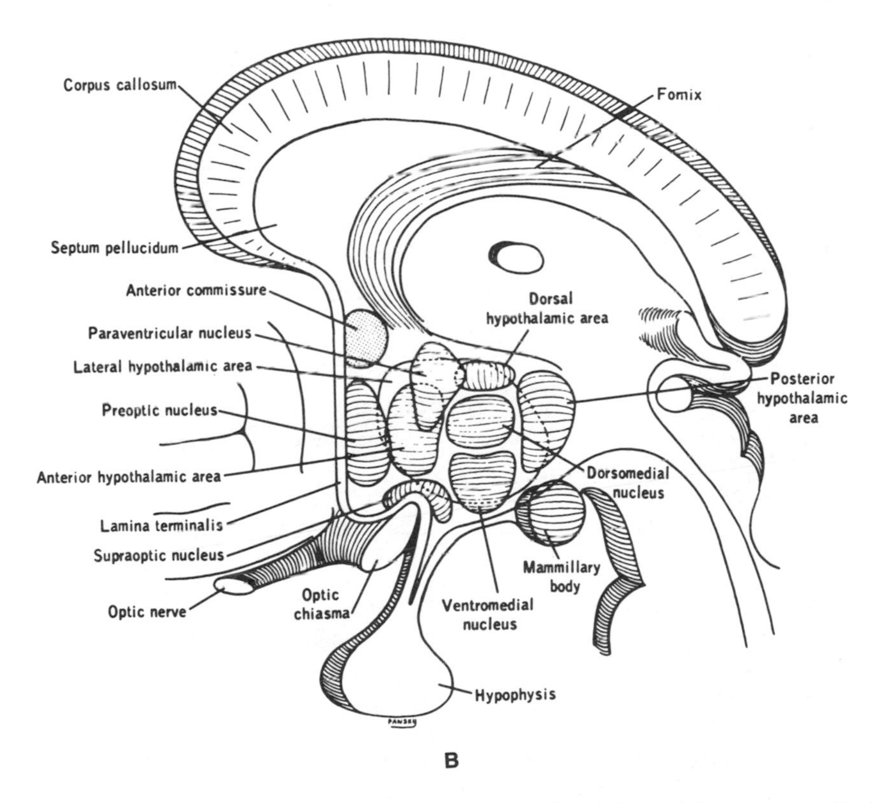

B

FIG. 2.4 The hypothalamus. A. Its location in the brain in relation to other structures. B. A three-dimensional diagram showing the location of various nuclei (groups of neurons). Although not shown on this diagram, there are many nerve fibers and pathways woven throughout this complex, which makes the hypothalamus the ''Grand Central Station'' of the brain. (Diagram B from House & Pansky, 1967.)

TABLE 2.2
Major Hormones in Man

Gland	Hormone	Target or function
Anterior pituitary	Growth hormone	Growth
	Trophic hormones:	
	ACTH	Adrenal cortex
	TSH	Thyroid
	FSH	Gonads
	LH	Gonads
	Prolactin	Mammary glands
Posterior pituitary	Vasopressin	Kidney, blood pressure
	Oxytocin	Mammary glands, uterus
Thyroid	Thyroxine	Development, metabolic rate
Parathyroid	Parathormone	Calcium, phosphorus metabolism
Adrenal cortex	Sex hormones	(See below)
	Glucocorticoids	Metabolism of carbohydrates, protein and fat
	Mineralocorticoids	Electrolyte, water balance
Adrenal medulla	Adrenalin	Circulatory system,
	Noradrenaline	glucose release
Pancreas:		
α cells	Glucagon	Glucose release
β cells	Insulin	Glucose transfer, utilization
Ovaries:		
Follicles	Estrogen	Development and maintenance
Corpus luteum	Progesterone	of sexual anatomy,
Testes	Testosterone	physiology, and behavior
Kidney	Angiotensin	Vasoconstriction, aldosterone release, water intake

behavioral responses for homeostatic regulation of the internal environment and for adaptation of the animal to the continuously changing external environment. These neural integrative activities ensure that the ''housekeeping chores'' are handled routinely in relation to the varying demands of the behaving animal for water, energy, oxygen, etc. Examples that illustrate these integrative functions of the hypothalamus are presented in the sections that follow.

The autonomic nervous system. The autonomic nervous system—consisting of two components, the sympathetic and parasympathetic divisions—connects the CNS with internal effectors such as cardiac muscle, blood vessels, urogenital tract, smooth muscles of the intestine and stomach, the liver, the sweat glands, and the medulla of the adrenal gland (Figure 2.5). Signals along sympa-

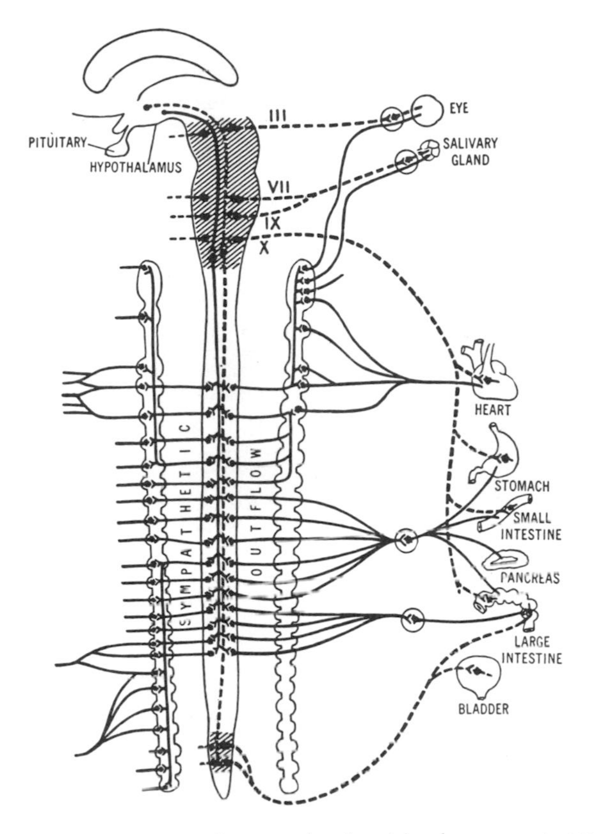

FIG. 2.5 The autonomic nervous system has central and peripheral components concerned with internal physiological regulations. The sympathetic division consists of short preganglionic fibers, which extend from the thoracic and lumbar regions of the spinal cord to the ganglia, that form the sympathetic chain and postganglionic fibers that innervate the heart, stomach, kidney, genitalia, and other internal organs. The parasympathetic division consists of long preganglionic fibers, which extend from the sacral region of the spinal cord and from the brain through cranial nerves, and short postganglionic fibers, which innervate these same structures. The dual innervation enables both sympathetic and parasympathetic effects to be exerted in cardiovascular, gastrointestinal, and other functions. Integrative centers for the autonomic nervous system are represented in the medulla and hypothalamus, and they are influenced by activity at other levels of the CNS, particularly the limbic system and reticular activating system. (From Gardner, 1968. Courtesy W. B. Saunders Co.)

thetic nerves increase during motivational arousal and in particular during stress and emotional excitement. For example, when an animal escapes from a predator or during aggressive behavior, there is an increase in heart rate and blood pressure, peripheral vasoconstriction, inhibition of activity in the gastrointestinal tract, and conversion of glycogen in the liver to elevate the level of blood glucose. In general, the sympathetic nervous system controls physiological responses that mobilize energy and enable the animal to engage in vigorous activity and to adapt to emergency situations.

The parasympathetic system, as shown in Figure 2.5, has connections with many of the same internal effectors but contributes to the storage rather than to the expenditure of energy and to the maintenance of body function. Signals along parasympathetic nerves slow the heart, reduce blood pressure, increase digestive secretions, and facilitate digestive movements. The parasympathetic nervous system is dominant, for example, when an animal is resting in a safe place after feeding.

Most effectors are innervated by both the sympathetic and parasympathetic systems, which typically exert opposite effects. The coordination of an appropriate balance between the two systems is determined by integrative activities in the medulla, hypothalamus, and other levels of the brain so that the internal environment is maintained relatively constant under a variety of environmental conditions. The anterior hypothalamus is associated with parasympathetic function and the posterior hypothalamus with sympathetic function.

The endocrine system and neuroendocrine integrations. The endocrine glands shown in Figure 2.6, are sources of chemical substances designated hormones that are transported by the circulatory system to produce effects on specific target structures (e.g., adrenal cortical hormone, ACTH) or throughout the body (e.g., growth hormone). A summary of the principal hormones and their sites of action and functions is presented in Table 2.2. In later chapters we

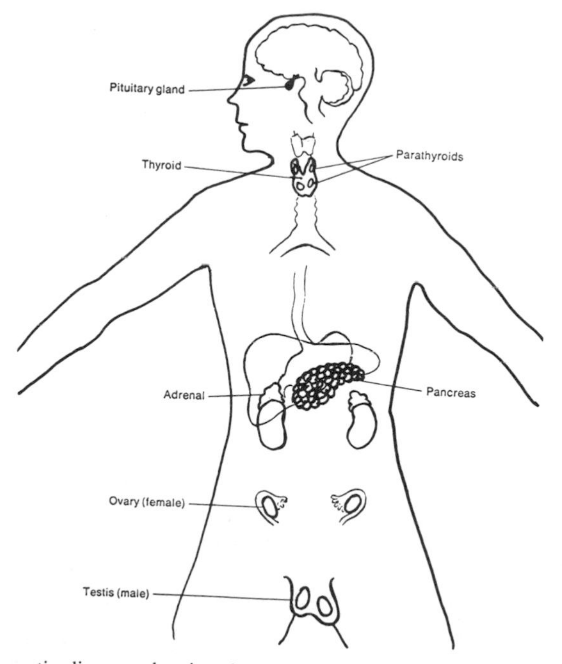

FIG. 2.6 A schematic diagram showing the principal endocrine glands. (Modified from Lewin, 1972, p. 12.)

TABLE 2.3
Functions of the Hypothalamus

Temperature regulation
Energy balance and exchange
Water balance and exchange
Sodium balance and exchange
Reproduction
Affective and emotional behavior
Control of autonomic nervous system
Neuroendocrine control

deal with a number of hormones related directly or indirectly to behaviors. For example, vasopressin or antidiuretic hormone (ADH) and angiotensin are involved in the control of water loss and water intake respectively and contribute to body-water homeostasis (see Chapter 5). ACTH is released from the anterior pituitary during stress to stimulate the production of adrenal cortical steroids (Chapter 9), and adrenaline and noradrenaline are released from the adrenal medulla in the cold (Chapter 3). Our main interest here is in the control of hormone release, particularly of hormones associated with behavior in which their release is integrated by the CNS in relation to events occurring in the external environment.

The hypothalamus is a functional link between the CNS and the endocrine system (Figure 2.7A). Two of the hormones—antidiuretic hormone and oxytocin—are produced by special neurosecretory cells in the hypothalamus and are transported along the hypothalamo-hypophyseal nerve tract to the posterior pituitary (Figure 2.7B). The release of ADH from the posterior pituitary is

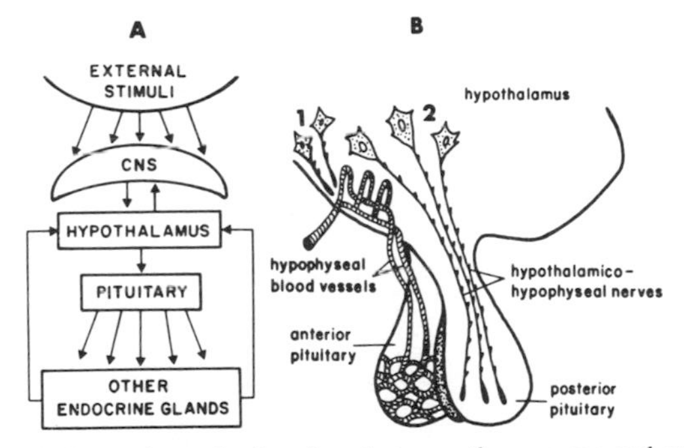

FIG. 2.7 A. The hypothalamus is at the interface between the nervous and endocrine systems. (From Lewin, 1972, p. 18.) B. Hormones released by the posterior pituitary are produced in the hypothalamus by neurosecretory neurons and transported along the fibers of these neurons to the posterior pituitary (2). Hormones released from the anterior pituitary are produced by the gland itself, and their release is controlled by signals from the hypothalamus. The signals are provided by releasing factors or releasing hormones transported by small blood vessels, the hypophyseal portal vessels, from the hypothalamus to the pituitary gland (1). (After Lewin, 1972, p. 18.)

initiated by the loss of too much water from the body, which increases the concentration of sodium in the blood and extracellular fluid. Osmoreceptors near the neurosecretory cells in the region of the supraoptic nucleus of the hypothalamus monitor the concentration of Na^+ and other electrolytes (osmolality) and trigger the release of ADH into the blood, which is carried to the kidney to facilitate the reabsorption of water and to reduce the production of urine (Figure 2.8). The release of ACTH and other hormones from the anterior pituitary is controlled by releasing hormones produced in the hypothalamus and is transported along the hypophyseal portal blood vessels (Figure 2.7B). There are one or more releasing hormones for each anterior pituitary hormone (Figure 2.9A).

The releasing hormones are under negative feedback (Figure 2.9B). For example, a high level of cortisol in the blood suppresses corticotrophic releasing factor (CRF) and thus the output of ACTH, whereas a low level of cortisol increases CRF and ACTH release. Similarly, thyrotropin releasing hormones are controlled by a negative-feedback mechanism and depend on the level of thyroid hormone in the blood. The releasing hormones, and therefore the anterior pituitary hormones, may also be influenced by external stimuli and other factors. For example, during emergency situations when more cortisol is needed to cope with the stress, CRF production is increased; neural structures, such as those in the limbic forebrain concerned with emotion and stress, influence the hypothalamic mechanisms that produce CRF. Another example is sexual receptivity in the rat, which occurs in a periodic fashion in relation to the levels of the gonadotropins controlled by the releasing hormones, FSHRF and LRF.

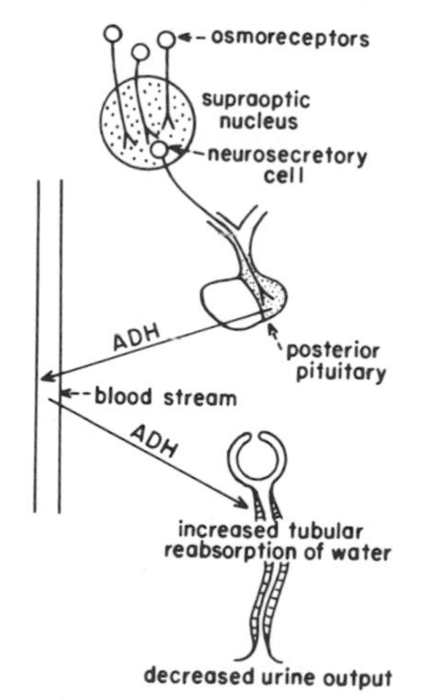

FIG. 2.8 Osmoreceptors in the anterior part of the hypothalamus respond to blood osmolality and regulate production and release of antidiuretic hormone (ADH), which descends via neurosecretory fibers into the posterior pituitary. Capillaries in the posterior pituitary gland pass the ADH into the blood stream and hence to the kidney, where it increases reabsorption of water by the kidney and reduces urine output. (From Mogenson, 1975b.)

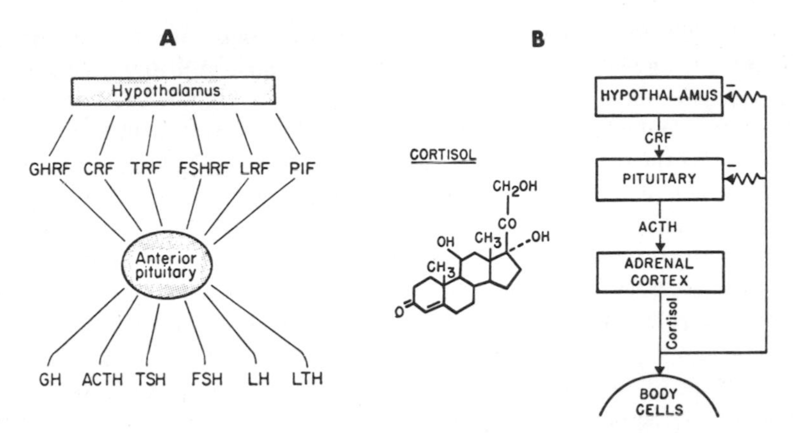

FIG. 2.9 A. The anterior pituitary gland, controlled by the hypothalamus and associated neural structures, produces several trophic hormones that act on other endocrine glands to control the hormones they produce. Releasing factors from the hypothalamus reach the pituitary by means of small blood vessels, and the trophic hormones reach target glands by means of the general circulation. The releasing factors, one for each trophic hormone, are called *thyrotrophin* releasing factor (TRF), *corticotrophin* releasing factor (CRF), *growth hormone* releasing factor (GH-RF), *luteinizing hormone* releasing factor (LH-RF), and *follicle stimulating hormone* releasing factor (FSH-RF). (From Van Sommers, 1972, p. 61.) B. The negative-feedback control of cortisol from the adrenal cortex. (Modified from Lewin, 1972, p. 27.)

The neural and endocrine systems work together closely as a ''partnership'' (Lewin, 1972, p. 22). First, as indicated in previous paragraphs, neural influences from limbic forebrain structures are funneled through the hypothalamus to control the pituitary gland. Second, hormones act back on the CNS for the negative-feedback control of the releasing hormones. Third, hormones act on the CNS to facilitate certain physiological and behavioral responses. For example, thyroid hormones and adrenal cortical hormones increase the excitability of the CNS, and testosterone and estrogen act on the hypothalamus to produce sexual arousal and sexual receptivity. We consider these examples in more detail in later chapters.

Integrative Systems and Adaptation to the External Environment

The evolution of specialized systems includes the skeletomotor system, complex neural mechanisms for its control, and the sensory systems. In higher mammals, especially in primates and man, the motor system, vision, hearing, and the other sensory systems become highly developed. Motor capabilities, such as locomotion or speech, and sensory capabilities, such as detecting faint sounds or recognizing complex visual patterns, are highly developed. Specialized motor and sensory systems enable the animal to make appropriate behavioral responses ("actions") to adapt to continuously changing circumstances of the external environment (see Figure 1.9). For example, an animal utilizes its sensory and motor systems to locate food or water, to construct a nest, or to escape from a predator. It is the sophisticated development of these systems, especially those for processing, storing, and utilizing sensory information (designated in later chapters as *perceptual* and *cognitive processes*), that are the basis of the uniquely human characteristics—the so-called higher mental functions.

Complex neural integrative mechanisms, involving a number of CNS structures, have evolved for the coordination of motor responses for adaptive and goal-directed behaviors (Figure 2.10). The investigation of the functional interactions of these neural structures for the control of movements is an important and vigorous field of research, although the precise mechanisms are still poorly understood (Henneman, 1974). This is not surprising if one considers the complex nature of the circuits and the interrelationships of key structures shown in Figure 2.11.

It is not our concern, and indeed beyond the scope of this book, to consider in detail the mechanisms for the execution of motor movements. The simplified

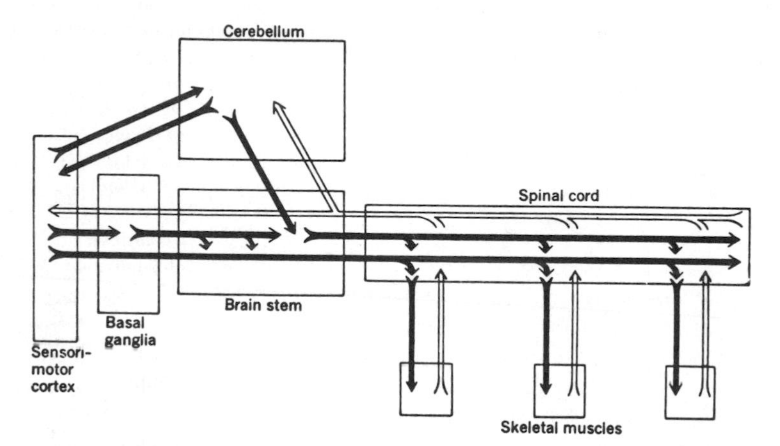

FIG. 2.10 This is a block diagram of the motor system from Henneman (1974). Sensory connections are shown by unshaded arrows. The basic control mechanism, shown at the bottom right, is a closed loop whereby nerve impulses are transmitted from a given muscle along sensory neurons to the motor neurons that control that muscle. This closed loop circuit can function autonomously for simple reflexes. There are also descending tracts to the lower motor neuron from the sensory–motor cortex (the corticospinal tract) and from the brainstem. The basal ganglia, which make up a large volume of the brain and pathology of which results in prominent motor disturbances, make an important but not well-understood contribution to motor function. They receive connections from the cortex and project to the brainstem. The cerebellum, "a large, highly organized outgrowth of the brainstem," receives many inputs (from muscle receptors, tendons, joints, skin, vision, hearing, and from the cerebral cortex), which are integrated, and signals sent to the sensory–motor cortex and brainstem (red nucleus, vestibular nuclei, reticular formation) "coordinat[e] the activity of motor circuits at all levels of the central nervous system." (From Henneman, E., Organization of the motor systems—a preview. In V.B. Mountcastle, Ed., *Medical Physiology,* 13th Ed., 1974; The C.V. Mosby Co., St. Louis.)

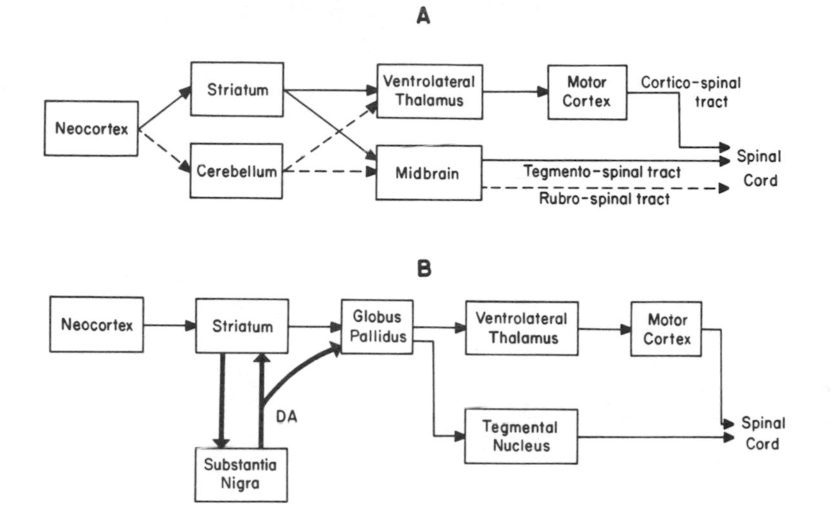

FIG. 2.11 Neural systems for the motor control of behavior. A. Parallel systems project from the neocortex to the lower motor neurons of the spinal cord via the striatum and cerebellum. The major contributions of the cerebellar route are to provide subroutines for the intended movement and to contribute to error detection through projections to the motor cortex. The striatum route contributes to the initiation of movement. B. The striatum, which samples the activity of several areas of the brain, is believed to be one of the key structures in the motor control of behavior (Figure 2.10). In addition to the topographical projections from sensory and association areas of the cerebral cortex, it receives inputs from the motor cortex and cerebellum, the latter two via the intralaminar nuclei of the thalamus. It also receives projections from limbic forebrain structures directly as well as indirectly via frontal association cortex. The striatum is in a good position, therefore, to contribute to the translation of the ''intention to respond'' into the appropriate ''command signals'' for initiation and control of movements. (After Mogenson and Phillips, 1976.)

models shown in Figures 2.10 and 2.11 are sufficient for the present discussion. Some aspects of Figure 2.11 are more relevant in later chapters, especially in Chapter 10 in which we deal with the activation of these complex circuits by the neural systems that subserve the initiating factors of behavior. However, it is premature at this stage to attempt to deal with the interface between the neural substrates of motivation and those for the motor control of behavior—we take up this interesting topic on p. 280.

There is also extensive literature on the visual, auditory, and other sensory systems and on the mechanisms for information processing at various levels of the CNS (Figure 2.12). These systems provide important information in guiding the behavioral responses of the animal in its environment. Sensory stimuli also initiate and enhance behavior either because of their intrinsic sensory characteristics, as with certain gustatory and olfactory stimuli that may facilitate ingestive behaviors, or because they have acquired biological significance through previous experience and thus have rewarding or punishing properties. The chemical senses, especially olfaction, make an important contribution to the behavior of

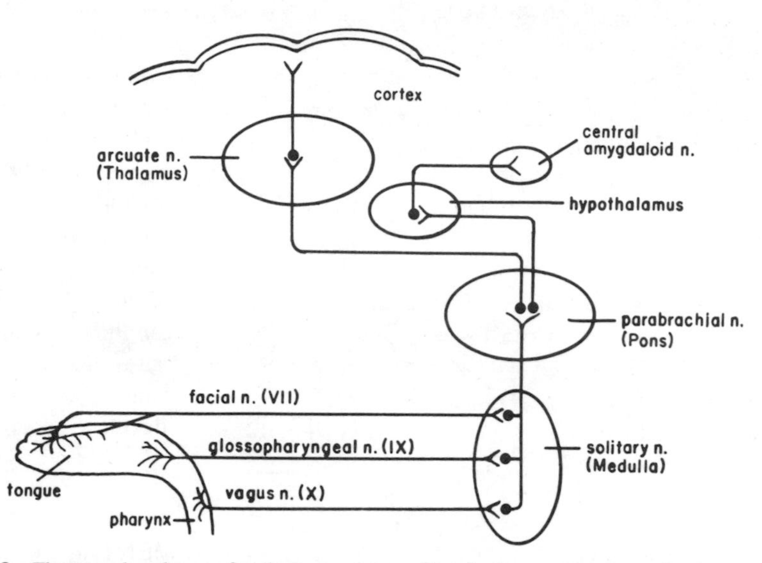

FIG. 2.12 The neural pathways for the taste system. Signals from receptors on the tongue are transmitted along the seventh, ninth, and tenth cranial nerves to the solitary nucleus of the medulla in the lower brainstem. Pathways extend rostrally through the pons to the thalamus and cerebral cortex and to the hypothalamus and amygdala. Neural projections to the hypothalamus, amygdala, and other forebrain structures are likely the route by which taste signals influence feeding, drinking, and brain-stimulation reward.

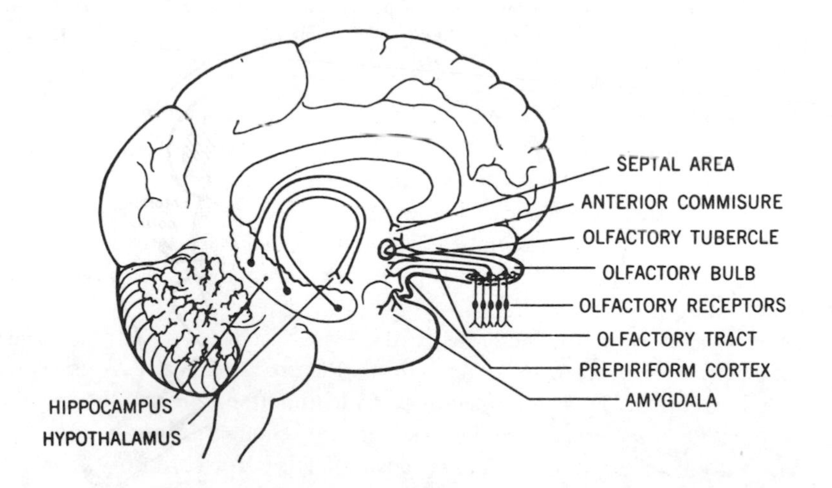

FIG. 2.13 A saggital section of the human brain showing a diagrammatic representation of the olfactory system. Direct connections reach the olfactory tubercle and amygdala and indirect connections reach several other limbic structures (see Figure 2.14). (From Rubenstein, H.S., *The Study of the Brain,* 1953; Grune & Stratton, Inc., Publishers, New York. Reprinted by permission.)

many animals, and a large amount of the brain is concerned with olfactory function (Figure 2.13).

The structures of the brain that receive olfactory inputs—comprising what was classically designated the *rhinencephalon* but in recent years called the *limbic system*—are involved in a number of important functions in addition to olfaction (Figure 2.14). They contribute to feeding, sexual, aggressive, and other behaviors and to internal physiological regulations controlled by the endocrine and autonomic nervous systems. For man and primates, in which olfaction is much less important than in rodents and other lower animals, limbic forebrain structures have as their major function a vital role in motivated and emotional behaviors. These structures have strong neural connections with the hypothalamus and represent a higher level of neural integration, which ensures that physiological regulations under the control of the endocrine and autonomic nervous systems are coordinated with behavioral responses. This concept is developed more fully in Chapter 9.

RESEARCH STRATEGIES AND EXPERIMENTAL TECHNIQUES FOR THE STUDY OF THE NEURAL SUBSTRATES OF MOTIVATED BEHAVIORS

The study of the neural substrates of drinking, thermoregulatory, aggressive, and the other behaviors dealt with in this book is not the prerogative of a single scientific discipline. Contributions are being made by scientists in various fields

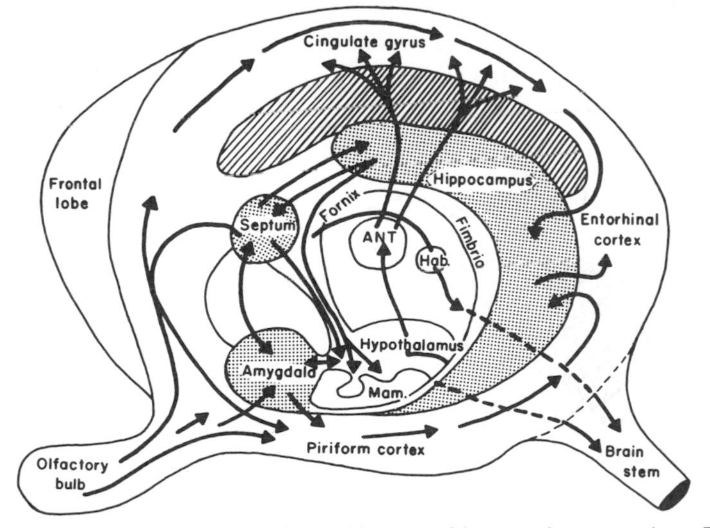

FIG. 2.14 A diagram of the limbic system with some of its many interconnections. The main structures (amygdala, hippocampus, septum) are connected to the hypothalamus and contribute to many of the functions in which the hypothalamus is involved. In general the limbic system is concerned with behavior and the preservation of the species. For additional details see the text.

including neuroanatomy, neurophysiology, neuropharmacology, and physiological psychology. Investigators currently concerned with brain mechanisms and behavior are building on the contributions of previous workers. As indicated earlier in this chapter and in Chapter 1, such contributions have come from biology, regulatory physiology, endocrinology, ethology, and experimental psychology, as well as from neuroanatomy and neurophysiology.

The techniques used by investigators interested in the neural substrates of motivated behaviors are similar to those used by other neuroscientists interested in functional questions. They are used in relation to three main research strategies: (1) to block or disrupt the functioning of some part of the CNS (e.g., by ablation, lesions, cooling, or chemical block) to observe what happens to the animal's behavior; (2) to excite some part of the CNS with electrical stimulation or by applying an appropriate chemical compound or drug to observe behavioral changes; and (3) to record electrical changes in the brain in relation to particular behaviors. For neural substrates of motivated behavior, the distinctive feature of any of these research strategies is that experiments are typically performed in unanesthetized animals observed in situations in which the behavior being investigated can occur.

It should be noted that lesion, stimulation, and electrophysiological recording techniques were not developed specifically for the study of the neural substrates of motivated behaviors. In fact, the application of these techniques in neurophysiology and other neuroscience disciplines for many years was primarily in the anesthetized animal. Although the initial observations of the effects of lesions and electrical stimulation of the brains of unanesthetized animals were made a century ago, interest was in sensory and motor functions and not complex behaviors of the sort considered in this book. For many years experiments in chronic, unanesthetized animals were mainly limited to the more superficial structures of the brain, and lesions and stimulation typically did not influence motivated behaviors in any striking way (Doty, 1969). It was not until the stereotaxic technique was introduced, making possible discrete lesioning and stimulation of the hypothalamus and other deep structures of the brain, that the serious study of the neural substrates of motivated behaviors began (Figure 2.15). This occurred about 1940, and during the next 10 or 15 years, a number of dramatic observations were made that implicated the hypothalamus and associated limbic structures in motivated behaviors. Examples of these pioneering experiments selected for presentation in later sections of this chapter are concerned primarily with feeding behavior.

Electrophysiological recording techniques, introduced into neurophysiology in the 1930s, were not used to any extent in unanesthetized, behaving animals until quite recently. The stereotaxic procedure provided the means for accurately positioning the recording electrodes, and developments in electronics and construction of microelectrodes made it possible to record from single neurons in the hypothalamus and limbic system. However, microelectrode recording in unanes-

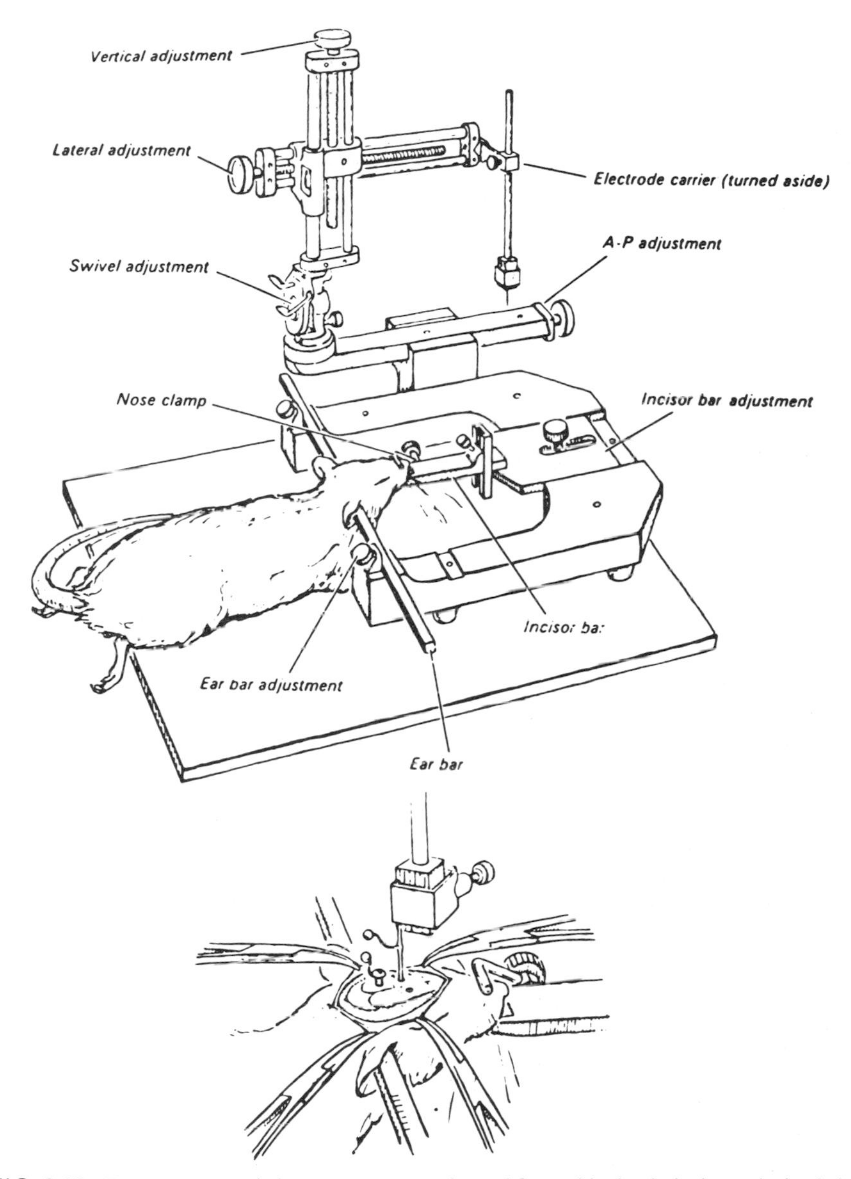

FIG. 2.15 *Top*—a stereotaxic instrument commonly used for positioning lesioning and stimulating electrodes in the rat. The head of the anesthetized animal is held firmly in position by means of the ear bars, incisor bars, and nose clamp. The electrode carrier can be moved in any one of three directions—anterior—posterior, medial—lateral, and dorsal—ventral—to position the electrode according to coordinates established by use of a stereotaxic atlas. *Bottom*—Stereotaxic surgery for implanting an electrode. The procedure involves exposing the skull, drilling holes in the calculated position, inserting several screws in the skull to which to anchor the dental acrylic, inserting the electrode to the desired position, and then building up a mound of dental acrylic around the electrode to secure it in place. (From *Experimental Neuropsychology*, by B. L. Hart. W. H. Freeman & Co. ©1969.)

thetized, behaving animals is not easy, and successful experiments of this sort became possible only during the last few years.

Disruption or functional block of CNS structure or pathway by surgical ablation. A classical technique to study the functions of the brain has been to remove by surgical ablation portions of the brain to observe the loss or disruption of function. The ablation technique used so effectively in pioneering studies by Flourens (1842) provided the first experimental evidence for motor and sensory areas of the cerebral cortex. However, more than 75 years later the first observations were made that implicated parts of the cerebrum in motivated and emotional behaviors: Klüver and Bucy (1937) observed hyperphagia, hypersexuality, and reduced fearfulness in monkeys following bilateral surgical ablation of the temporal lobes. About the same time, Jacobsen (1936) reported an attenuation of emotional reactions to ''frustrating'' test situations in monkeys with frontal lobes ablated. These classical studies implicated the temporal and frontal lobes in motivated and emotional behaviors.

Electrolytic lesions. Surgical ablation cannot be used to investigate the functions of deep brain structures because of extensive damage to the overlying neural tissue. The introduction of the stereotaxic procedure in the late 1930s for accurately positioning electrodes made it possible to produce discrete lesions in the hypothalamus and other deep structures of the brain. The classical experiments with this technique demonstrated that the regulation of body temperature was disrupted by lesions of the hypothalamus (for details, see Chapter 3 and Figure 3.11). About the same time, it was observed that lesions of the ventromedial nucleus of the hypothalamus resulted in a substantial increase in food intake and a disruption of body-energy homeostasis. This phenomenon of hypothalamic hyperphagia and obesity, shown in Figures 1.7D and 2.16A, has been reproduced by many investigators.

The technique of making electrolytic lesions with the aid of the stereotaxic technique has been used extensively and has implicated the hypothalamus and limbic structures in a number of the behaviors dealt with in this book.

Brain transections and knife cuts. Another classical approach in the investigation of brain function has been to transect the CNS at various levels in order to observe what happens to particular functions (the ''salomi'' technique). It was with this technique that the hypothalamus was implicated in temperature regulation and emotional behaviors. When the brain was transected rostral to the hypothalamus, temperature regulation was normal, but a transection caudal to the hypothalamus disrupted temperature regulation (for details, see Chapter 3). Similarly, a transection rostral to the hypothalamus did not disrupt aggressive behavior; in fact, it was exaggerated, but transections caudal to the hypothalamus did (for details, see Chapter 9).

In recent years more limited transections or knife cuts have been made to disrupt particular neural pathways. Sectioning neural pathways near the ven-

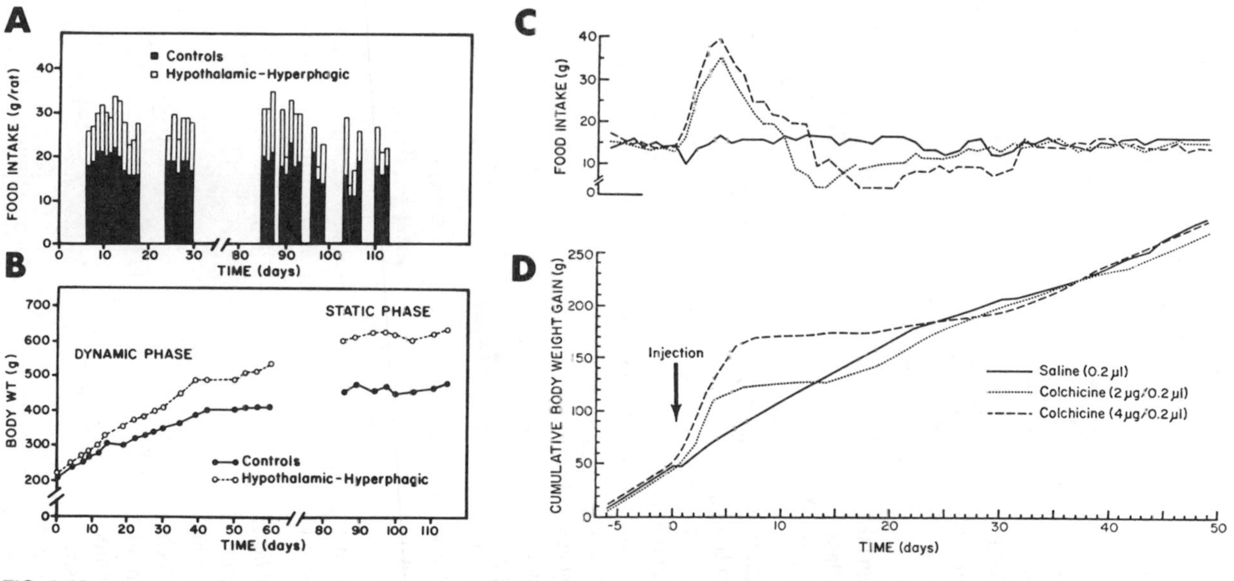

FIG. 2.16 Records of food intake (A) and body weight (B) of rats made hyperphagic with electrolytic lesions of the ventromedial hypothalamic nucleus (VMH) and their unoperated controls (after May and Beaton, 1966). (C) and (D) show similar records of rats made temporarily hyperphagic by an injection of colchicine, a drug which produces a reversible neural block, into the VMH. (From Avrith and Mogenson, unpublished observations.)

tromedial hypothalamus (VMH) has been reported to result in hyperphagia and obesity similar to but less pronounced than that observed in the classical experiments involving VMH lesions (Albert & Storlien, 1969; Gold, 1970). As indicated elsewhere experiments of this sort have implicated noradrenergic pathways in feeding behavior.

Reversible neural block. The transmission of action potentials along nerve fibers can be blocked using a cooling probe. The technique has the important advantage that the disruption of a neural pathway or structure is reversible. Although this technique has been very fruitful in the study of neural circuits involved in the control of limb movements (Brooks, 1975), it has not been used to investigate the neural substrates of motivated behaviors.

The administration of certain chemical compounds to a neural structure or pathway has been used to produce a reversible block of neural pathways involved in feeding behavior. This was first done by administering procaine, a local anesthetic agent, to the VMH (Epstein, 1960). An increase in food intake was observed, but the effect was small because the neural block lasts only a short time. Another procedure is to administer colchicine, which disrupts axoplasmic transport in nerve fibers by combining with microtubule proteins thereby blocking the transmission of nerve impulses at synapses. Because the combination of colchicine with microtubule proteins is not permanent, lasting only a few days, the neural block is reversible. Using this procedure a reversible hyperphagia and increased body-weight gain has been observed when colchicine was administered bilaterally to the VMH (Figure 2.16B).

Excitation of CNS Structure or Pathway

Electrical stimulation. Another classical technique to investigate the functions of the brain has been to administer electrical stimulation to a neural structure in order to observe physiological and behavioral responses. Using this procedure Fritsch and Hitzig (1870) demonstrated the motor cortex (Figure 2.17). More recently Hess (1954), the first investigator to stimulate deep structures of the brain, observed feeding, aggressive, and other behaviors when electrical stimulation was administered to the hypothalamus and related structures in chronically prepared cats. Subsequent investigators used the stereotaxic procedure to position stimulating electrodes accurately in these deep structures of the brain and confirmed and extended these pioneering observations. Electrical stimulation of the hypothalamus and forebrain structures was reported to elicit feeding, brain stimulation reward, drinking, copulatory, and attack behavior (for details, see Chapters 4, 5, 8, 9).

An example of feeding elicited by electrical stimulation of the lateral hypothalamus is shown in Figure 2.18. When the hypothalamic stimulation is administered daily, it has been reported that rats become obese (Steffens, 1975).

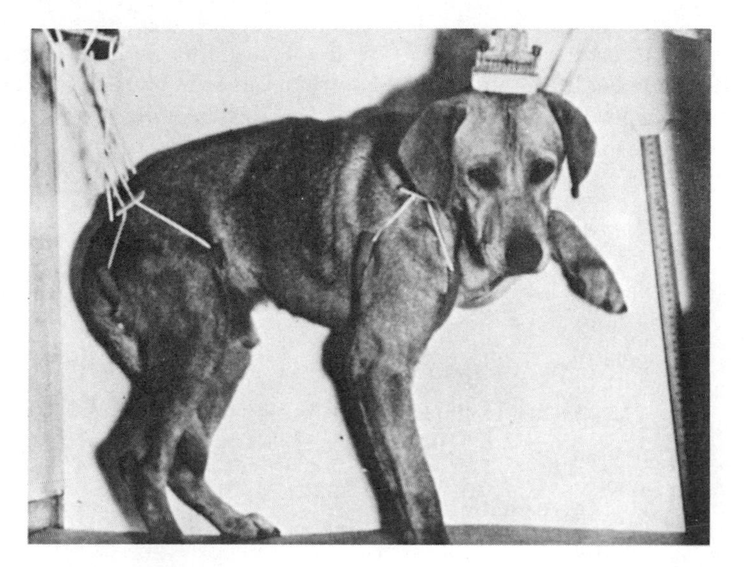

FIG. 2.17 Stimulation of the motor cortex of the dog on the right side of the brain elicits leg flexion of the left front leg. ''Command'' signals are transmitted along the pyramidal tract, which crosses over to the other side in the lower brainstem. In addition to the pyramidal system originating in the motor cortex, there are a number of other structures that contribute to motor function—caudate nucleus, putamen, globus pallidus (designated basal ganglia), cerebellum, thalamic nuclei, red nucleus (see Figures 2.10 and 2.11). The pyramidal system is important for the control of finer movements–of the fingers, for example. (From Doty, 1961. Courtesy University of Texas Press.)

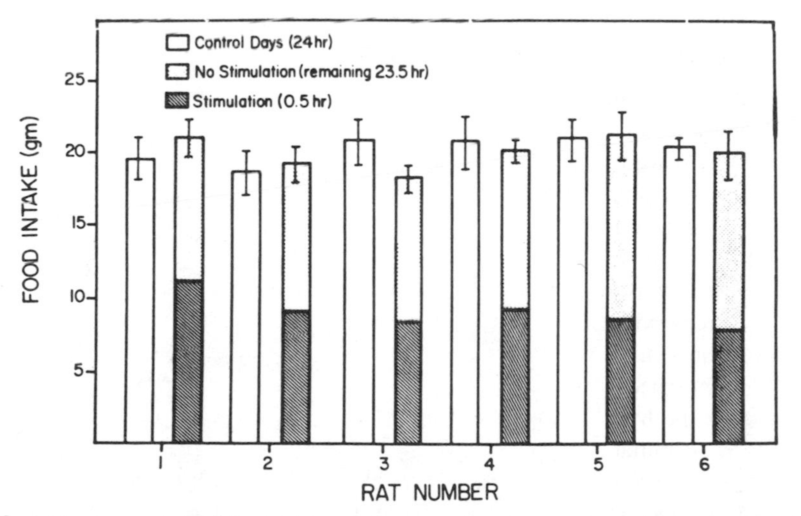

FIG. 2.18 Comparison of food intake of six rats with permanent electrodes implanted in lateral hypothalamus on control days without stimulation (12 sessions) versus days with stimulation (18 sessions). Food was available ad libitum during 30-minute stimulation periods and during the remainder of the day. Vertical lines represent ±SEM. (From Steffens, Mogenson, & Stevenson, 1972.)

Chemicals. A widely used research strategy has been to infuse drugs and other chemical compounds directly into the brain usually by means of chronic intracerebral cannulae (Figure 2.19). In the first experiments of this type, hypertonic NaCl was infused into the brains of goats to induce thirst, apparently by activating central osmoreceptors (Andersson, 1953). Subsequently, this procedure for "chemical stimulation" of the brain has been used for the central administration of noradrenaline and acetylcholine (or carbachol) to initiate feeding and drinking (Grossman, 1962; Fisher, 1964) and of noradrenaline, acetylcholine, and serotonin to induce thermoregulatory responses (Myers, 1970). A few years ago there was a good deal of enthusiasm about the possibility that the central administration of drugs that were neurotransmitters or that acted at synapses could be used to elucidate neural systems that subserve feeding, drinking, and other behaviors (Miller, 1965). Unfortunately, it is frequently difficult to exclude the possibility that behaviors result from pharmacological effects of the drugs rather than to the physiological activation or blocking of a neural system.

Hormones. Hormones have also been administered to specific brain sites of unanesthetized, unrestrained animals. The first study was by Fisher (1956) who observed copulatory behavior in male rats when testosterone was administered to the anterior hypothalamus. More recently it has been shown that drinking is elicited by administering Angiotensin II, a kidney hormone, to the anterior hypothalamus-preoptic region and other brain sites (Epstein et al., 1970). For further details of such studies, see Chapters 4 and 5.

Recording Electrical Changes from the Brain

Electrical changes may be recorded from sensory and motor pathways and from various other sites in the brain; action potentials are recorded from nerve fibers, and slow potentials from dendrites and cell bodies (Figure 2.20). Recordings

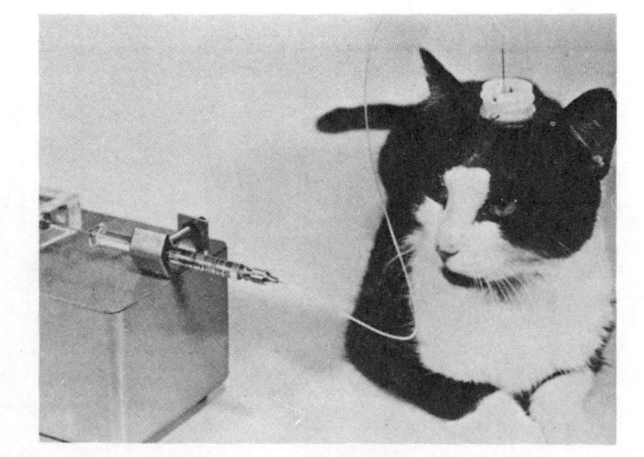

FIG. 2.19 Cat with a chronic intracerebral cannula for the microinjection of chemical compounds. (From Myers, 1971.)

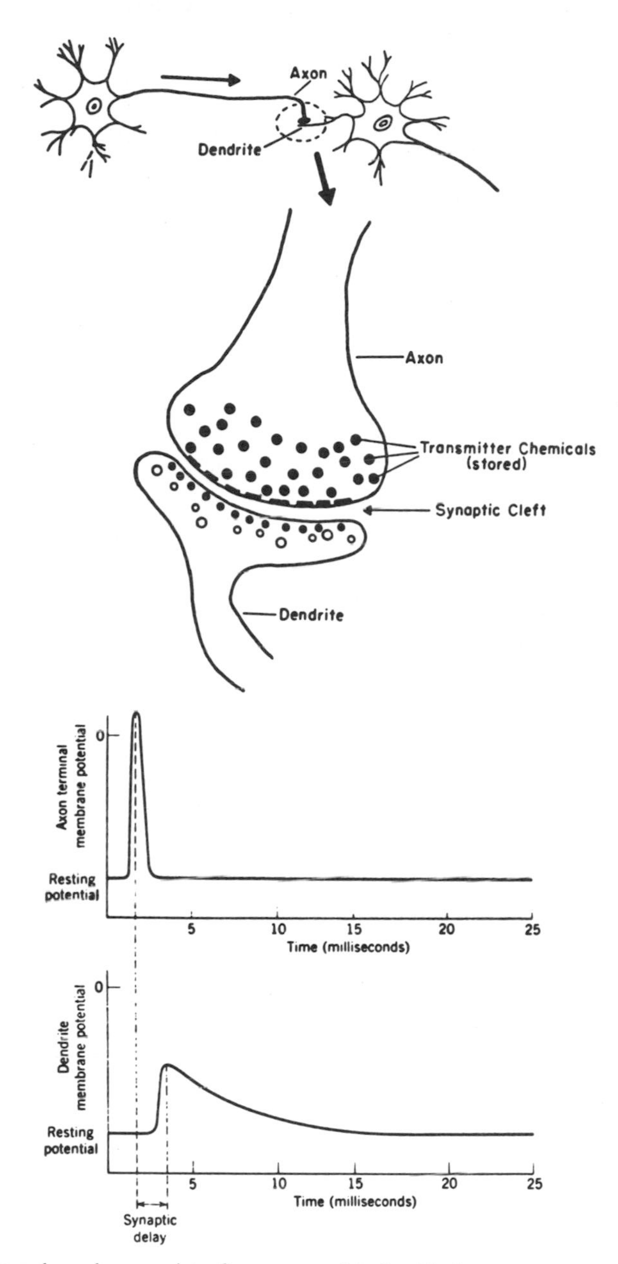

FIG. 2.20 Neuron shown here consists of a soma or cell body with short processes or dendrites and a longer process called the axon which makes contact with a dendrite of a second neuron. The region of contact of the axon with the dendrite or cell body is called the synapse. When action potentials travel along the axon and reach the axon terminal a chemical transmitter is released which depolarizes the postsynaptic or dendritic membrane to initiate an action potential in the neighboring neuron. An example of a chemical transmitter is acetylcholine at the neuro-muscular junction. Action or spike potentials occur in axons or fibers whereas in dendrites the potentials are smaller and of longer duration (see bottom of figures). Action potentials are propagated at speeds up to 100 m/sec depending on the diameter of the fibers whereas dendritic potentials are local and decremental. (From Stevens, 1966.)

were first made from the scalp of human subjects, and the electroencephalogram (EEG), as it was called, was found to be correlated with the level of consciousness; the "brain waves" were of low amplitude and high frequency during wakefulness and of high amplitude and low frequency during sleep (see Figure 7.2). The basis of the EEG is not very well understood, but it is thought to reflect the summation of dendritic potentials rather than action potentials.

When electrodes with small tips (1—4 microns) are used, designated *microelectrodes,* it is possible to record the electrical activity of single neurons in the brain. The microelectrodes may be made of very fine glass micropipettes filled with sodium chloride solution or of tungsten or stainless steel etched to a fine tip with the shaft insulated except at the tip. It is possible to record with the electrode near a neuron (extracellular recording) or in some cases with the electrode penetrating the neuron (intracellular recording). The microelectrode recording technique has been used by many investigators to study sensory systems and the neurophysiological basis of information processing.

Microelectrode recording techniques have also been used to record from single neurons in the hypothalamus and associated structures. Results of experiments by Oomura and co-workers are shown in Figure 2.21. The discharge of neurons in the lateral hypothalamus and ventromedial hypothalamus were observed to be reciprocally related in support of the dual-mechanism model (Figure 1.8).

An interesting and productive application of microelectrode recording has been to investigate the neural control of body temperature. Recordings have been made from neurons in the anterior hypothalamus-preoptic region that respond to changes in ambient temperature and/or to warming or cooling the preoptic region with a thermode (Figure 2.22). These are designated *temperature-sensitive neurons.* Some are central temperature receptors, and others integrate temperature signals from the CNS and the skin and are part of the neural mechanism for the control of thermoregulatory responses. For further details about this mechanism, see Chapter 3, p. 80.

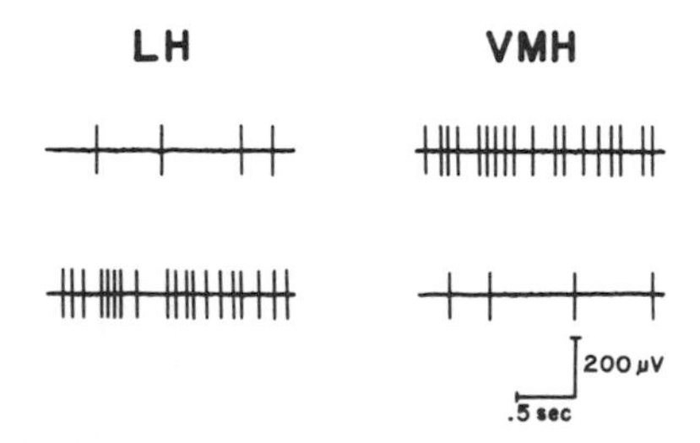

FIG. 2.21 Action or spike potentials (see Fig. 2.20) are recorded from single neurons using glass tungsten or stainless steel microelectrodes with small tips. Oomura and co-workers (1967) have demonstrated that when neurons in the lateral hypothalamus (LH) of the rat are discharging slowly, neurons in the ventromedial nucleus are discharging rapidly (top records) and vice versa (bottom records). As indicated in more detail in Chapter 4, this has been considered support for the dual-mechanism model for the control of food intake (see Figure 1.8a).

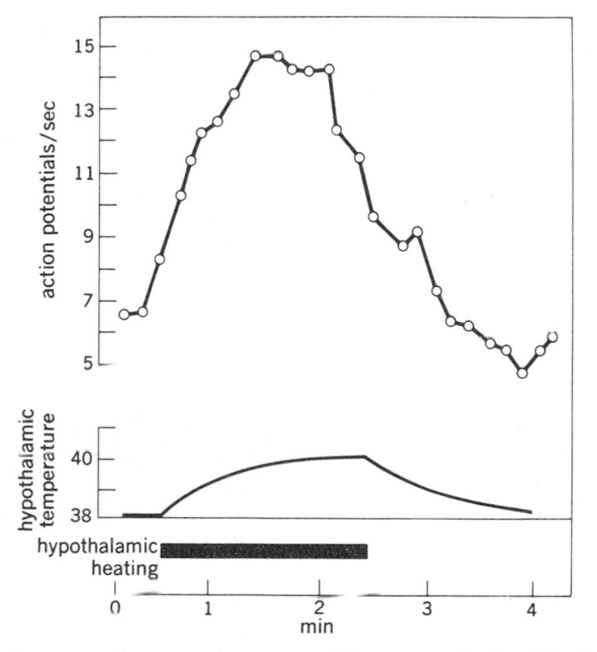

FIG. 2.22 Discharge rates from a temperature sensitive neuron after local heating of a discrete hypothalamic area. (After Nakayama, et al., 1963.)

Electrophysiological recording of the activity of a single neuron can be combined with the administration of chemical compounds, often putative neurotransmitters, to the same neuron (Figure 2.23). This technique is known as *microiontophoresis*. Multibarrel glass micropipettes are used, one filled with sodium chloride solution (2–3 Molar) for recording from the neuron and the others filled with solutions of the compounds whose effects on the neuron are to be investigated.

The chemical compound is ejected from the micropipettes by iontophoresis. Using this technique, neurons that respond to the application of glucose or hypertonic saline have been identified in the hypothalamus (Oomura et al., 1969). The possibility that these are glucoreceptors and osmoreceptors for the initiation of feeding and drinking is considered in Chapters 4 and 5.

For a number of years, experiments were performed only in anesthetized animals, and the results could not be easily generalized to the freely behaving animal. In recent years, there has been increasing use of electrophysiological recording of the activity of single neurons in the awake animal trained to make certain behavioral responses. The strategy of combining behavioral control and electrophysiological recording in the unrestrained or partially restrained animal has been used with considerable success to investigate the central mechanisms

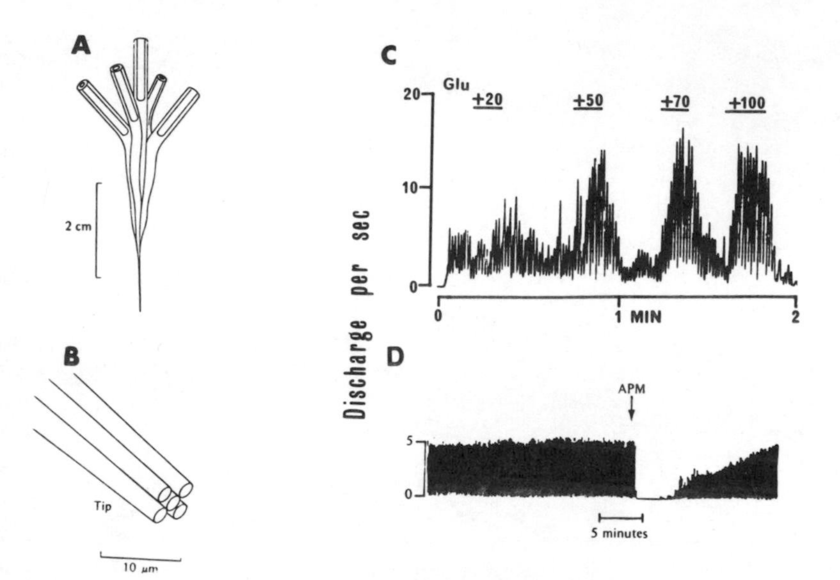

FIG. 2.23 A. A multibarreled microelectrode assembly for the application of micro amounts of chemicals to the brain in electrophysiological recording experiments. Application of drugs is by iontophoretic ejection of charged drug molecules from the tip (see B) of the finely drawn glass micropipettes by electrical current application. The rate of ejection of the drug can be controlled by regulating the current flowing through the electrode tip. One of the micropipettes, filled with 3 molar NaCl, is used to record action potentials from a single neuron (after Iversen & Iversen, 1975. Copyright 1975, Oxford University Press.). The recording in C (top) shows the response of a neuron in the hypothalamus to the iontophoretic application of increasing amounts of glucose (after Oomura et al., 1975). This is a glucosensitive neuron. The lower recording shows the response of a dopaminergic neuron in the pars compacta of the substantia nigra to the iontophoretic application of apomorphine (APM).

subserving limb movements (Evarts, 1975) and to investigate the neural substrates of feeding and other behaviors (Olds, 1970; Ranck, 1975; Rolls, 1976). As indicated by Mountcastle (1976), "it is one of the most productive methods of experimentation now available in brain physiology and . . . results obtained with it will revolutionize our ideas about many central neural mechanisms previously studied only in anesthetized animals [p. 2]."

Some Newer Techniques Used in the Investigation of the Neural Substrates of Motivated Behaviors

In recent years the classical techniques for studying the neural substrates of behavior—lesions, stimulation, recording electrical activity—have been supplemented by other techniques that have been introduced for the investigation of brain function. Of particular importance are developments in neuropharmacology, histochemistry, and neurochemistry.

Drugs that act on the CNS have been used for many years to activate or block neural systems. However, this approach has become more sophisticated and powerful following the visualizing and mapping of monoamine-containing neurons using the technique of histofluorescence (Figure 2.24). The new chemical anatomy of the brain, which has emerged from studies utilizing histofluorescence and neurochemical techniques, and the elucidation of the biosynthesis and synaptic action of neurotransmitters, has made it possible to relate drug action to known anatomical systems. Research utilizing neuropharmacological techniques and interpreted in relation to these new advances in chemical neuroanatomy has had a major impact in recent years on the investigation of the neural substrates of motivated behaviors and has led to a reassessment of older views and theoretical models (e.g., Mogenson & Phillips, 1976).

There are many drugs that influence the functioning of the CNS and thereby the behavior of animals and man. Some drugs block the transmission of nerve impulses at synaptic junctions; other drugs facilitate transmission at certain synapses. For example, haloperidol is a dopaminergic blocker or antagonist that blocks transmission at dopaminergic synapses by competing with dopamine for receptors on the postsynaptic membrane. On the other hand, apomorphine acts on dopamine receptors to enhance synaptic transmission and is called a *dopamine receptor agonist*. Drugs may also be used to release the neurotransmitter from axon terminal, to prevent the re-uptake of the released neurotransmitter by the presynaptic membrane, or to block the biosynthesis of the neurotransmitter. These properties enable neuropharmacologists and other investigators to use drugs as experimental tools to investigate brain function and behavior.

For purposes of illustration, the effects of certain drugs on one of the dopaminergic systems of the brain—the nigrostriatal pathway, which projects from the substantia nigra to the neostriatum—is now considered. The nigrostriatal pathway is part of the extrapyramidal motor system, and if it is lesioned unilaterally, the animal assumes an asymmetrical posture: The limbs ipsilateral to the lesion are kept close to the body, and the contralateral limbs are extended; the head and tail deviate toward the side of the lesion (Ungerstedt, 1971). When amphetamine is administered to the unilaterally lesioned animal by intraperitoneal injection, it circles toward the side of the lesion (Figure 2.25). This rotational behavior has been attributed to the release by amphetamine of dopamine from the axon terminals of the unlesioned nigrostriatal pathway (Figure 2.25). The administration of haloperidol, a dopamine receptor antagonist or blocker, prevents the occurrence of the rotational behavior following the injection of amphetamine. If apomorphine, the dopamine receptor agonist, is administered, the animal rotates toward the nonlesioned side (Figure 2.25). This has been attributed to the "supersensitivity" of the dopamine receptors on the postsynaptic membrane to apomorphine—the "supersensitivity" being the consequence of damage to the dopaminergic fibers and subsequent disuse of dopaminergic synapses.

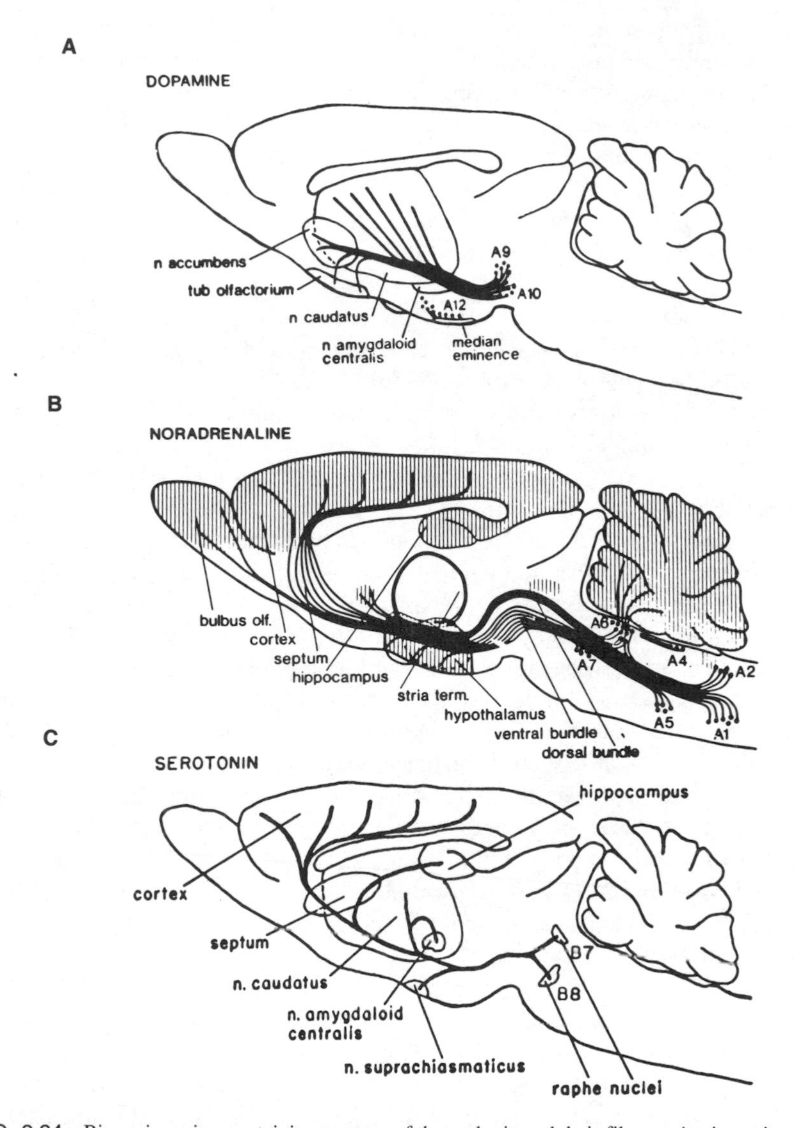

FIG. 2.24 Biogenic amine-containing neurons of the rat brain and their fiber projections visualized using the techniques of histofluorescence. A. Dopamine neurons in the substantia nigra (A_9) project to the caudate nucleus and putamen (neostriatum). Dopamine neurons in the ventral tegmental area (A_{10}) project to limbic forebrain structures (nucleus accumbens, olfactory tubercle) and to the prefrontal cortex. B. Noradrenergic neurons in the pons (locus coeruleus, A_6) project to the hippocampus and cerebral cortex. Noradrenergic neurons in the pons and medulla (A_1, A_2, A_5, A_7) project to the hypothalamus, septum, and basal forebrain. C. Serotonergic neurons project from the raphe nuclei of the midbrain to the hypothalamus, septum, amygdala, caudate nucleus, hippocampus, and cerebral cortex. (Modified from Ungerstedt, 1971.)

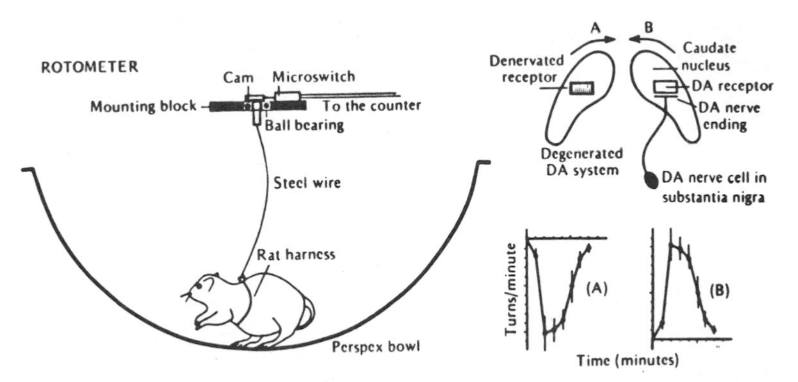

FIG. 2.25 *Left*—A schematic drawing of the rotometer. The movements of the rat are transferred by the steel wire to the microswitch arrangement. *Top right*—The principal outline of the experimental situation shown in a horizontal projection of the nigrostriatal DA system. When stimulation of the denervated receptor dominates, the animal rotates in direction B. *Bottom right*—The rotational behavior is presented as turns per minute over time. The curves are given negative *y*-values when stimulation of the denervated receptor dominates and positive *y*-values when stimulation of the innervated receptor dominates. Each point represents the mean of several animals. (After Ungerstedt, 1974 as modified by Iversen & Iversen, 1975, p. 281.)

A further example of the use of neuropharmacological techniques is in the study of the neural mechanisms that control food intake; in recent years such studies have pointed to the importance of monoamine systems in feeding behavior. Some of the pertinent evidence is as follows. The administration of 6-hydroxydopamine, a neurotoxin that selectively damages dopaminergic and noradrenergic neurons, into the cerebral ventricles or ventral midbrain produces aphagia (Ungerstedt, 1971). The administration of noradrenalin directly to the hypothalamus, as indicated earlier, induces feeding (Grossman, 1962). The reduction of food intake produced by the administration of amphetamine has been reported not to occur when one of the noradrenergic pathways of the brain (the ventral noradrenergic pathway) is damaged (Ahlskog & Hoebel, 1973), implicating this noradrenergic pathway in the anorectic effects of amphetamine. Similarly, fenfluramine does not exert its anorectic effect when serotonergic neurons are damaged or blocked, suggesting that fenfluramine acts via serotonin systems (Blundell, 1975).

Finally, brief mention should be made of another approach—the measurement or assay of levels of noradrenaline, dopamine, or other neurotransmitters in the brain in relation to behavior. For example, it has been reported that there is an association between noradrenaline levels and depression in humans and, in experimental animals, an increase in the synthesis and turnover of serotonin following sleep deprivation (for details, see Chapters 7 and 9). By using the push–pull cannulae technique (Figure 2.26), it is possible to measure the release of neurotransmitters in localized regions of the brain. This procedure has been used to

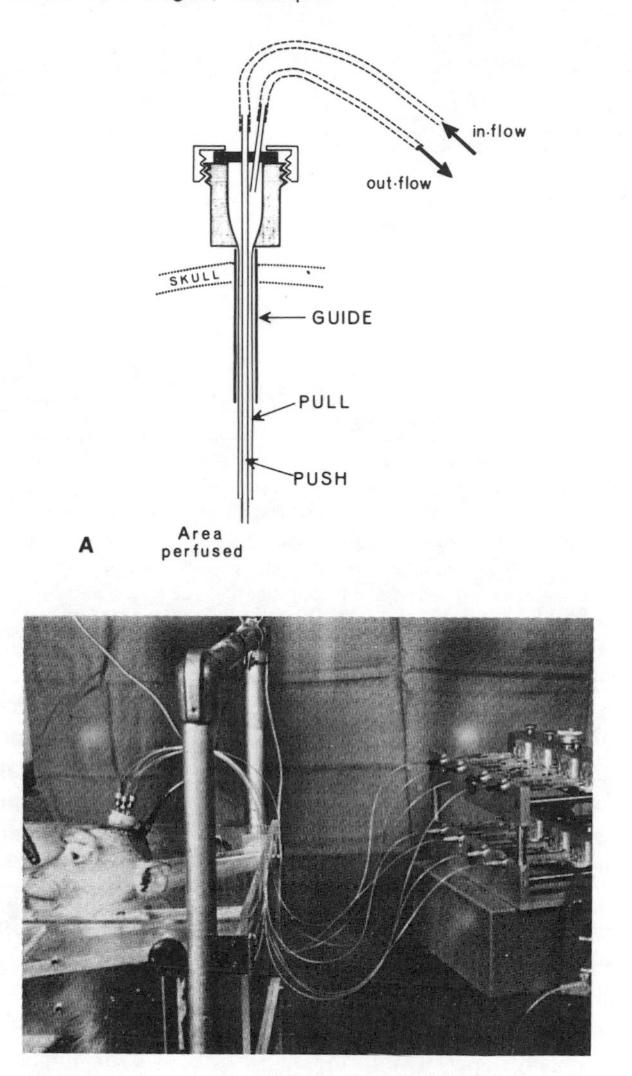

FIG. 2.26 A. A push–pull cannulae for collecting chemicals released in the brain. A perfusion fluid is infused through the inner cannula (push) and withdrawn through the outer cannula (pull). The fluid can then be analyzed using a biochemical assay for norepinephrine, acetylcholine, or some other neurotransmitter released into the perfused area. B. A monkey with a push–pull cannulae in an experiment in which the release of norepinephrine in the hypothalamus is being investigated. (From Myers, 1974a.)

demonstrate the release of norepinephrine in the amygdala during self-stimulation of the lateral hypothalamus (Stein & Wise, 1969) and the release of norepinephrine from the hypothalamus during feeding (Martin & Myers, 1975). The endogenous stores of norepinephrine are initially labeled by microinjecting (^{14}C) norepinephrine into the cerebral ventricles or into the site of the push–pull cannula. The release of labeled norepinephrine, which has been taken up by the axon terminals, is determined by perfusing artificial cerebrospinal fluid into the push–pull cannulae, collecting the perfusate and using a scintillation counter to measure the amount of (^{14}C) norepinephrine released during the experimental procedure. In such experiments, as in those in which neuropharmacological techniques are used, neurochemical techniques can be used with greatest advantage in the study of neural systems whose chemical anatomy is known.

SUMMARY

The study of brain function and behavior is an interdisciplinary enterprise. Progress in studying the neural substrates of feeding, drinking, sexual, thermoregulatory, and aggressive behaviors have relied heavily on advances in the neurosciences as well as in regulatory physiology, endocrinology, and other fields of biology and of psychology. The central nervous system controls the behaviors or ''actions'' of the animal by integrating various sensory inputs and sending ''command signals'' to the motor system. It also coordinates, in conjunction with the endocrine system, the other specialized systems of the body that have a vital role in maintaining the internal environment (''housekeeping chores'').

Earlier studies of the brain mechanisms that subserve thermoregulatory feeding, and the other behaviors dealt with in this book—as with the study of other brain functions—typically began with the use of lesioning and stimulation techniques. These techniques implicated the hypothalamus and later limbic forebrain structures in motivated and emotional behaviors (e.g., Figures 1.7 and 1.8). In recent years other techniques have been utilized in the study of the neural mechanisms of behavior, and the experimental results have enriched the field. Electrophysiological recording techniques are particularly promising when used in chronically prepared, freely moving animals. Neuropharmacological and neurochemical techniques are also providing important new findings and are particularly valuable in studying the contribution to behavior of neural systems whose chemical neuroanatomy has been elucidated using histofluorescence and other histochemical techniques.

For further readings, the students should consult the following:

Curtis, B. A., Jacobson, S., & Marcus, E. M. *An introduction to the neurosciences*. Philadelphia: Saunders, 1972.

Guyton, A. C. *Textbook of medical physiology*. Philadelphia: Saunders, 1976.

Milner, P. M. *Physiological psychology*. New York: Holt, Rinehart & Winston, 1970, chaps. 3 and 4.

Myers, R. D. (Ed.). *Methods in psychobiology*. New York: Academic Press, 1971.

Skinner, J. E. *Neuroscience: A laboratory manual*. Philadelphia: Saunders, 1971.

3
Thermoregulatory Behavior

Thermal homeostasis in the animal world is achieved by two principal systems: behavioral and autonomic regulation [Benzinger, 1964, p. 831].

An important biological principle is that physiological and behavioral adaptations occur in response to variations of the animal's environment. One of the important environmental variations is temperature, and mechanisms have evolved that enable warm-blooded mammals to maintain a relatively constant body temperature in the face of wide fluctuations of air temperature from one time of the year to another and from one part of the day to another. The moose or buffalo of the North American plains experiences extremely cold weather in the winter and rather high temperatures during certain days in the summer. The temperature during periods of the year may differ by 15°–20°C between midday and midnight. The lion and tiger living in tropical zones experience ambient temperatures of 40°C or higher, and when captured and placed in a zoo in northern Europe or Canada, they may be exposed to −10° to −20°C or lower during the winter. At the other extreme, the polar bear survives subarctic weather but can also adjust to temperatures of the temperate zone. The regulation of body temperature in animals living in environments with such wide fluctuations of air temperature is, as indicated by the quotation from Benzinger, the result of both behavioral and physiological thermoregulatory responses.

These two kinds of responses of an animal to changes in the temperature of its environment have a complementary role in maintaining body-temperature homeostasis, and they frequently occur concurrently. For example, a rat exposed to the cold may build a nest or in the laboratory may press a lever to turn on a heat lamp; at the same time, there will be peripheral vasoconstriction to reduce heat loss and increased discharge of the sympathetic nervous system and increased output of thyroid hormones to increase metabolic heat production. It seems appropriate to begin, therefore, with a consideration of these two kinds of thermoregulatory responses before directing our attention later in the chapter exclusively to behavioral thermoregulatory responses. In the later sections, we deal first with the characteristics and determinants of thermoregulatory behaviors as a perspective for considering what is known about the neural substrates.

Thermoregulatory behavior has been selected as the first motivated behavior to be considered in detail, because it is closely associated with homeostasis. It is the example par excellence of a ''homeostatic drive,'' occurring in response to a deviation in core body temperature.

BODY-TEMPERATURE REGULATION[1]

There is an optimal temperature for the functioning of body cells—regulation of body temperature is one of the conditions of homeostasis. In man, body temperature is maintained at 37°–38°C with minor variations. A deep body temperature above 43°–44°C is lethal in man and many mammals, and a body temperature below 25°C is incompatible with life.

The regulation of body temperature within rather narrow limits means that heat being continuously produced by body cells must be balanced by heat loss. There are physiological mechanisms regulated by hormones and the autonomic nervous system that control both heat production and heat loss (Figure 3.1). These are considered briefly in the first part of this section. Behavioral, or what Richter (1943) called ''total organism,'' responses also contribute to body-temperature regulation, and they are considered later.

Physiological Mechanisms for Body-Temperature Regulation

Heat gain. Heat is generated continuously in the body of a mammal as a consequence of metabolic activity (see Figure 3.1). Heat production depends on: (1) the basal rate of metabolism of cells; (2) increased metabolic rate resulting from activity of muscles, including shivering; (3) increased metabolic rate due to the effects of thyroid hormones, noradrenalin, and sympathetic stimulation on body cells (noradrenalin produces a short-term increase in heat production, whereas thyroid hormones produce an increase that develops slowly but is prolonged); and (4) the specific dynamic action of ingested protein causing an increase in the metabolic rate.

Besides heat from cellular metabolism, the external environment may also be a source of heat gain. If surrounding objects, such as desert sand, are at a higher temperature than that of the animal's body, it will gain heat by radiation. The ingestion of hot foods or fluids will also raise body temperature at least to a small extent.

[1]This section contains only a brief presentation of the physiology of temperature regulation. For further details, consult a textbook in mammalian physiology (e.g., Guyton, 1976, chap. 72).

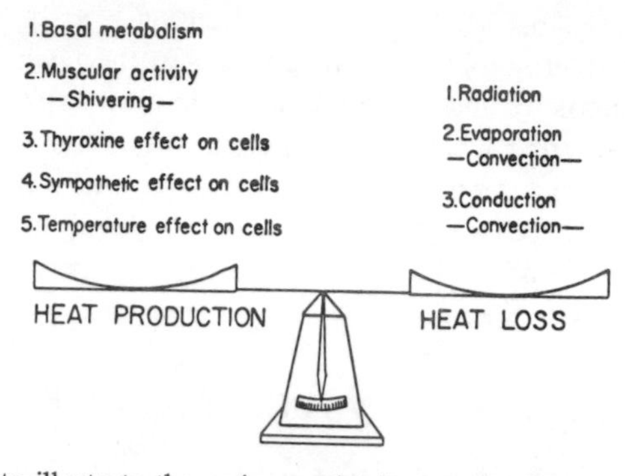

FIG. 3.1 Diagram to illustrate the major mechanisms responsible for heat production and heat loss to regulate body temperature. (From Guyton, 1976, p. 956. Courtesy W. B. Saunders Co.)

Heat loss. As shown in Figure 3.1, heat is lost from the body by radiation, conduction, and evaporation. The amount of heat exchanged with the external environment depends on the difference in temperature (temperature gradient) between the body and the surrounding air (ambient temperature), the humidity and movement of the air, and the amount of body insulation (Figure 3.2). Loss of heat via these three routes in a nude human subject for several ambient temperatures is shown in Table 3.1. At lower ambient temperatures, a relatively larger proportion of heat is lost by radiation and conduction. At higher ambient temperatures, the amount of sweating increases, and as a consequence, evaporation becomes the major route of heat loss. As much as 4–5 liters of water may be lost per day, and this increases with heat acclimation. There will also be peripheral

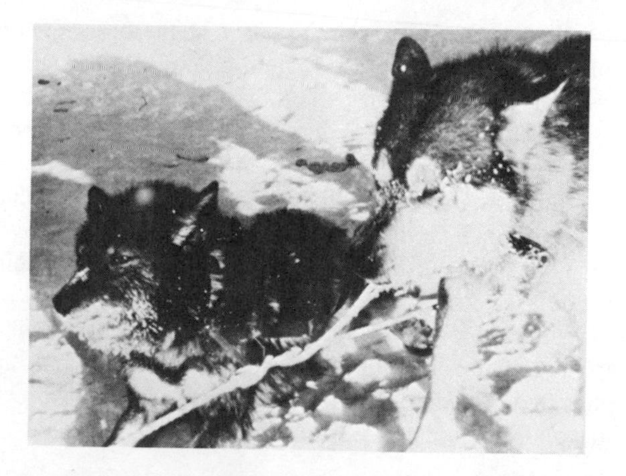

FIG. 3.2 Husky dogs after a 30-mile run in extremely cold weather (−33°C). The thick fur serving as an insulation is an important factor in regulating their body temperature in the cold. (From Folk, 1974.)

TABLE 3.1
Heat Loss from the Body via Various Routes Varies with Ambient Temperature

Room Temperature	Radiation	Convection	Evaporation
25°C	66%	10%	24%
30°C	41%	34%	25%
35°C	4%	7%	89%
40°C	0%	0%	100%

vasodilatation resulting in the movement of heat from the core of the body to the skin so that it can be more readily dissipated.

Some animals such as the rat do not sweat, but they utilize evaporation for heat loss by spreading saliva over their fur (Hainsworth & Epstein, 1966). As much as 15 ml of water per day may be lost in this way when a rat is in the heat. Dogs pant and thereby dissipate heat, with water evaporating from the tongue and oral cavity.

Physiological responses to cold. When a rat or other animal is exposed to a cold ambient temperature, physiological responses occur that increase heat production and that reduce heat loss to the external environment. An increase in heat production results from increased secretions of thyroid hormones and noradrenalin and increased discharge of the sympathetic nervous system. Shivering may occur as well as increased voluntary activity. In the cold-acclimated animal, there is a greater output of thyroid hormones and increased sensitivity of tissues to the thermogenic effects of thyroxine and noradrenaline. Shivering is less likely to occur, therefore, because of greater heat production by means of nonshivering thermogenesis. When the ambient temperature of a rat is reduced from 24°C to 5°C, its heat production doubles, and if the rat has been shaved, heat production will increase threefold.

A number of physiological responses occur in the cold-exposed rat that have the effect of reducing heat loss from the body: for example, constriction of the blood vessels in the skin, resulting in a reduced loss of heat by radiation and convection; and piloerection of hair and, in the case of birds, fluffing of the feathers. These responses increase the insulation of the body from the cold ambient air.

Behavioral thermoregulatory responses also occur in response to cold, and they are considered later.

Physiological responses to heat. With a high ambient temperature, the rate of metabolic activity and heat production decrease, and heat loss mechanisms are activated. There is peripheral vasodilatation, and thereby the transfer of heat to the surface of the body. As skin temperature increases, so does the temperature gradient between the body and the air, providing the ambient temperature is not too high. This increases net heat loss by radiation and conduction as long as the body temperature is greater than ambient temperature. When air temperature is higher than skin temperature, the only means of losing heat from the body is by the evaporation of sweat. The loss of large amounts of water and salt as sweat in turn initiates other physiological responses (e.g., increased output of ADH and aldosterone for the conservation of water and sodium loss by the kidney) and the ingestion of water (and salt).

Systems Analysis of Temperature Regulation

In temperature regulation the core and skin temperatures are monitored, and signals are provided that control thermoregulatory responses. Because these thermoregulatory mechanisms operate in a feedback manner, it has been useful to use control theory and systems analysis to assist with the study of temperature regulation. Systems analysis helps to clarify what is known about the control mechanisms and to guide research in dealing with unanswered questions.

Body temperature is regulated around a certain level, as shown in Figure 3.3; this is assumed to be due to a set-point mechanism. Any deviations above or below the set or ideal body temperature are detected by a comparator in the CNS that receives signals from temperature receptors in the skin and in the brain. If there is a discrepancy between the actual temperature and the set-point temperature, then "command signals" elicit a thermoregulatory response, such as shivering, shown in Figure 3.3. Shivering results in increased heat production so that body temperature gradually increases. As a result, the feedback signals from the temperature receptors are reduced, and when there is no longer a discrepancy between actual temperature and set temperature (no error), the shivering response

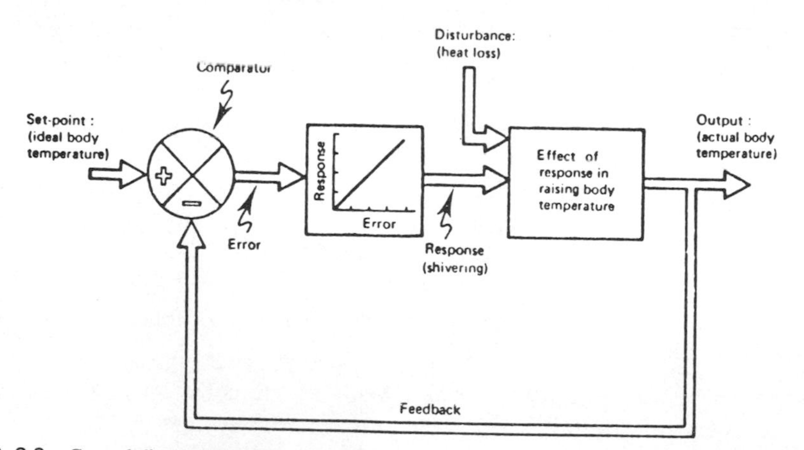

FIG. 3.3 Control diagram showing postulated mechanisms for thermoregulation. Shivering occurs when a "discrepancy" or error occurs between the actual body temperature and the set-point for ideal body temperature. (From Van Sommers, 1972.)

ceases. This is a negative feedback mechanism that controls body temperature in much the same way that room temperature is controlled by adjusting the thermostat.

Under certain conditions body temperature is elevated, for example, during fever and during exercise. This is considered to be the result of resetting the set-point to higher levels. There is evidence that pyrogens act on the preoptic area, the assumed locus of the comparator, or set-point mechanism, to raise body temperature (Veale & Cooper, 1974). Body temperature also varies throughout the day, being lowest in the early morning (at about 4:00 a.m.) and highest in the afternoon (about 4:00 p.m.). The circadian rhythm is thought to be controlled by altering the set-point mechanism.

Other thermoregulatory responses besides shivering, such as increased metabolic rate, peripheral vasoconstriction and vasodilatation, piloerection of the hair, and sweating, are also controlled by negative feedback mechanisms.

The Contribution of Behavior to Body-Temperature Regulation

Body-temperature homeostasis, as indicated earlier, depends on behavioral responses as well as on physiological thermoregulatory responses. Not only is there peripheral vasoconstriction and increased metabolic heat production when a rat is placed in the cold but also an increase in activity and nest building. Huddling together occurs in a number of species when exposed to a low ambient temperature, thereby reducing heat loss (Figure 3.4). Some animals seek shade from the sun or go into cool burrows in the ground (Figure 3.5); others will cool off by going into a lake or stream. In hot weather the elephant may spray water over the surface of its body (Figure 3.6). According to Bligh (1973), thermoregulatory

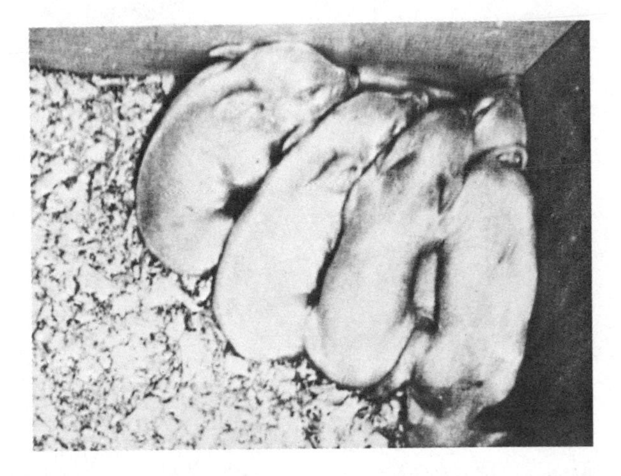

FIG. 3.4 Piglets huddling in the corner of a box to keep warm. (From Mount, 1968.)

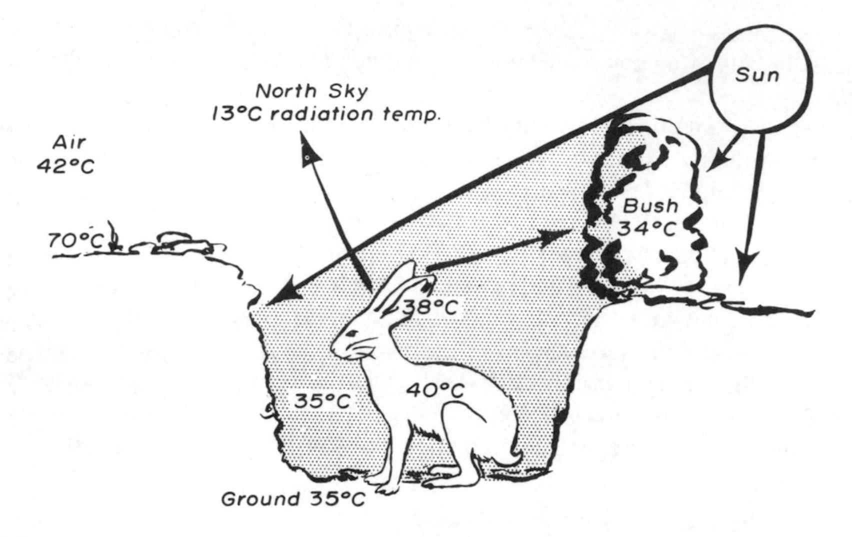

FIG. 3.5 By remaining in a shaded depression during the heat of the day, the desert jack rabbit maintains thermal balance. (From Folk, 1974.)

FIG. 3.6 The elephant uses a thermoregulatory behavior—spraying water over its body—to assist in regulating body temperature in the heat. (Photo courtesy B. Box.)

behaviors, defined as "complex patterns of response to environmental and other circumstances by which the organism varies the transfer of heat between itself and the environment," are of three types: "i) changes in surface area–to mass relations (huddling or limb extension; ii) voluntary variation in the extent of external insulation (choice of clothing or bedding); and iii) selection or creation of a less thermally stressful habitat (e.g., shade seeking, nest building) [p. 211]."[2]

Behavioral thermoregulation has been frequently considered of secondary importance in comparison to physiological regulations. This seems to have been because most of the earlier laboratory studies were in rather artificial circumstances in which behavioral thermoregulatory responses were minimal or inoperative (Bligh, 1973). In many circumstances behavioral thermoregulatory responses have a more important role than physiological thermoregulatory responses. According to Bligh (1973):

> This may be particularly true of man who does not shiver immediately in response to cold, nor sweat immediately in response to heat. It is the use of shelter and clothing rather than shivering and sweating that enables man to inhabit a wide range of climatic conditions. Thus it is behavioral rather than autonomic thermoregulation which determines the environmental limits in which the homeothermy of man can be maintained, but some have argued that it is the autonomic functions (shivering, sweating and/or panting, and peripheral vasomotor tone) which effect the fine control of body temperature. However, behavioral responses to heat and cold also play an important role in the fine balance of body temperature so long as there is freedom to use them. Given the freedom to do so, we open and shut windows, and put on and take off jackets rather than sweat and shiver [pp. 192–193].

The importance of behavioral responses for body-temperature homeostasis was clearly demonstrated some years ago in a series of experiments by Curt Richter. Because of the significance of this work, which had a major impact on the formulation of theoretical models for dealing with motivated behaviors from the biological point of view, one of these experiments is presented here in detail. You should note, in particular, Richter's (1943) research strategy, which was to demonstrate a behavioral regulatory response by disrupting an important physiological regulatory mechanism:

> The individual cages used for these experiments were each equipped with a roll of soft paper ½ inch wide and 500 feet long, with the free end readily accessible to the rat within the cage. [Figure 3.7A] shows a cross-sectional view of one of these cages. By means of a cyclometer and a scale to compensate for the progressively decreasing diameter of the roll, the amount of

[2]The distinction between behavioral and physiological thermoregulatory responses may not be entirely appropriate, because the behavioral responses do have a physiological basis. For this reason Bligh (1973) has designated the so-called physiological responses as "*autonomic response*." He suggests that the behavioral thermoregulatory responses:

> are physiological in the sense that they depend on a complex central nervous response to a thermal stimulus, but the actual changes in the thermal relations between the organism and environment are achieved by behavioral adjustments, and not by adjustments of autonomic effector processes such as shivering, panting or the blood supply to the skin [p. 211].

paper used each day was measured, and interpreted as an effort made by the rat to conserve heat by covering itself. All used paper was removed each day at noon. It was found that normal male and female rats used approximately equal amounts of paper to build nests which varied in size with changing external temperatures, for example, a drop in room temperature from 80 to 45 degrees increased the amount of paper used daily from 500 to 6000 centimeters. With this method we were also able to show that hypophysectomized rats built much larger nests than normal animals as a result of their inability to produce adequate amounts of heat, which consequently threatened them with a fatal reduction in body temperature. [Figure 3.7B] shows the effect produced on nest building activity of a rat by hypophysectomy. The length of paper used daily increased from 700 to 3500 centimeters. When nest building paper was no longer made available, the rat died after 35 days, with a body temperature more than 15 degrees below normal. Thyroidectomized rats, which likewise have lost their ability to produce adequate amounts of heat, also built very large nests in an effort to cover themselves and thus to conserve heat. Both thyroidectomized and normal rats treated with large amounts of thyroid extract stopped building nests altogether. Some of the hypophysectomized and thyroidectomized rats used the entire roll of 15,000 centimeters (500 feet) of paper in 24 hours. Thus we have another instance in which, after removal of the physiological regulators, homeostasis was maintained by a total organism response [pp. 69–70].

CHARACTERISTICS OF THERMOREGULATORY BEHAVIOR

Thermoregulatory behavior has the same characteristics—purposiveness, persistence, periodicity, and priorities—as other motivated behaviors. We have selected as an example to illustrate these characteristics the experiment by Richter quoted in the previous section.

Richter (1943) studied nest building as a thermoregulatory behavior both in the normal, intact rat and following surgical interventions that disrupted physiological regulations that contribute to body-temperature regulation (e.g., removing pituitary or thyroid gland). As indicated previously, he observed nest-building behavior of rats in a cold ambient environment and used as a quantitative index of this behavior the amount of paper from a roll utilized by the animal in a certain time period. It is clear from his experiments that thermoregulatory behavior has the characteristics of motivated behavior outlined in Chapter 1. Nest-building behavior was purposive and persistent. Rats engaged in this activity for considerable periods of time, using large amounts of paper and building very large nests. However, nest-building behavior was discontinued at intervals while the animal ate, rested, or slept. Like many other motivated behaviors, nest building was periodic. Frequently nest building terminated when the animal began to engage in some other motivated behavior that for a time had a higher priority (Adolph, 1947). ''Motivational time-sharing'' is a prominent characteristic of motivated behaviors (McFarland, 1974).

In more recent laboratory studies of thermoregulatory behavior, animals have been required to make an arbitrary learned response (operant behavioral response) to obtain heat or cold. For the rat this is usually pressing down on a

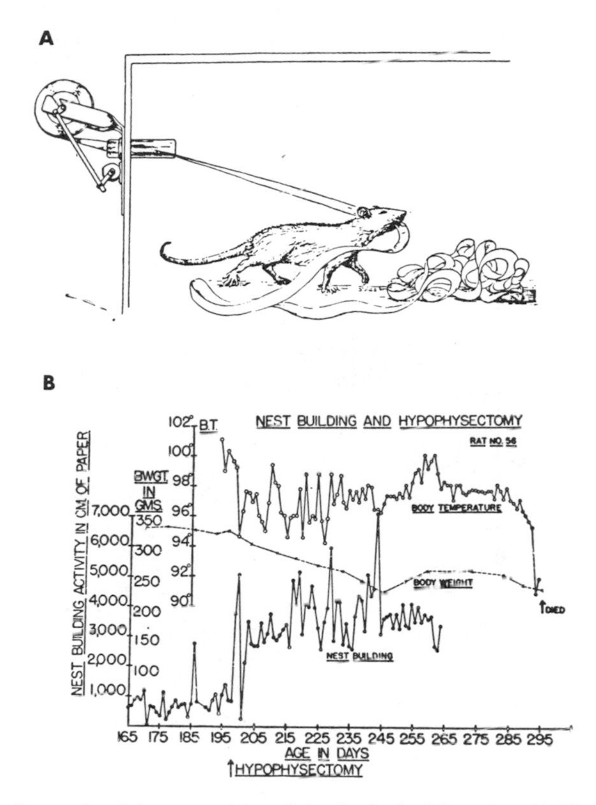

FIG. 3.7 The first study of thermoregulatory behavior in the laboratory. A. The amount of paper from a roll used by the rat to build a nest in the cold was measured. B. Records of nest-building activity, body weight, and body temperature for rat No. 56 after removal of the pituitary gland. (From Richter, 1943.)

lever, and monkeys have been trained to pull a chain in order to activate a fan that blows warm or cold air into the test chamber (Figure 3.8). The characteristics of purposiveness, persistence, and periodicity of the thermoregulatory behavior are obvious by observing the animal directly or the record of its responses on a cumulative recorder. If feeding, drinking, and other behaviors are recorded concurrently, the ''motivational time-sharing'' of thermoregulatory behavior with these other motivated behaviors is clearly demonstrated.

FACTORS THAT CONTRIBUTE TO THERMOREGULATORY BEHAVIOR

The several factors that initiate and contribute to behavioral thermoregulation are discussed in this section.

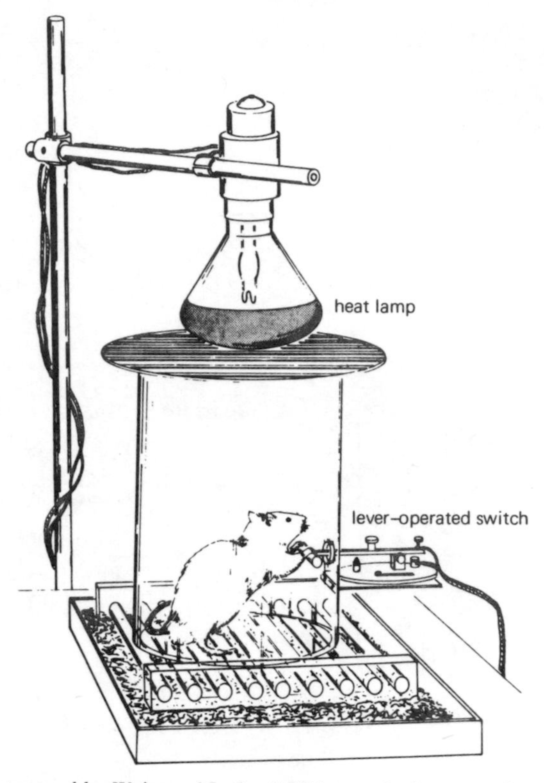

FIG. 3.8 The apparatus used by Weiss and Laties (1961) to study thermoregulatory behavior of a rat in the cold. The animal turned on a heat lamp by pressing a lever in its plastic cage. (From Milner, 1970.)

Homeostatic Signals

When a rat or other laboratory animal is placed in an apparatus of the sort shown in Figure 3.8, it makes an appropriate behavioral operant response to turn on a heat lamp or to activate a mechanism providing a momentary flow of warm air under two kinds of conditions. The first condition is when the ambient temperature is reduced below the level appropriate for the animal's "thermal comfort." When the ambient temperature is about 5°C, rats reliably press the lever to obtain heat, and if the rats are shaved, the rate of behavioral responding is increased (Weiss & Laties, 1961). The stimulus that initiates the behavior is a reduction in the temperature of the body detected by receptors in the skin and receptors in the central nervous system. The second condition for initiating thermoregulatory behavior is to lower the temperature of the preoptic area or some other region of the brain (e.g., medulla) that contains central temperature receptors using a thermode, a slender probe insulated except at the tip, through which a cold fluid is circulated. Using this procedure, it has been demonstrated that the goat, dog, and other animals make physiological and behavioral thermoregulatory responses when the preoptic region is cooled, and body temperature rises. On the other hand, when the temperature of the preoptic region is raised by circulating a warm fluid through the thermode, thermoregulatory responses that reduce heat production and increase heat loss are observed, and body temperature falls.

We discuss these experiments in more detail later, but the point is that the brain itself, and especially the preoptic region, has temperature receptors, and when these receptors are activated, thermoregulatory responses occur.

Hormones

Hormones, particularly thyroxine, noradrenalin, and adrenal cortical hormones, contribute to heat production by controlling metabolic activities in the body. Although these hormones do not initiate thermoregulatory behaviors directly, they do have an influence on such behaviors in that their reduction or absence necessitates a greater reliance on thermoregulatory behaviors in order to regulate body temperature. This was first demonstrated experimentally by Richter (1943) when he showed that rats engaged in more nest-building behavior following removal of the thyroid gland or the pituitary (see Figure 3.7). More recently, Laties and Weiss (1959) have reported that the thermoregulatory behavioral response of turning on a heat lamp is significantly increased in thyroidectomized rats.

When animals are exposed to a low ambient temperature for a long period of time, the levels of thyroxine and metabolic heat production increase as cold acclimation occurs. In the rat kept at 3°–5°C, complete cold adaptation occurs in 5 or 6 weeks (Sellers et al., 1957). When cold-acclimated, the rat does not engage in as much thermoregulatory behavior as initially. This was demonstrated in an experiment in which rats, kept at a room temperature of 22°C or 5°C, were permitted to press a lever to turn on a heat lamp during a 2½-hour test period at −10°C (Figure 3.9). As shown at the left side of the figure, there was no difference in the number of lever presses for the two groups at the beginning before cold-acclimation. However, when the test for behavioral thermoregulation was repeated 40, 41, and 42 days later, the rats kept at a room temperature of 5°C now pressed the lever to turn on the heat lamp significantly fewer times than the rats maintained at 22°C. That cold-acclimation had occurred in the rats kept at 5°C is indicated by the measures of oxygen consumption, an indicator of metabolic activity, shown at the right of Figure 3.9.

These observations illustrate further the "trade-off" between physiological and behavioral thermoregulatory responses, which was shown initially in the pioneering experiments of Richter. Physiological and behavioral thermoregulatory responses are complementary in body-temperature homeostasis; metabolic heat production is greater in the cold-acclimated rat because of hormonal and

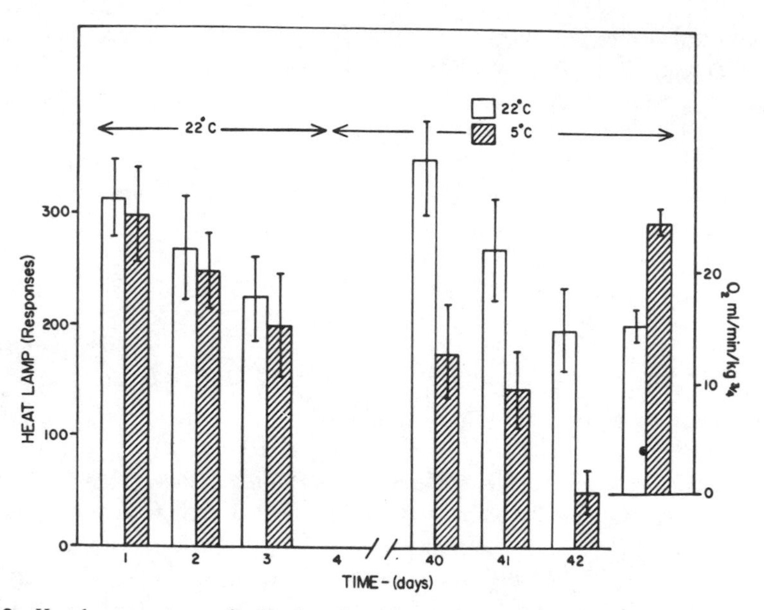

FIG. 3.9 Heat-lamp responses of cold adapted rats (hatched columns) and their nonadapted controls (open columns) when exposed to extreme cold (−10°C) for 2½ hours. Pressing a lever turned on a heat lamp for 3 seconds. Days 1 to 3 represent the training period before cold adaptation. Metabolic rate of the two groups during the sixth week is shown on the right. (From Mogenson et al., 1971.)

other adaptations, and as a consequence, the animal relies less on behavioral thermoregulatory responses.

Circadian and Biological Rhythms

Body temperature varies systematically throughout the day (Figure 3.10). In the rat, a nocturnal animal, body temperature is highest at night and lowest in the daytime. Body temperature tends to be highest when the animal is more active and lowest during sleep. However, there is more to the systematic fluctuation of body temperature than an influence of activity. It is the set-point for body temperature that varies as determined by a "biological clock" entrained to the light–dark cycle. A reduction of the temperature set-point during sleep is biologically adaptive, because sleep reduces the demands on metabolic heat production and on physiological regulatory mechanisms for body-temperature homeostasis, and thermoregulatory behavior is thus minimal or absent.

For some animals the set-point for body-temperature regulation may be lowered for an extended period of time—for several months in certain hibernating animals. During hibernation the animal is asleep or minimally active (Figure 3.11). Because the body temperature is regulated at a lower level, the expendi-

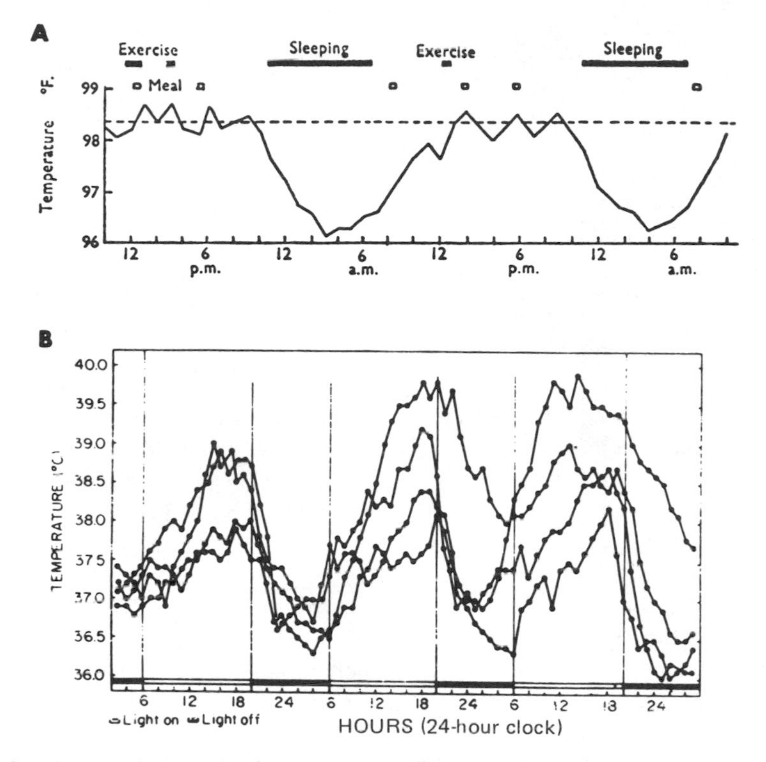

FIG. 3.10 Circadian rhythms of body temperature are illustrated. A. Continuous body-temperature records of man taken during exercise and sleep during a 2-day period. (From Green, 1964. Copyright 1964 by the Oxford University Press.) B. Continuous record of rectal temperatures of four baboons during a 3-day period. (From Sundsten, 1969. Reprinted by permission of the author and the New York Academy of Sciences.)

ture of energy for metabolic heat production is greatly reduced. As a result the limited supply of energy as stored fat lasts for a longer period of time.

Experiential and Cognitive Factors

Stimuli in the external environment serve as cues to guide the thermoregulatory behavior of the animal. For example, cattle may seek the shelter of trees during a windy day in winter, and birds may search for sticks, reeds, etc., to construct nests. External stimuli may also be utilized by animals to make behavioral responses prior to, or in anticipation of, homeostatic signals. According to Bligh (1973), thermoregulatory behaviors are frequently initiated by thermal discomfort rather than by a deviation of the core temperature of the body, and, particularly in man, behaviors that contribute to thermoregulation often occur in anticipation of thermal discomfort. Animals may not leave their shelters on a cold day,

FIG. 3.11 Ground squirrel in hibernation. Electrodes were previously implanted in its brain for electrophysiological studies. (From Kayser, 1961.)

and man may drive to work in his car rather than walk when he looks out the window and sees that it is stormy. Such behaviors, which enable you to avoid or reduce thermal discomfort and which contribute to body-temperature regulation, utilize learning, memory, and other cognitive processes. Benzinger (1970) emphasized the importance of experiential and cognitive factors for behavioral thermoregulation in man when he stated: "Shivering indicates a failure of intelligent behavior [p. 853]."

Driving to work in the car and putting on warmer clothing are thermoregulatory behaviors that depend not on homeostatic signals and physiological mechanisms operating according to negative feedback but rather on neural processes that permit the person to anticipate homeostatic signals and to act in a way to prevent their occurrence. Oatley (1973) has suggested that this depends on "representations of the world" in the brain that "is mediated and embedded, as it were, in a structure of feed-forward [p. 214]." Benzinger (1964) has noted that man is unique in thermoregulatory behavior because of his more sophisticated development of cognitive processes as compared to lower animals:

> In behavior, the artificial means of man for locomotion from one environment or climate into another are exceptionally fast and far-reaching. Moreover, human protective clothing, different from the furs or feathers of animals, may be applied, removed or changed at will. The shelters of man are more elaborate; the use of fire and other means of heating with external sources of energy are a privilege of the human race. Man alone bathes for comfort in artificially heated water. Less than one century ago, discoveries in physics added another dimension: cooling. This has enabled man to refrigerate or air-condition his dwellings and working spaces. Even more recently, heating or cooling, applied to fast-moving craft for land, sea, air and space travel have permitted the invasion of areas with the most adverse conditions, including polar ice-caps, ocean-depths, high altitudes and terrestrial orbits. These means will in the future permit man to inhabit remote celestial bodies [p. 49].

ONTOGENY OF THERMOREGULATORY BEHAVIORS

The newborn of some mammalian species are capable of maintaining a stable body temperature. A remarkable example is the cariboo calf, born in arctic weather, which shortly after birth is able to follow the mother and the rest of the herd, and its activity increases heat production. At the other extreme are rat and mouse pups, which are relatively helpless, cannot increase metabolic heat production in response to cold, and rely on the warmth of the mother's body and the nest she builds. Because of lack of neural and motor development, newborn rats and mice cannot increase skeletomotor activity or make other behavioral thermoregulatory responses. They are further disadvantaged by inadequate development of thermal insulation (e.g., hair), by the limited availability of stored energy for heat production, and by their small size. Body size is an important factor in temperature regulation, because the young animal has a larger surface area to body mass ratio than the adult and therefore loses more heat per gram of body weight.[3]

The first behavioral thermoregulatory response observed in the rat is huddling at 6–7 days of age. At about the same age, rat pups are able to move to a warm compartment in a choice apparatus (Fowler & Kellogg, 1975). By Days 8 or 9, they become more active, and the resulting increase in heat production together with increased body weight and amount of hair result in greater stability of body temperature. By the end of the second week of life, the rat is able to regulate its body temperature fairly well. This is due in part to increased motor capacity associated with maturation of the CNS and motor apparatus, but it may also be related to the development at that age of central monoaminergic and cholinergic neurons concerned with the control of body temperature (Fowler & Kellogg, 1975).

Body-temperature regulation is also poor in the human newborn, and both physiological and behavioral thermoregulatory responses appear during ontogenetic development. The period of relative helplessness and reliance on the mother or other human adult is much longer than in the rat or mouse. Thermoregulatory behaviors, which require sensory systems, motor systems, and sensory–motor integrations, appear as the CNS and the sensory and motor apparatus reach maturity. The period of ontogenetic development is also extended in man, in comparison to the rat or mouse or even the primate, because of man's capacity to make behavioral thermoregulatory responses in the absence of thermal discomfort or of deviations in body core temperature. These behaviors depend on memory, learning, and other cognitive processes that take several years to develop.

[3]Hull (1973) states that:

The thermoregulatory capacities of all mammals whether young or old are limited by their thermal insulation and the area of the exposed surface; the smaller the mammal the greater is the exposed surface relative to its body weight. These two factors obviously impose limits on the maximum and minimum size of mammals living in different climates if they are to achieve homeothermy. Only the larger mammals can live on the earth's surface in arctic climates, smaller mammals would be immobilized by the fur necessary to retain their body heat. Pearson (1948) calculated that mammals weighing below 2.5 gm could not exist in any climate for they could not eat sufficient food to support the necessary rate of heat production to maintain homeothermy. These considerations emphasize the problems of newborn mammals which are obviously much smaller than the adult, and they are often poorly covered with fur [p. 169].

In the previous section, the various factors that initiate thermoregulatory behaviors were considered. It is clear from what has just been said that the contribution of these factors varies among species as well as during the ontogenetic development of a particular species.

NEURAL INTEGRATIVE SYSTEMS SUBSERVING PHYSIOLOGICAL AND BEHAVIORAL THERMOREGULATORY RESPONSES

Our major interest in this section is the neural substrates of thermoregulatory behaviors. However, this subject must be considered from a more general perspective, because body-temperature regulation depends on physiological as well as behavioral thermoregulatory responses. Physiological and behavioral responses are complementary for temperature regulation, as indicated in an earlier section, and it is therefore necessary to consider the possibility that there are neural integrative mechanisms for the overall coordination of physiological and behavioral thermoregulatory responses.

The neural substrates of thermoregulatory behavior, and of body-temperature regulation in general, are considered in relation to the factors that initiate and contribute to thermoregulatory responses. Because temperature regulation is important for homeostasis, a deviation in body temperature is followed by thermoregulatory responses. The deviation is detected by receptors on the surface of the body and in the CNS. Most of what is known is concerned with the hypothalamic and other neural mechanisms for the integration of these signals and for sending command signals to autonomic, endocrine, and skeletomotor systems that control the responses. Hormones, one of the factors identified earlier as influencing body-temperature regulation, is not considered here in detail, because they do not have a direct effect on thermoregulatory behavior. A number of hormones have an indirect effect, because they control energy metabolism and thereby contribute to heat production. If the production of these hormones is disrupted (e.g., surgical removal of thyroid or adrenal gland), body temperature is likely to drop in the cold, and as a consequence, thermoregulatory behaviors occur in response to thermal signals.

The neural substrates of the experiential and cognitive determinants of thermoregulatory behavior are also not dealt with separately. This is not because experiential and cognitive factors are unimportant, especially in man and higher mammals, but because of the dearth of evidence. Indirect evidence implicates higher brain structures (e.g., cerebral cortex and limbic forebrain structures) in memory, learning, and other cognitive processes, but the precise neural mechanisms involved when experiential and cognitive factors influence behavioral thermoregulation are unknown. As indicated in Chapter 10, the study of the neural substrates of cognitive processes is an important field of future investigation if we are to eventually have an adequate neurology of motivated behaviors.

Neural Substrates of Thermoregulatory Responses Initiated by Thermal Stimuli

Investigations of the neural mechanisms controlling body temperature began in the 1930s, with the demonstration that surgical transections of the brain caudal to the hypothalamus disrupted temperature regulation, whereas such transections rostral to the hypothalamus did not (Keller & Hare, 1932). Using the newly introduced stereotaxic procedure to place discrete lesions, further evidence was obtained that implicated the hypothalamus in temperature regulation. When the anterior hypothalamus-preoptic region was lesioned, animals could not maintain body temperature against a high ambient temperature, and when the posterior hypothalamus was lesioned, they could not maintain it against the cold (Ranson & Ingram, 1935). These observations suggested that heat-loss mechanisms were represented in the anterior hypopthalamus and heat-gain mechanisms were represented in the posterior hypothalamus. When the temperature of the anterior hypothalamus-preoptic region was elevated by local heating, responses such as panting and sweating were observed. In subsequent experiments the temperature of the preoptic region was lowered as well as raised using the intracranial thermode technique, and both behavioral and physiological thermoregulatory responses were observed. From these experiments it was concluded that the preoptic region integrates signals from receptors in the skin and in the preoptic region itself and functions as the "controller" of thermoregulatory responses for body-temperature regulation (Figure 3.12). Hypothermia from lesions of the posterior hypothalamus was attributed to the disruption of mechanisms for heat production that received "command signals" from the "controller" in the preoptic region.

Physiological responses to cooling and warming the preoptic region of the goat have been studied extensively by Andersson and his co-workers (1962) in Stockholm (Figure 3.13). The goat is an ideal animal for such experiments because of its size and ease of handling, making it easier to record appropriate physiological

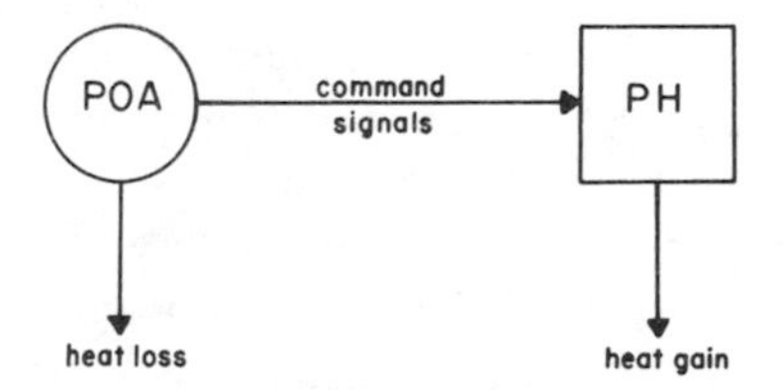

FIG. 3.12 Schematic diagram showing postulated central mechanisms for the control of body temperature. Experimental evidence from lesion and transection experiments suggested that heat loss was controlled by the anterior hypothalamus and heat gain by the posterior hypothalamus, with the preoptic region acting as a controller.

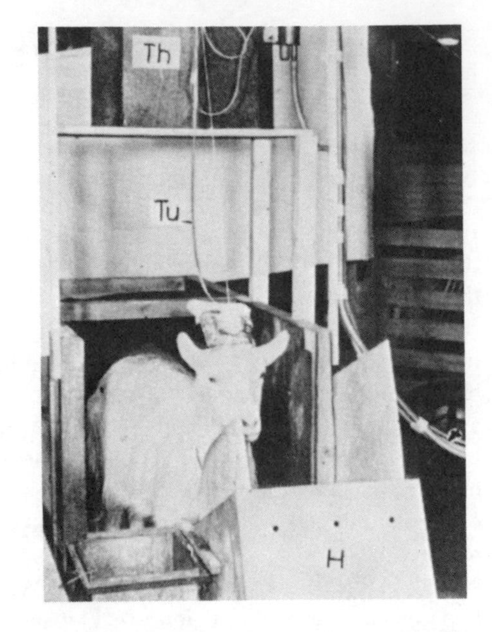

FIG. 3.13 The experimental apparatus of Andersson and co-workers (1962) for heating or cooling the brain of a goat. Th = thermos for containing water to perfuse the permanently implanted thermode through rubber tubing (Tu). H = Hayrack. (From Andersson, Gale, & Sundsten, 1962.)

parameters and to alter the temperature of the preoptic region without influencing more distance structures of the brain. When the preoptic region was cooled, these investigators observed an increase in the output of thyroxine, adrenalin, and noradrenalin, increased O_2 consumption, peripheral vasoconstriction, and shivering (Figure 3.14). These responses increase heat production and reduce heat loss so that body temperature, measured by monitoring the rectal temperature of the goat, increased. Warming the preoptic region, on the other hand, increased heat-loss responses (e.g., panting) and reduced heat-gain responses (e.g., peripheral vasodilatation). The results for a similar experiment in the baboon are shown in Figure 3.15, and it can be seen that when the preoptic region was warmed, there was peripheral vasodilatation and reduced urinary epinephrine. Body temperature monitored from the midbrain was decreased.

Cooling or warming the preoptic region also initiates behavioral thermoregulatory responses as well as physiological thermoregulatory responses of the sort just described. This was first demonstrated by Satinoff (1964), who reported that rats pressed a lever to turn on a heat lamp when the preoptic area was cooled by means of a thermode chronically placed in this region of the brain. Similar results, showing that animals make an operant response to obtain heat when the preoptic area is cooled, have been obtained in the pig (Baldwin & Ingram, 1967), the squirrel monkey (Adair, Casby, & Stolwijk, 1970), and the baboon (Gale, Matthews, & Young, 1970). In other experiments the preoptic region was warmed, and rats made operant responses to cool themselves (Corbit, 1970; Murgatroyd & Hardy, 1970).

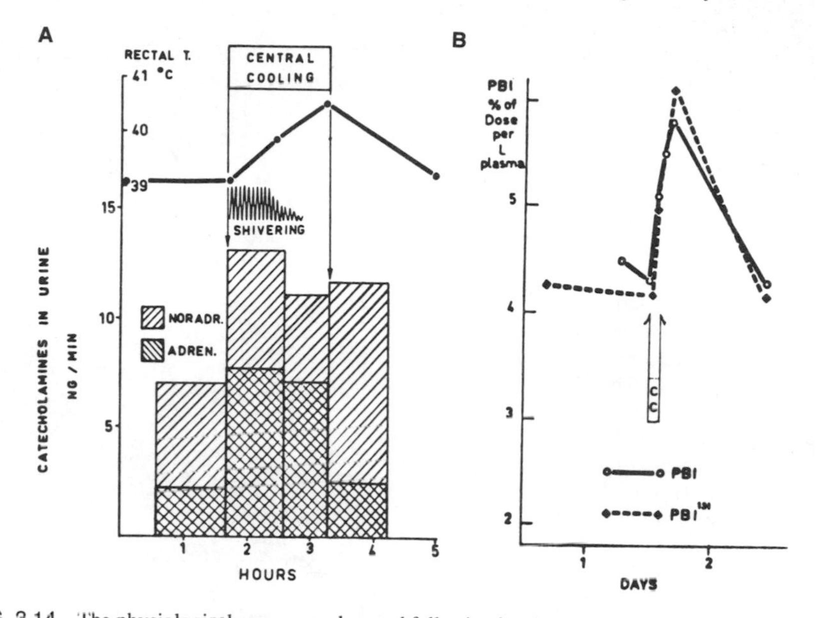

FIG. 3.14 The physiological responses observed following local cooling of the preoptic area of the goat. A. Adrenalin output increases as the animal begins to shiver. B. Thyroid hormone secretion rises, as indicated by a similarity between the curves for serum PBI (solid line) and plasma PBI^{131} (broken line). (From Andersson et al., 1964; 1965.)

Behavioral thermoregulatory responses have also been initiated by cooling or heating other regions of the CNS with thermodes. Lipton (1971) has reported that cooling or warming the region of the caudal pons and rostral medulla in rats elicited an operant response to obtain warm or cool air respectively. Roberts and Mooney (1974) demonstrated lever pressing for cool air as well as other thermoregulatory responses when the medulla was cooled. Cabanac (1972) observed that warming the spinal cord of dogs elicited operant responses to obtain cool air, but the spinal cord was much less sensitive than the preoptic region in eliciting behavioral thermoregulatory responses. The medulla may also be less sensitive, and according to Lipton (1971), it has a secondary role in temperature regulation compared to the preoptic region. Most investigators assume that the medulla, pons, and spinal cord are additional sources of information about body temperature and, like skin temperature receptors, send signals to the preoptic region, which functions as the integrative and control mechanism for temperature regulation (see Figure 3.12).

The crucial importance of the preoptic region in temperature regulation has been demonstrated with other experimental procedures as well. When bacterial pyrogens, responsible for the elevation of body temperature in fever are placed directly into the preoptic region using cannulae, fever is produced (Cooper et al., 1967; Villablanca & Myers, 1965). Further evidence that this is the "fever

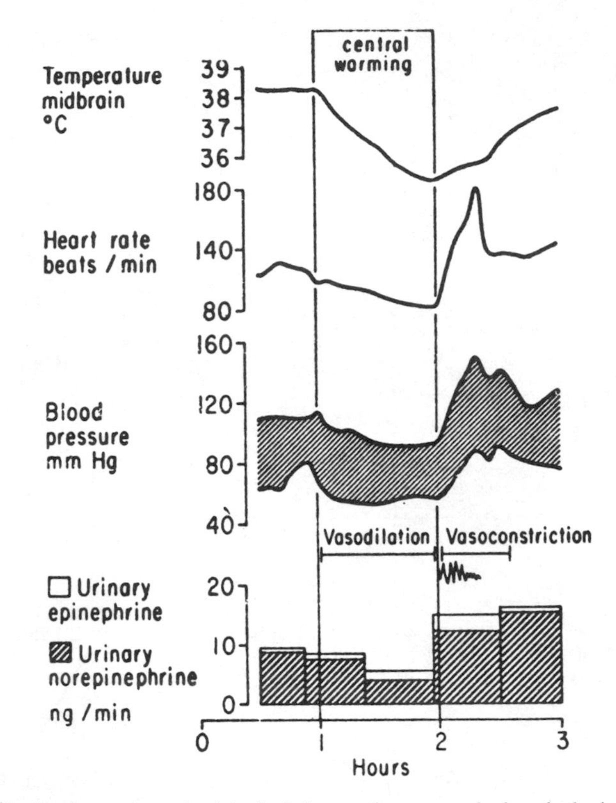

FIG. 3.15 Graph showing various physiological changes that occurred when the brain of a baboon was warmed. (From Morishima & Gale, 1972.)

center'' has been obtained by lesioning the preoptic region and demonstrating that fever is no longer produced by the systemic administration of pyrogens. Also it has been reported that prostaglandins serve as mediators of pyrogen-induced fever (Veale & Cooper, 1974).

Noradrenalin, serotonin, and other compounds known to be CNS neurotransmitters, have been injected into the anterior hypothalamus-preoptic region or cerebral ventricles by means of cannulae while recording body temperature. The initial experiments were in the cat, and it was observed that infusing serotonin resulted in a rise in body temperature and infusing noradrenalin resulted in a fall in body temperature (Feldberg & Myers, 1963). Feldberg's and Myers' pioneering experiments suggested a possible role for noradrenergic, serotinergic, cholinergic, and other transmitter-specific neurons, in the control of body temperature. These monoaminergic neurons, which project to the hypothalamus and various forebrain structures, are located in the midbrain and pons (see Figure 2.24), and the drugs presumably act at the synapses of their fiber projections in the preoptic region. In later experiments Myers and Veale (1970) observed that an excess of sodium ions in the posterior hypothalamus caused a rise in body temperature and an excess of calcium ions, a fall in body temperature. They hypothesized that body temperature is regulated around a set-point determined by the ratio of sodium and calcium ions in the posterior hypothalamus, which sends a reference input to the controller in the preoptic region. Myers (1974b) has formulated a model that attempts to account for the effects of centrally administering these compounds (Figure 3.16). Further research is needed to investigate the species differences in the effects of these compounds and to elucidate the neural mechanisms. It will be important to relate these mechanisms to what is known about the chemical anatomy of the brain based on visualizing monoaminergic neurons and the fiber projections using histofluorescence (Figure 2.24).

The presence of thermosensitive neurons in the preoptic region has been demonstrated with electrophysiological recording techniques. The action potentials

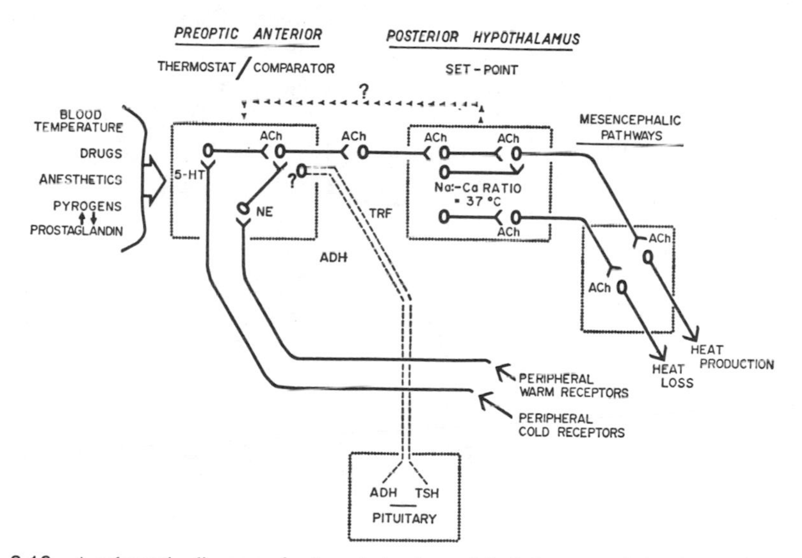

FIG. 3.16 A schematic diagram of a hypothalamic model of thermoregulation based mainly on experimental findings in the monkey and cat. The anterior hypothalamic preoptic area contains neurons that are thermosensitive as well as a comparator mechanism that contrasts the set-point with the local temperature. The region is also sensitive to a number of substances, as indicated, including pyrogens and prostaglandins, which may interact functionally. The release of 5-hydroxytryptamine (5-HT) acts through a cholinergic pathway to signal heat production, whereas the release of norepinephrine (NE) blocks the cholinergic heat production pathway that traverses the posterior hypothalamus. The set-point of 37°C depends upon the intrinsic ratio of Na+ to Ca++ ions, an aberration of which will evoke a shift in the set-point presumably via independent cholinergic pathways. The regulation of body water and thyroid activity may also be partially mediated by synapses arising in the anterior hypothalamus. ACh = acetylcholine; TRF = thyroid releasing factor; ADH = antidiuretic hormone; TSH = thyroid stimulating hormone. (Figure and legend from Myers, 1974b.)

recorded from single neurons increase when the preoptic region is warmed with a thermode (see Figure 2.22). These are designated warm-sensitive neurons, and some of them have been shown to be warm receptors. Neurons that increase their activity when the preoptic region is cooled have also been demonstrated, and some are cold receptors (Figure 3.17A). Experimental evidence from electrophysiological recording experiments suggests that it is the action of pyrogens on these neurons that is responsible for the elevation of the set-point for body-temperature regulation in fever. As illustrated in Figure 3.17B, pyrogens reduce the responsiveness of warm-sensitive neurons, and as a result, there is reduced drive on heat-loss mechanisms and body-temperature rises. There is also evidence that salicylates (e.g., aspirin) exert antipyretic effects by restoring the responsiveness of warm-sensitive neurons of the preoptic thermostat to the warm blood flowing through the preoptic region.

Warm and cold sensitive neurons in the preoptic region have been shown, using electrophysiological recording techniques, to respond to changes both in the ambient temperature and in central temperature. This suggests that they integrate signals from temperature receptors in the skin and in the CNS and are part of the "controller" that generates "command signals" for appropriate thermoregulatory responses—autonomic, endocrine, and behavioral. However, everyone does not agree that there is a single integrative mechanism for all

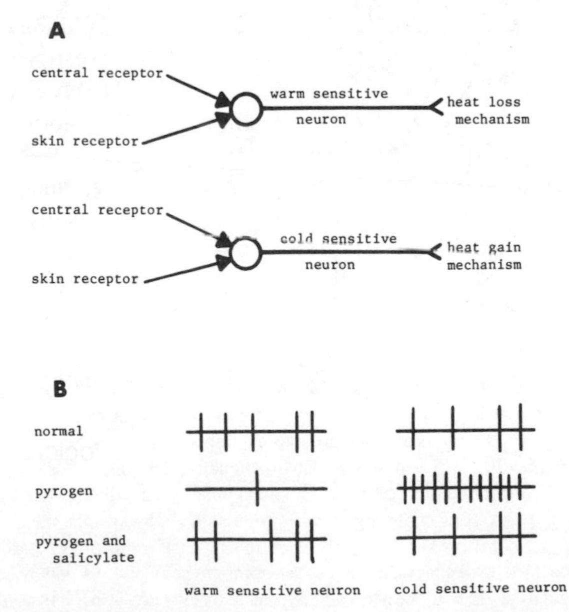

FIG. 3.17 A. Warm and cold sensitive neurons in the preoptic region receive inputs from central and peripheral temperature receptors (for electrophysiological procedures, see Figure 2.22). B. The administration of bacterial pyrogens reduces discharge of the warm sensitive neuron, and this effect is reversed by salicylates.

thermoregulatory responses (Bligh, 1973; Satinoff, 1974). We return to this issue later.

Biological Rhythms and Temperature Regulation: Neural Substrates

There are circadian fluctuations in body temperature in a number of mammals (see Figure 3.10A). This circadian rhythm is not the result of some defect in temperature regulation but seems to be the result of a change in the set-point. In some species the set-point for body-temperature regulation is markedly reduced for weeks and even months during hibernation (see Figure 3.11).

It has been suggested that circadian rhythms in activity, ingestive behavior, and sleep–waking are controlled by a "biological clock" entrained to the photoperiod (Rusak & Zucker, 1975). These rhythms, as well as the estrous cycle of sexual receptivity, have been disrupted by lesions of the suprachiasmatic nucleus in rats (Stephan & Zucker, 1972) and in hamsters (Stetson & Watson–Whitmyre, 1976), suggesting that this hypothalamic nucleus is associated with the "biological clock." Stetson and Watson–Whitmyre (1976) tentatively conclude that "the nucleus suprachiasmatic is a biological clock in the hamster [p. 199]," but Rusak and Zucker (1975) are more cautious and comment that "whether the SCN serves as a master clock, a central coupler of rhythms, or as a complex oscillator in a multi-oscillator system is unknown [p. 148]."

The role of the suprachiasmatic nucleus in circadian and ultradian (e.g., hibernation) fluctuations of body temperature has not been investigated. It may be involved in adjusting the set-point for body temperature in relation to the photoperiod and in controlling the occurrence of thermoregulatory behaviors associated with the circadian rhythm of sleep–waking and the more general rhythms of rest–activity. For example, nest building, increased activity (skeletomotor responses), and feeding occur at night in rodents when they are awake. On the other hand, huddling in some species may occur to a greater extent during the periods of sleep. These observations suggest that the suprachiasmatic nucleus (and the biological clock) may contribute to thermoregulatory behaviors, but experimental studies are needed to determine the neural mechanisms by which it influences behavioral thermoregulation.

Is There a Single Neural Integrative System for Physiological and Behavioral Thermoregulatory Responses?

According to Bligh (1973), "whether behavioral and autonomic thermoregulatory processes are separate but complementary, or whether they are integrated components of physiological temperature regulation [p. 193]" is a major issue in the neural control of body-temperature homeostasis.

There are two points of view. The first view, accepted by most investigators of temperature regulation, assumes that the preoptic region integrates thermal signals from the skin as well as from temperature receptors in the CNS and is the source of "command signals" for various physiological thermoregulatory responses as well as behavioral responses (Figure 3.18). An increase in metabolic rate, peripheral vasoconstriction or vasodilatation, sweating, and other physiological thermoregulatory responses are mediated by the endocrine and autonomic nervous systems under the control of the hypothalamus. The hypothalamus activates neural systems in the limbic forebrain and midbrain that interface with neural systems for the motor control of behavioral thermoregulatory responses. According to this view, the hypothalamus has an overall integrative function for physiological and behavioral thermoregulatory responses.

An alternative view, that the neural mechanisms controlling physiological and behavioral thermoregulatory responses are separate, has been proposed by Bligh (1973). Bligh suggested that the two kinds of responses are initiated by different stimuli, have different neural substrates, and have a different role in body-temperature homeostasis. Behavioral thermoregulatory responses, says Bligh (1973), are initiated by thermal discomfort detected by temperature receptors in the skin and are concerned with "the maintenance of thermal comfort [p. 193]." On the other hand, physiological (or what he designates *autonomic*) ther-

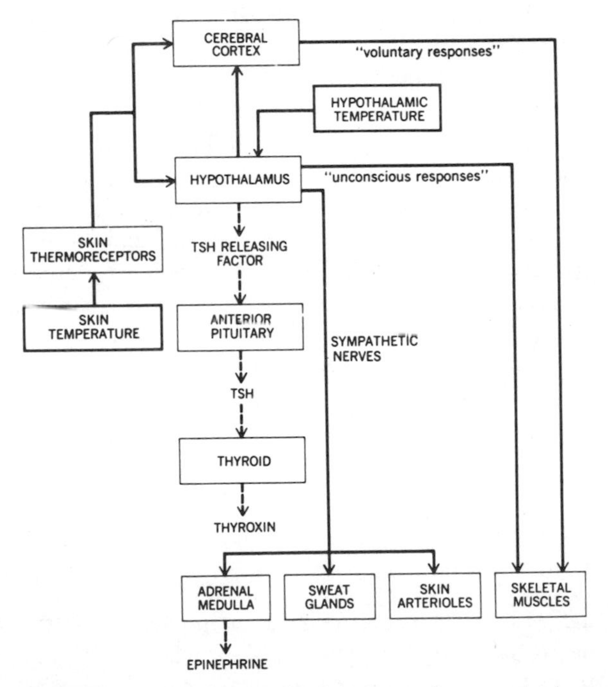

FIG. 3.18 Control diagram showing postulated thermoregulatory mechanisms under the control of the nervous and endocrine systems. Broken lines indicate hormonal pathways, and solid lines indicate neural pathways. (From Vander, Sherman, & Luciano, 1970.)

moregulatory responses are "concerned with the maintenance of a stable core temperature [p. 193]" and are initiated by central temperature receptors that detect deviations in core body temperature. Experimental evidence consistent with this view that neural mechanisms for behavioral and physiological or autonomic thermoregulatory responses may be dissociated have been reported by Adair and Hardy (1971). They observed that when the posterior hypothalamus of the squirrel monkey was cooled, the animal made the operant response of pulling a chain to obtain warm air, but autonomic thermoregulatory responses did not occur.

Satinoff (1974) has gone a step further by suggesting that not only are behavioral and physiological thermoregulatory responses under the control of separate neural mechanisms but that the various physiological thermoregulatory responses themselves are subserved by separate and independent neural mechanisms. Satinoff's (1974) proposal is as follows:

> Do thermoregulatory responses occur together for reasons other than simply that the thermal stimulus excites many independent mechanisms at the same time? The reason, then, that the preoptic area has been considered as a major integrative center may have nothing to do with integration *per se,* but simply may be an artifact of the functionally separate but anatomically overlapping passage of individual pathways through that part of the brain [p. 73].

In support of this view, Satinoff refers to the results of experiments in which recovery of body-temperature regulation was studied in rats after lesions of the anterior hypothalamus-preoptic region. It was observed that shivering, peripheral vasoconstriction, and increased metabolic rate in response to a cold ambient temperature, recovered independently and in different sequences for various animals. Satinoff attributes the differential recovery to differential amounts of damage to separate neural pathways associated with the three thermoregulatory responses and concludes that there is not a single integrative mechanism for body-temperature regulation.

The evidence seems to favor the views of Bligh (1973) and Satinoff (1974) that there is not a single integrative mechanism for the responses that contribute to body-temperature homeostasis. However, there may be an overall neural integration, to ensure "biological fitness" and survival, which determines the relative contributions of physiological and behavioral thermoregulatory responses in various circumstances. For example, a squirrel or rat placed in a cold room containing no materials for constructing a nest will shiver and increase metabolic rate and food intake, whereas these animals at the same ambient temperature in their natural habitat would rely on nest building and other behavioral thermoregulatory responses and not utilize physiological thermoregulatory responses. A squirrel kept from its nest and store of food in midwinter by a predator might rely primarily on physiological thermoregulatory responses and even tolerate a drop in core body temperature. Behavioral thermoregulatory responses do contribute to body-temperature homeostasis, and in many circumstances, they are the major adjustment, but behavioral thermoregulatory responses (like other motivated behaviors) compete with other behaviors in accordance with "motivational time-sharing" to ensure "biological fitness" and survival.

SUMMARY

Body temperature, like body water, sodium, pH, and other aspects of the internal environment, is regulated within narrow limits. Body-temperature homeostasis depends on thermoregulatory behaviors as well as internal physiological regulations, and, as shown by the classical experiments of Richter, behavioral and physiological thermoregulatory responses are complementary. These responses are initiated by thermal signals from the skin and/or from receptors in the CNS that respond when there is a deviation in core-body temperature from normal. Thermoregulatory behaviors have the same characteristics—being purposive (or goal-directed), persistent, and periodic—as other motivated behaviors considered in later chapters. Because of their close association with homeostasis, thermoregulatory behaviors have a high priority in the "motivational time-sharing" of most mammals.

The contribution of behavioral thermoregulatory responses to body-temperature homeostasis has been minimized and even ignored in the past. A major reason for this seems to have been that much of the earlier research on temperature regulation was done under artificial circumstances in which behavioral thermoregulatory responses were minimal or inoperative. Frequently behavioral responses have a more important role than physiological thermoregulatory responses, and in higher mammals, especially in man, behavioral thermoregulation not only reduces the necessity to utilize the physiological responses but also increases the range of thermal conditions in which man and higher animals can survive.

Studies of the central control of body temperature have focused on the hypothalamus, and there is a good deal of evidence to indicate that this small but strategically located neural structure makes an important contribution. The anterior hypothalamus-preoptic region contains temperature-sensitive neurons and is believed to be involved in the integration of signals from these and other central thermal sensors and of signals from temperature receptors in the skin. It has been proposed that the preoptic region is the "thermostat" or "central controller" for body-temperature regulation—integrating central and peripheral thermal signals and sending "command signals" to effector systems that control thermoregulatory responses. Some of these effector systems, represented in ventral and caudal regions of the hypothalamus, control the release of pituitary hormones that regulate endocrine mechanisms involved in energy metabolism and heat production and that control the peripheral autonomic nervous system. Experiments utilizing local warming and cooling of the preoptic region with a thermode indicate that the hypothalamus can also activate neural systems that subserve behavioral thermoregulatory responses.

A major issue is whether hypothalamic and associated neural mechanisms provide an overall integration for the control of physiological and behavioral thermoregulatory responses that are frequently complementary. The widely accepted view is that the preoptic region (designated the *temperature "controller"*) is the integrator of thermal signals and the source of "command signals" for both physiological and behavioral thermoregulatory responses. An alternative view, and the one favored in this chapter, is the proposal of Bligh that the two kinds of responses are initiated by different thermal signals, have different neural substrates, and have a different although complementary role in body-temperature homeostasis. Physiological thermoregulatory responses are initiated primarily by central temperature receptors that detect deviations in core body temperature and that maintain a stable core temperature. Behavioral thermoregulatory responses, on the other hand, are usually initiated by thermal discomfort detected by temperature receptors in the skin, and they contribute to the maintenance of thermal comfort. The neural substrates of behavioral thermoregulatory responses involve limbic structures and the cerebral cortex as well as the hypothalamus. Thermoregulatory behaviors are controlled by some of the same neural structures that subserve other motivated behaviors and compete with these behaviors ("motivational time-sharing") to ensure "biological fitness" and survival.

4
Feeding Behavior

Feeding is a commonplace yet essential behavior in man and other animals. Food is the source of energy for work and heat production and of ''building blocks'' for growth and tissue maintenance. Man living in primitive cultures and many animals devote a good deal of effort to procuring food, and other animals, such as cattle, spend much of their waking hours feeding. Squirrels and chipmunks store food at the end of the summer in preparation for winter when food is scarce; other animals, such as the bear, increase their food intake substantially in the late summer and fall, storing a good deal of energy as body fat prior to hibernating for the winter. The scarcity of food in certain parts of the world is the cause of social and political as well as health problems for people living in those regions. On the other hand, in affluent western countries, overeating and obesity is a major health hazard. It is important, therefore, to investigate feeding behavior and to understand how the brain controls food intake.

Feeding has been selected as the second motivated behavior for detailed consideration mainly because food intake is closely related to homeostasis in that it contributes to energy balance—to the regulation of body energy content. First we consider food intake from this point of view, and it is seen that energy deficit signals are only one of the determinants of feeding behavior: The palatability of food, external stimuli, circadian rhythms, and experiential and cognitive factors also contribute; these are discussed in relation to the characteristics of feeding as a motivated behavior. Consideration of the characteristics and determinants of feeding provide a background to presentation, in the final section of this chapter, of what is known about the neural substrates of feeding behavior.

CHARACTERISTICS OF FEEDING BEHAVIOR

A rat, cat, or other laboratory animal that has not had any food since the previous day will feed promptly when food is made available. A rat trained previously to traverse a straight alley or a complex maze to obtain food will run promptly along the alley or through the maze to the food and begin to eat. It will then feed for several minutes with occasional pauses that gradually become longer until feeding finally ceases completely. The behavior shows purposiveness (it is goal-directed) and persistence, like a number of other motivated behaviors (see Chapter 1).

The temporal pattern of feeding can be recorded by attaching an electronic device to the food dish (foodometer) or by having the rat press a lever to obtain food pellets. The record of feeding by a rat living in its home cage with food and water continuously available is shown in Figure 4.1. It is clear that feeding is periodic, another characteristic of motivated behaviors. The rat eats meals spaced a few hours apart with most of the meals occurring during the night. Feeding behavior follows a circadian rhythm with 80–90% of the total food intake occurring at night. The circadian rhythm of food intake is considered in more detail later.

Feeding behavior is periodic, in part, because the signals for its onset and termination wax and wane with the availability of body energy and, in part, because of events occurring in the external environment. Feeding competes with other behaviors that under certain circumstances have a higher priority ("motivational time-sharing").

FOOD INTAKE AND BODY-ENERGY HOMEOSTASIS

As indicated in Chapter 1, homeostasis, physiological regulations, and behavior are related. This is the case for feeding behavior, because body-energy content as reflected by body weight is relatively constant; energy intake as food equals energy expenditure, at least in the long term. Many investigators have considered the control of food intake from the viewpoint of regulatory physiology—of energy regulation—and have regarded feeding as a behavioral adaptation that contributes to body-energy homeostasis.

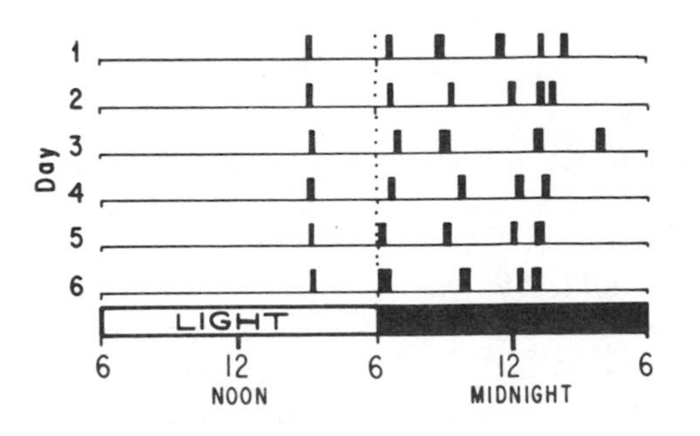

FIG. 4.1. Records of meals in the rat showing a circadian pattern of food intake; most of the feeding occurs in the dark period. (From Folk, 1974.)

Food Intake in Relation to Energy Balance

The basal energy requirements in man are approximately 1800 calories per day or 1.2 calories per kilogram of body weight per hour; for the rat, used extensively in laboratory studies of food intake and energy balance, the basal energy requirements are higher (5 calories per kilogram of body weight per hour), because it has a higher metabolic rate. In the normally active man, energy requirements are about 2500—3000 calories per day, and these requirements are considerably higher during intense exercise and physical work and in a cold environment. They are also increased during rapid growth and during gestation and lactation. In order to provide for and replenish such expenditures, energy is taken into the body as food.

It is widely known that depriving an animal of food for several hours is a reliable way to ensure that feeding behavior will occur. Feeding has been considered a response to energy deficits, and it has been widely assumed that animals eat to obtain calories. When the caloric content of the diet is altered, as in the classical diet dilution experiments (Adolph, 1947), rats readily adjust their food intake to maintain a caloric intake appropriate to energy expenditure. Similarly, when energy requirements are increased, as in the lactating female, during exercise or exposure to a cold environment, food intake increases in relation to increased energy expenditure (Stevenson, 1969). These observations suggest that an animal monitors energy expenditure (energy deficits) and responds by ingesting an appropriate number of calories of food. In the next section, we consider some of the signals and receptors associated with energy depletion and repletion that contribute to the initiation and termination of feeding behavior.

We should emphasize at this point that although the intake of energy as food matches energy expenditure, this is not done in a simple and direct manner. There are three complicating factors in attempting to account for food intake in terms of a simple feedback mechanism that reflects current energy expenditure. The first complication is that there is a delay between the ingestion of food and the subsequent availability of energy to the cells and tissues of the body. The delay occurs because time is required for digestion, absorption, and transport of the energy constituents. If feeding was regulated by a simple feedback control from energy depletion and repletion, this time lag would result in an excess or overshooting of food intake. Apparently there are other signals and feedback loops that enable an animal to turn feeding off before the food is digested and the energy made available for utilization by the tissues (for further details, see the next section and Figure 4.2). The second complicating factor is that the body can store appreciable amounts of energy. In man there are sufficient glycogen reserves, as a readily utilizable short-term energy source, for about a day and enough fat stores for about a month. Although energy expenditure is continuous, as a consequence of the capacity to store energy in the body, feeding may be and is usually periodic; man and higher animals are meal eaters. Energy stored in the

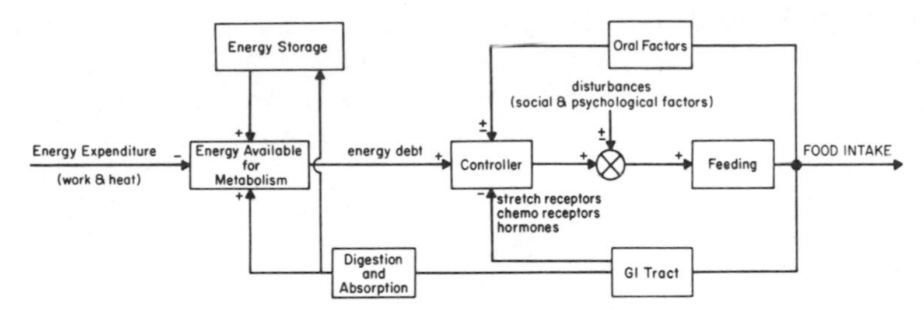

FIG. 4.2. A simplified model illustrating some of the signals and feedback loops that contribute to the control of food intake and the regulation of energy balance in the body. (From Mogenson & Calaresu, 1977.) This control theory and system analysis is a formal and rigorous approach to the biological principles formulated by Richter (1943), which associated behaviors such as feeding with physiological regulations and homeostasis. This has been stated very succinctly by Collier et al. (1976) as follows:

> In the strict sense (energy) homeostasis implies negative feedback, specialized receptors, and moment-to-moment monitoring of energy balance. In the case of feeding, the process is assumed to consist of successive depletion and repletion phases (De Ruiter, 1967). Depletion occurs as a result of metabolism. When energy stores are depleted below a threshold or critical value, feeding behavior (search, seizure, ingestion) preempts other ongoing activities. Ingestion leads to repletion, and, when an upper threshold of the energy stores or some surrogate of these stores is exceeded, ingestive behavior ceases. This depletion–repletion cycle differs from the more usual homeostatic systems since the item being controlled (e.g., food or water) is discontinuously present in most environments in contrast to an item such as oxygen which is continuously present (cf. Cannon, 1932). Thus feeding can only occur in episodes rather than continuously [p. 31].''

body can be made available in the intervals between meals so that the animal is free to engage in other activities.

Energy intake does not need to precisely equal energy expenditure in the short term, although in the long term, energy expenditure is balanced by energy intake as reflected by the fact that body weight (that is body-energy content) remains relatively constant; there is a long time constant for energy balance (Hervey, 1971). There are a number of cases in which energy intake does not match energy expenditure in the short term. Edholm and co-workers (1955) reported that there was no or only a low correlation between energy intake and energy expenditure of individual military cadets as determined for 24-hour periods but a good correlation when these measures were averaged over periods of 2 weeks. Rats take in more calories than they expend at night (and fat synthesis or lipogenesis occurs) (Le Magnen, 1975). Man may overeat on weekends and holidays but takes fewer calories than expended during part of the periods between.

The third complicating factor in attempting to account for food intake in terms of a simple feedback mechanism that reflects current energy expenditures is that the animal lives in a complex, variable, and even hostile external environment. Feeding competes with other behaviors (e.g., motivational time-sharing). Under certain circumstances food intake is postponed even in the presence of strong "hunger signals" because of the higher priority of another activity (e.g., territorial defense). On the other hand, feeding may occur in the absence of "hunger signals," especially when the food is highly palatable, or in man because of the influence of social or cultural factors (see pp. 95–96).

These considerations have led many investigators to suggest that food intake is influenced by a number of signals and that feeding is under multifactor control (Adolph, 1947; Stevenson, 1964). In the short term, some of these signals may result in an intake of food-energy that either exceeds or is less than current energy expenditure. However, because body weight remains relatively constant, there must be mechanisms that bring food intake in line with energy expenditure in the long term.

The next section deals with signals, feedback loops, and other mechanisms that control food intake in relation to energy balance.

Systems Analysis of the Mechanisms for the Regulation of Food Intake

"Feeding behavior is under multifactor control and the multiplicity of signals means that there are a number of feedback loops for the initiation and termination of food intake" (Mogenson & Calaresu, 1977, p. 10). For this reason systems analysis and a control diagram of the sort shown in Figure 4.2 are useful in specifying the signals, in identifying receptors and feedback loops, and in suggesting integrative mechanisms that should be investigated. For a contemporary discussion of this approach see Booth (1977) and Booth et al. (1976).

Glucose and fats are major sources of energy for work and heat production (see left side of Figure 4.2). The body also requires proteins (amino acids) for growth and the maintenance of tissue. There is evidence that the availability of glucose, fats, and amino acids is monitored and that glucostatic, lipostatic, and aminostatic signals are utilized in the control of food intake. Following the ingestion of food, the blood levels of these nutrients gradually increase as digestion and absorption from the gastrointestinal tract occurs, and, as shown in Figure 4.2, one or more of these signals that initially induced feeding behavior is then reduced.

Digestion and absorption of food require time, as indicated earlier, and feeding usually terminates before the nutrient-deficit signals disappear. Other signals are generated from the oral cavity (e.g., chemoreceptors in the mouth) and from the gastrointestinal tract (e.g., stomach distention; release of cholecystokinin, a possible "satiety" hormone), and these signals may contribute to preabsorptive satiety and to the termination of feeding behavior before there is any appreciable change in blood nutrient levels. These signals keep animals from eating in excess of their energy requirements and becoming fat. McFarland (1970) has emphasized that the relative contribution of signals and factors for the cessation of feeding vary from one species to another. He suggests that although the delays involved in digestion and absorption entail some sort of short-term satiety mechanism, different species have solved the problem in different ways (McFarland, 1970).

The feedback loops for the control of food intake are complex and, according to Van Sommers (1972), are organized in a hieararchy; they comprise:

> a system of superimposed controls operating in the mouth and stomach during eating, in the bloodstream as food is absorbed, and throughout the body as food is stored as fat or as body weight increases. Each successive level of control involves detection of changes over a longer time period and each serves to compensate for errors in the adjustment of the others. Finally, the further dimension of palatability must be added to these controls [p. 82].

Palatability, associated with taste, and oropharyngeal sensation facilitate food intake via positive feedback loops so that when feeding behavior begins, it is more likely to continue (DeRuiter et al., 1974; Wiepkema, 1971). This explains why you eat an attractive and tasty dessert after being satisfied by a big meal. Similar effects have been demonstrated in laboratory animals; for example, when the taste of food was changed every few minutes during a 2-hour feeding period, rats increased their food intake (Le Magnen, 1971). Positive feedback effects of taste and palatability are shown by a positive sign (oral factors) in Figure 4.2 (see figure legend).

Oral, gastrointestinal, and other postingestional signals occur together when an animal feeds, but there are experimental procedures that make it possible to separate their effects on food intake. When the stomach is removed surgically and the esophagus attached directly to the intestine, animals eat more frequent, smaller meals, but total food intake is typically unchanged. The influence of oral factors on food intake in the absence of postingestional factors has been studied in animals prepared with chronic esophageal or gastric fistulae. On the other hand, the influence of oral factors can be studied in the absence of postingestional factors by infusing liquid food through a tube that is inserted into the nostrils, extending through the nasal passage, pharynx, and esophagus to the stomach (Figure 4.3). When rats are trained to feed using this "electronic

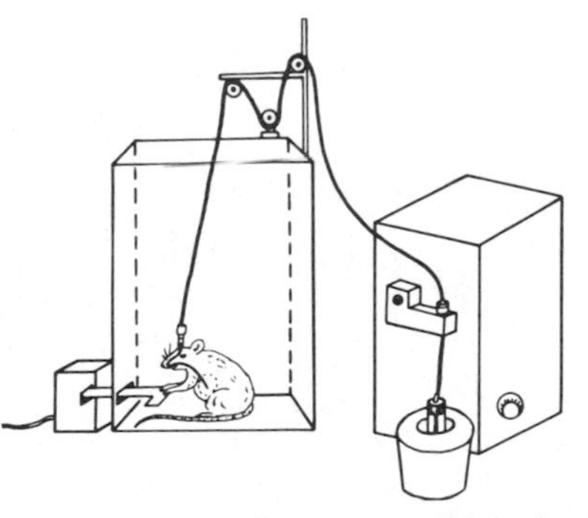

FIG. 4.3 Diagram of the apparatus for intragastric self-injection by the rat. The pipetting machine on the right delivers liquid diet directly into the rat's stomach via the chronic oropharyngeal gastric tube, when the rat depresses the bar. (Adapted from Epstein & Teitel baum, 1962.)

esophagus,'' they maintain themselves although at a somewhat lower body weight than normal; they adjust caloric intakes appropriately when the diet is diluted or made more dense (Epstein & Teitelbaum, 1962). Under these experimental conditions, smell, taste, and other oral stimuli are not necessary for the regulation of food intake. However, when these stimuli are available, they do influence food intake, and in the natural habitat of the animal, they may have an important role in discriminating and procuring food.

Another experimental strategy for eliminating the effects of oral factors on food intake is to employ chronic intravenous infusion of nutrients. Nicolaidis and Rowland (1976) used this procedure in rats to infuse a balanced liquid diet varying from 50 to 140% of normal daily intake. The results for one animal are shown in Figure 4.4. When the infusion equaled or was greater than normal intake, there was still an appreciable oral intake of the liquid diet. These investigators attribute the oral intake to a ''nonhomeostatic'' contribution to normal feeding behavior and suggest that there is an ''oral need'' independent of caloric or energy need.

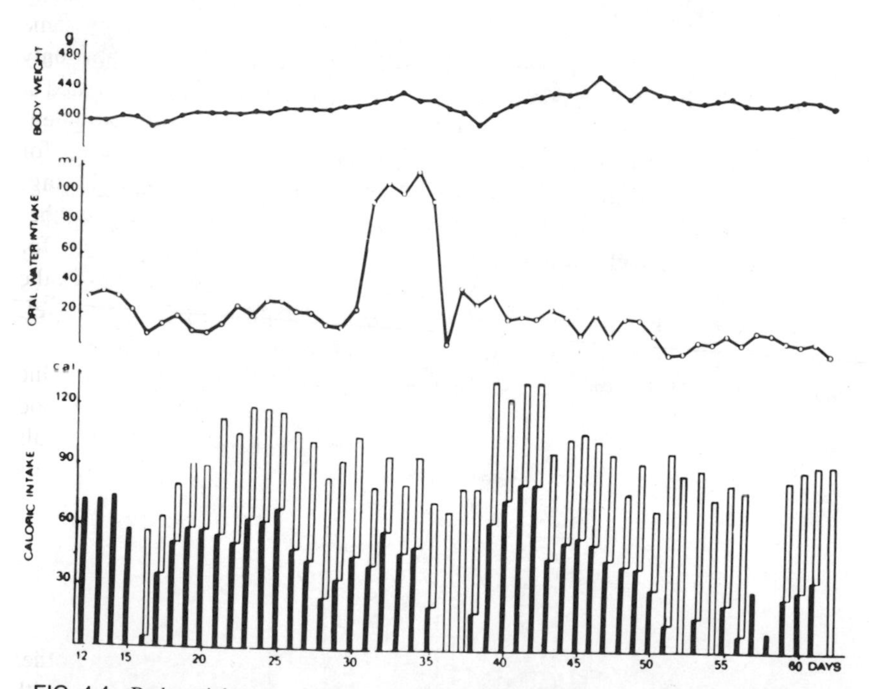

FIG. 4.4 Body weight, water intake, and caloric intake during long-term intravenous infusions of liquid diet. Black bars show the oral intake of liquid diet, and white bars show the amount infused (expressed in KCal). Note that there is a substantial oral intake of food, which varies with the amount infused. Note also the increase in water intake without any disruption of energy regulation during Days 31 to 35, when the diet was made hypotonic by the addition of NaCl. (From Nicolaidis & Rowland, 1976.)

In summary, a number of signals have been implicated in the initiation and termination of feeding behavior from laboratory studies, but much remains to be shown about their role, if any, in spontaneous, day-to-day feeding behavior.

FACTORS THAT INITIATE AND CONTRIBUTE TO FEEDING BEHAVIOR

The several determinants of food intake are discussed in this section in order to provide a perspective for considering later the neural substrates of feeding behavior.

Energy-Deficit or Homeostatic Signals

As indicated previously a reliable way to elicit feeding behavior is to deprive an animal of food. When the body weight of a rat or other experimental animal is reduced by 10 or 15%, there is a negative energy balance and strong energy deficit signals that initiate feeding. The most readily utilized, and at the same time the least abundant, source of energy is glucose. The body can store only enough glucose (as glycogen) for about 10–12 hours, so it must be replenished at relatively short intervals. It is not surprising, therefore, that considerable attention has been given to the possibility that the availability of blood glucose for body cells could serve as a signal for the initiation and termination of feeding.

The largest source of energy in the body is fat stores. Man, as well as a number of lower animals, typically have enough energy stored as fat to survive for several weeks. There is evidence that signals reflecting the amount of fat in the body influence food intake so that fat stores are regulated at an ''ideal'' level, around a ''set-point'' (Powley & Keesey, 1970).

There may be other homeostatic signals for the initiation of food intake. One possibility is the level or concentration of certain amino acids in the blood (Mellinkoff et al., 1956; Harper et al., 1970). Another possibility is that signals from dietary deficiencies (e.g., thiamine) act as the basis for specific hungers (Rozin, 1967).

The Influence of Palatability and External Stimuli on Food Intake

Internal signals related to the availability of glucose, free fatty acids, and other nutrients are not the only initiators of feeding behavior. When dogs prepared with gastric fistula had an amount of food equivalent to their normal daily intake placed into their stomachs prior to a feeding period, they still ate food by mouth when it was presented (Janowitz & Hollander, 1955). Appreciable food intake was observed even after a preload equivalent to 175% of their daily intake.

The smell and taste of food can initiate feeding, and if they increase palatability, total food intake may be increased (Young, 1967). A diet high in fat is very palatable for rats, and they increase their caloric intake and become very obese (Hamilton, 1964; Mickelsen et al., 1955); similar results have been reported for mice (Fenton & Dowling, 1953). Le Magnen (1971) has demonstrated in rats that following the same period of deprivation, food intake varies considerably depending on the oropharyngeal and olfactory stimuli associated with the food.[1] Food deprivation and palatability interact; hunger increases the range of foods that are accepted, and it increases palatability. Rats will accept a bland diet after being deprived and a bitter quinine-adulterated diet only after prolonged food deprivation. On the other hand, a highly palatable food will be eaten when the animal has just finished a meal of Purina chow, just as humans eat a rich, high-calorie desert after being satiated by the main course. As satiety or supersatiety occur, palatability decreases, as shown by the rating by human subjects of certain foods, initially judged pleasant, as unpleasant following a stomach load of 100 grams of glucose (Cabanac, 1971).

Cabanac (1971) suggests that the palatability of food and thus the amount of food eaten is determined by energy stores in relation to the set-point for body-weight (energy) regulation. He makes the speculative suggestion that it may be difficult for obese people, in whom the set-point for body-weight regulation is abnormally high, to remain lean after losing weight because their body weight is now well below the set-point; the palatability of the food is thereby increased, and body weight gradually drifts up again as they begin to increase their food intake.

Food intake may also be initiated by the sight of food, particularly if it is a highly preferred food, and in man and higher animals by the sight of other individuals eating. This depends on past experience in acquiring preferences for certain foods and, as indicated in the next section, on social learning whereby custom, cultural constraints, cuisine, etc., influence the choice of food and eating habits.

Schachter (1967) and his co-workers have called attention to the importance of external stimuli by comparing obese people with individuals of normal weight. In one study it was shown that obese subjects ate more cashews that had a light focused on them than cashews in normal lighting. It was also reported that obese

[1]In fact, as pointed out by Le Magnen (1971), taste usually contributes less to palatability than odor:

> Discriminative performances and thus the possibility of specific feeding responses offered by the two chemosensory systems are very different. Gustatory properties of food stimuli and the gustatory apparatus permit only a rough separation of four categories of stimuli: sweet, sour, salty and bitter. . . . On the contrary, olfactory action of a food in combination with the mechanosensory (proprioceptive touch) one offers the possibility of unlimited sensory-specific responses of palatability [pp. 222–223].

When the odor of food was changed every half hour during the 2-hour feeding period, rats ate 270% as much as compared to the standard diet (Le Magnen, 1971).

subjects ate more sandwiches if they were nearby on a table than if they were in a refrigerator out of sight. In both cases normal-weight subjects ate the same amount whether the cashews or sandwiches were visually prominent or not. Schachter proposed that obese subjects were more responsive than normal-weight individuals to external food-related stimuli (e.g., smell, sight, taste, other people eating) and that they have a deficiency in responding to internal signals reflecting the energy state of the body. This hypothesis of heightened external and reduced internal responsiveness of obese people is supported by a number of studies (e.g., Rodin & Slochower, 1976). For example, one of Schacter's students has shown that overweight subjects consumed excessive amounts of a chocolate milkshake through a tube when the feeding machine was in another room but adjusted intake to deprivation when the machine was in the same room and visual cues concerning intake were available.

Experiential and Cognitive Factors

The influence of food-related stimuli on feeding behavior may be direct, as in the case of food intake being enhanced by a sweet taste, or indirect, as in the case of acquired aversions to foods associated with poisoning (Rozin & Kalat, 1971) or food preferences and aversions related to being raised in certain cultures (Rozin, 1976). The role of experiential and cultural factors has not been emphasized in previous laboratory studies, which mainly used the laboratory rat, usually food-deprived and fed laboratory chow. However, even in the rat, the role of associative and other cognitive processes in the control of food intake has been demonstrated. Le Magnen (1971) fed rats three 1-hour meals each day spaced 7 hours apart. Following the deletion of one of the meals, the rats initially ate more during the meal that followed 15 hours later. Within a few days, however, they ate more during the preceding meal. It has been suggested by a number of investigators of feeding behavior that food intake also occurs prior to energy-deficit signals for a number of species in their natural habit. Very good arguments for this view are given in a recent article by Collier et al. (1976), who summarize the evidence as follows:

> The basic problem for the animal is to partition his time and energy between the many different activities which insure his reproductive success. An animal which did not consider the density, cost, and character of the item to be procured but waited to initiate feeding when it was "hungry" in the physiological sense would simply be unlikely to survive in any but the most permissive environment. It seems more likely that animals must, in some sense, feed in *anticipation* of, rather than in response to, their needs. Obvious examples of such anticipating behavior are premigration hyperphagia (Odum, 1960) and hibernation (Mrosovsky, 1971) [p. 46].

According to Oatley (1973), this anticipatory feeding (food intake that occurs prior to or in anticipation of energy deficits) depends on the animal utilizing neural representational processes of the external environment "deployed not in the feedback mode so that deficits are corrected, but in a feed-forward mode so that they can be anticipated [p. 221].''

In higher mammals and especially in man, there is a greater utilization of experiential and cognitive processes for the initiation of various behaviors including feeding. For man, what is eaten, when, and how much, is strongly influenced by social and cultural factors (Figure 4.5). For example, Italians, as well as Italian immigrants to North America, prefer certain foods and have favorite dishes; the food choice and eating habits of the Japanese are quite different. The role of social and cultural factors has been studied by Rozin (1976), and we quote a relevant section from his recent review:

> Human food selection is largely determined by cultural constraints or cuisine, rather than individual experiences or decisions. Cuisines are sets of practices concerning the basic foods eaten, the flavors added to these foods, the preparations of foods, and special constraints such as taboos. They are the produce of a complex interaction of food-related and social factors [p. 285].

> The analysis of cuisines is quite complicated, since food is more than something to eat for humans. Religious and social factors combine food-related factors in determining a cuisine. Food is a characteristic cultural expression, which like custume, can serve to provide a group with a distinct identity. The universe of potentially eatable foods is usually highly restricted: it is determined in large part by availability and the cuisine of the culture in question. Humans are virtually alone in the world in having a body of rules or limitations about what to eat and how to eat it [p. 297].

FIG. 4.5 This is an example of food intake not induced by ''hunger'' deficit signals. The young man in the photos is showing obvious enjoyment while eating a banana split. (Photo by B. Box.)

Hormones

There are a number of examples of marked changes in food intake when hormones are administered or endocrine glands removed. The administration of insulin increases food intake, and if the injections continue for some days, the animals become obese (MacKay, Calloway, & Barnes, 1940; Hoebel & Teitelbaum, 1966). Following hypophesectomy food intake is reduced but is restored by the administration of growth hormone (Bray, 1976). In the female rat, food intake, as well as the rate of body-weight gain, are reduced after puberty as compared to the male. Subsequently, food intake varies inversely with the plasma level of estradiol in the female rat during estrous cycles. Following ovariectomy there is a marked increase in food intake and body weight and a reversal of these effects when estradiol is administered (Mook et al., 1972; Wade & Zucker, 1970a), suggesting that estrogens have an inhibitory effect on feeding behavior.

It is necessary to consider whether these changes in food intake are the result of direct effects of hormones on neural systems that control feeding behavior or are secondary to changes in blood nutrient levels. In most cases it is the latter. For example, in the cold-exposed animal, increased food intake is associated with increased output of thyroid and adrenal cortical hormones and of adrenal catecholamines. These hormonal responses increase metabolic rate and energy expenditure, and the increase in food intake is secondary to increased energy expenditure for heat production rather than to a direct effect of thyroid or adrenal cortical hormones on neural mechanisms that control feeding behavior (Bray & Campfield, 1975).

Even the inhibitory effects of estrogens on food intake may not be a direct influence on feeding behavior. Several investigators have suggested that estrogens act by reducing the set-point for body-weight regulation (e.g., Fishman, 1976). This possibility is considered in a later section dealing with the neural substrates of feeding behavior.

Insulin administration increases food intake, as indicated previously, apparently because blood sugar levels are reduced and central or peripheral glucoreceptors are activated. However, this is not the normal physiological response to insulin. Steffens (1970) has reported a large endogeneous release of insulin during and following a meal in the rat, suggesting that insulin is associated with the termination of feeding rather than its initiation. Results of electrophysiological experiments, discussed in a later section, indicate that insulin has a direct effect on the activity of hypothalamic neurons implicated in the termination of food intake (see p. 106).

Circadian Rhythms and Food Intake

As indicated earlier, feeding behavior occurs in a periodic manner and in many species is clearly circadian. Richter (1943), one of the first investigators to study the periodicity of feeding, demonstrated that rats maintained on a 12-hour dark–12-hour light cycle ingest 80–90% of their daily food intake during the dark period (Figure 4.1). It is now widely accepted that the rhythmical occurrence of feeding, like other circadian rhythms, is endogenously controlled by a "biological clock" and that the rhythms become entrained to light (Rusak & Zucker, 1975).

Although circadian rhythms of food intake may be modified, they persist under a variety of conditions. They continue during water deprivation but with a reduced total intake of food. During lactation, when energy demands and food intake double or triple, the circadian rhythms of food intake are observed, but there is a proportionately greater increase of food intake in the light than in the dark (Levin & Stern, 1975). The pups do much of their suckling in the light so that energy expenditure by the mother is presumably high during this period. Similarly, when total food intake is increased by lesioning the ventromedial hypothalamus, there is a larger increase in food intake during the light than during the dark (Kakolewski et al., 1971). In the female rat, food intake is depressed during proestrus when there is a peak of running activity; the 4-day estrus cycle is "superimposed on a circadian rhythm of food intake similar to that of the male" (Ter Haar, 1972).

In infant rats the circadian rhythm of food intake is markedly influenced by the behavior of the mother. Recently it has been reported that from Day 4 to 17, rat pups gained more weight (and presumably consumed more mother's milk) during the day than at night. This is the period when maternal behavior is most prevalent and the pup receives more attention from the mother. After Day 19, when the pups are eating solid food, most of their feeding occurs at night. A similar pattern (diurnal intake of milk in the light period from Days 4 to 17 and nocturnal intake of solid food from Days 19 to 35) was also observed in blind rats raised by sighted mothers. When the blind pups were weaned at Day 21, Day 29, or Day 31, nocturnal feeding continued for only about 4 days after weaning. Subsequently, they ate about the same amounts of food during the day and night. It appears that maternal responses and feeding activities of the mother rather than a circadian light–dark cycle determined the food intake pattern of the young.

Although food intake is strongly influenced by circadian rhythms, this circadian periodicity has no biological utility in the relatively constant conditions of the laboratory. In the natural habitat, however, circadian rhythms of food intake do have biological significance and must be taken into account if we are to eventually understand the neural substrates of feeding behavior. Circadian rhythms of ingestive behavior and rest–activity, considered from the evolutionary point of view, ensure that energy expenditure and activity occur at times of the day optimal for the survival of the animal and the species (Rusak & Zucker, 1975). In hibernating species the annual rhythm of hyperphagia during the fall is also adaptive, because it enables the animal to have sufficient energy stores to survive during the winter when food is scarce; during hibernation energy expenditure is much reduced (Mrosovsky, 1971).

ONTOGENY OF FEEDING BEHAVIOR

For many animal species, feeding behavior does not appear in the "full-blown" adult pattern at birth. The capacity for sensory and motor functions and for sensory–motor integration develop as a function of maturity of the CNS and the opportunity to interact with the environment. Immaturity of sensory, motor, and integrative mechanisms in the newborn and young animal may particularly limit food-seeking and food-procuring responses. Many animals are relatively helpless for a period of time after birth, and the appetitive behaviors that precede the ingestion of food are very limited or even absent. Some of the signals utilized by the adult animal for the initiation and cessation of feeding behavior may not be responded to by the young animal, because the sensory and/or motor neural mechanisms have not matured sufficiently. A good deal can be learned from the ontogenetic approach; in particular, it demonstrates that the various factors that initiate and contribute to feeding behavior have a differential role depending on the age of the animal and the stage of development.

Suckling, a reflexive, consummatory response, is, however, present at birth in most infant mammals including the rat. Infant rats can also increase or decrease the intake of the mother's milk when the stomach is emptied by deprivation or filled by milk or kaolin (Houpt & Epstein, 1973; Satinoff & Stanley, 1963). According to Epstein (1976a):

> At birth, the fullness or emptiness of the upper gut controls feeding behavior. Evidently the pup roots and suckles to fill a gastric void and the nutrient quality of the fluid in the gut plays no detectable role. Sleep is the pup's dominant behavioral state. The pup seems to be aroused by signals from an empty stomach. Then, aided by the dam, it suckles, performing the only ingestive behavior for which it is fully competent, and having filled the upper GI tract it lapses back into sleep [p. 195–196].

Food intake relative to body weight is considerably greater in the infant rat than it is in the adult. According to Harper and Boyle (1976), "energy intake of the young rat varied as body weight to the 0.87 power over the wide range of growth rates and body sizes—for mature rats, food intake varied as body weight to the 0.35 power [p. 9]." It appears that the infant rat is a glutton, food intake ceasing only when the stomach is full. Gradually within the first few weeks of life, gastric distention is supplemented by other controls, and as a consequence, the ingestive behavior of mammals is changed dramatically. According to Epstein (1976a):

> In the second week of the suckling period physiological competence for thermoregulation, locomotion, sustained wakefulness and use of distance-sense improves and the animal begins to show some mastery of its habitat. Appetitive behaviors are now possible. By 14 to 15 days after birth the animal can, in fact, be forcibly weaned. But natural weaning is delayed for at least another week assuring maximal success in the transition to the more independent life of the free-foraging adult. Like birth, weaning is delayed well beyond the developmental age at which survival is merely possible [pp. 197–198].

About the time of weaning, the circadian rhythm of food intake observed in the adult first appears. As indicated previously, during the first 2 to 2½ weeks, when the mother's milk is the only or the major source of food, feeding and body weight gain is diurnal (occurring in the daytime). When the young rat has switched to eating solid food, the nocturnal pattern of food intake begins (at about the end of the third week of life) and continues throughout the life of the animal (Levin & Stern, 1975). It may be that diurnal suckling in the rat pup is the result of the greater frequency of maternal attention during the light period. The mother rat is also less active in the daytime, so suckling is not as frequently disturbed by her locomotor activity.

The feeding response to the administration of insulin, which reduces blood glucose levels (Lytle, Moorcroft, & Campbell, 1971), or 2-deoxy-D-glucose, which produces glucoprivation (Houpt & Epstein, 1973), does not occur until the fourth or fifth week after birth in the rat. "The glucoprivic control is clearly not necessary for normal ingestive behavior in early postnatal life and may play only a minor role thereafter, once it matures in early adulthood" (Houpt & Epstein, 1973, p. 65).

NEURAL SUBSTRATES OF FEEDING BEHAVIOR

Investigations of the neural substrates of motivated behaviors began primarily, as indicated in Chapter 1, with the study of the neural mechanisms of feeding, and an extensive literature on this subject has accumulated during the last 35 years. It is not possible in the space available for this chapter to deal with all of the experimental work that has been done. In any case, as indicated in detail later, recent observations using newer experimental techniques have led to a reinterpretation of some of the earlier findings, which may not have the same theoretical significance assumed a few years ago. It seems appropriate, as with the behaviors discussed in other chapters, to consider what is known about the neural substrates of feeding behavior in relation to the factors that contribute to feeding. However, in view of the major impact that earlier research on the neural mechanisms of feeding behavior has had on theoretical ideas and models, this section begins with a brief historical overview in order to provide a perspective for what follows.

Historical Developments

The foundations for the study of the neural substrates of feeding behavior were established in the 1940s and 1950s: Using the recently introduced stereotaxic procedure, lesions of the ventromedial hypothalamus (VMH) were reported to cause hyperphagia and obesity; a few years later, lesions of the lateral hypothalamus (LH) were observed to result in aphagia, weight loss, and death in many cases if the animals were not fed by stomach tube (Figure 4.6). The results of these classic experiments suggested the presence in the hypothalamus of a dual

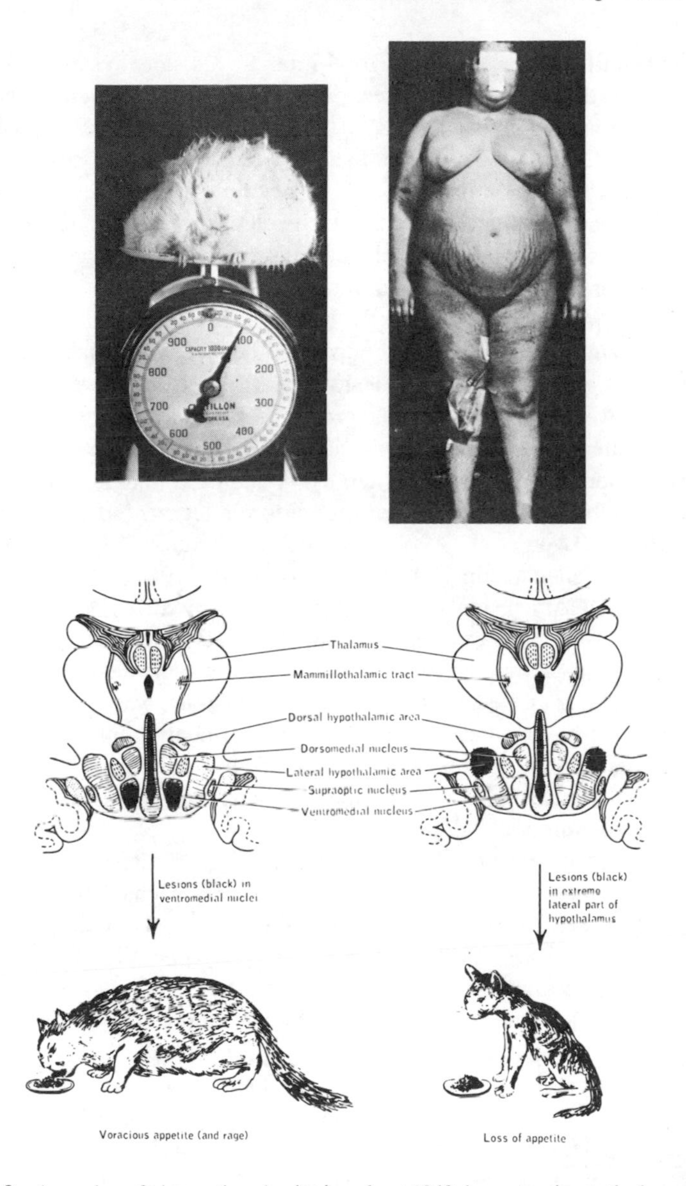

FIG. 4.6 A number of observations beginning about 1940 drew attention to the hypothalamus in the control of food intake. Top left: A rat with bilateral lesions of the ventromedial hypothalamus overeats and becomes obese (from J. A. F. Stevenson). Top right: A patient with a tumor of the ventromedial hypothalamus which produced overeating and obesity (from Reeves & Plum, 1969). At bottom: A diagram to illustrate the effects of hypothalamic lesions on food intake. The cat on the left with lesions of the ventromedial nucleus has a voracious appetite and the cat on the right with lesions of the lateral hypothalamus exhibits aphagia (from House & Pansky, 1967). These and other observations supported the dual-mechanism model which was influential and popular for a number of years.

neural mechanism for the control of food intake—a satiety system in the VMH and a feeding system in the LH (see Figure 1.8A). Subsequently, additional experimental observations supported this dual-mechanism model: Feeding was elicited by electrical stimulation of the LH (Miller, 1960); food intake was increased when the VMH was depressed by procaine, a local anesthetic (Epstein, 1960); in electrophysiological experiments the electrical activity of VMH and LH were shown to be reciprocally related (see Figure 2.21) and to contain glucosensitive and glucoreceptive neurons (Oomura et al., 1969).

About 15–20 years ago, the results of lesion and electrical stimulation experiments began to implicate limbic forebrain and other extrahypothalamic structures in feeding behavior. For example, lesions of the amygdala were observed to produce either hyperphagia or hypophagia, apparently depending on the locus of the lesions (Fuller et al., 1957; Green et al., 1957; Fonberg, 1969), and electrical stimulation of the amygdala was shown to reduce food intake (Fonberg & Delgado, 1961). Subsequently, lesions of the midbrain tegmentum were reported to produce aphagia (Lyon et al., 1968; Parker & Feldman, 1967), and stimulation of this region elicited feeding (Wyrwicka & Doty, 1966). As indicated previously, the dual-mechanism model was retained but modified by postulating that limbic forebrain structures exerted modulatory effects on the primary satiety and feeding systems in the hypothalamus, which in turn generated "command signals" that were transmitted to the midbrain (see Figure 1.8B). This expanded model continued to be widely accepted and influential through the 1960s, stimulating new research and providing a conceptual framework for interpreting the results obtained; during this period the model and the empirical findings were mutually reinforcing.

Toward the latter part of the 1960s, reservations and criticisms were beginning to be expressed about the appropriateness of the dual-mechanism model—about the specificity and stability of the LH systems that subserve feeding and drinking (Valenstein et al., 1968) and about the specificity of the VMH satiety system, which was assumed to terminate feeding (Grossman, 1966). Then in 1971 there was a crucial development that gave considerable impetus to a reassessment of the dual-mechanism model and to a reinterpretation of the lesion and stimulation experiments that had focused attention for so many years on the hypothalamus. Ungerstedt (1971) reported that aphagia and adipsia, similar to that produced by the classical LH lesions, occurred when the dopaminergic (DA) neurons of the nigrostriatal pathway were damaged. The fibers of these DA neurons project from the substantia nigra along the lateral edge of the LH to the caudate and putamen (see Figure 2.24). In subsequent years a number of investigators turned their attention to these DA projections and to other transmitter-specific pathways, and the hypothalamus was deemphasized. We deal with some of these experimental studies in a later section and then consider what role, if any, the hypothalamus has in feeding behavior.

In the meantime we return to the basic question: What initiates feeding behavior? From the recent developments just referred to, it does not seem to be a good idea to organize what is known about the neural substrates of feeding behavior in relation to the dual-mechanism model. Instead, the relevant experimental evidence is considered from the viewpoint of the various factors that initiate and contribute to food intake.

Homeostatic Deficit Signals and Feeding: Neural Substrates

As indicated in an earlier section, the body can detect energy deficits. For example, following a period of deprivation, a dog eats eagerly when food is presented, and a rat performs an acquired operant response, such as pressing a lever, to obtain food pellets. Feeding behavior is also initiated when blood glucose levels are reduced by the injection of insulin (Hoebel & Teitelbaum, 1966). According to classical theories, the availability of glucose (e.g., glucostatic theory; Mayer, 1952) and fats (e.g., lipostatic theory; Kennedy, 1953) are monitored and signals generated for the initiation and termination of feeding behavior. Evidence for glucoreceptors and for receptors that monitor other nutrients is considered in the following.

The first evidence for central glucoreceptors, assumed to monitor blood sugar levels, came from experiments in which the administration of the chemical compound goldthioglucose by intraperitoneal injection resulted in hyperphagia and obesity in mice (Mayer & Marshall, 1956). It was suggested that nerve cells, which showed a preferential uptake of goldthioglucose and which were damaged by the heavy metal, were glucoreceptors. Because there was extensive damage to the region of the VMH, and this nucleus had in earlier lesion studies been implicated as a "satiety" center, it was concluded that glucoreceptors in the VMH controlled the termination of feeding behavior. This view was subsequently supported by the results of electrophysiological recording experiments of Anand and co-workers (1962), who demonstrated that neurons in the hypothalamus changed their rate of discharge when glucose was injected into the circulation. Although these recording experiments showed that there are glucosensitive neurons in the hypothalamus, the results did not prove that these neurons are glucoreceptors, because the recording could have been from neurons in the neural circuit rather than from receptors, which might be located elsewhere in the brain or even in the periphery (Mogenson, 1975).

More recently Oomura and his co-workers have demonstrated glucosensitive neurons in the VMH, as well as in the LH, and have provided evidence that those in the VMH are glucoreceptive. Using the technique of iontophoresis, involving multibarrel pipettes so that action potentials can be recorded from a single neuron and small quantities of glucose (or other chemical compounds) administered directly to the neuron, these investigators showed that about one-third of the neurons in the VMH increased their discharge rate when glucose was applied (Figure 2.23). Neurons in the LH decreased their discharge rate when glucose was applied iontophoretically; that is, they fired more often when glucose was not present in their extracellular environment as would be expected of cells that would initiate feeding when blood glucose levels decreased. In both cases the response to glucose was insulin-sensitive. It seems clear, as concluded by Oomura (1976), that neurons (glucoreceptors) in the hypothalamus can respond to changes in the levels of glucose in the blood. However, it has not been established as yet whether these glucoreceptors contribute to the feeding behavior. Because feeding is not reliably reduced by the direct administration of glucose to the hypothalamus (Panksepp, 1975a) or initiated by the administration to this neural region of 2-deoxy-D-glucose (a glucose analogue and antimetabolite of glucose that causes glucoprivation and should activate glucoreceptors), it has been suggested that glucoreceptors in the hypothalamus may be concerned not with the initiation or termination of feeding behavior but rather with the autonomic and endocrine regulations for controlling blood glucose levels (Epstein et al., 1975; Panksepp, 1975a). It should be noted, however, that feeding is initiated by administering 2-deoxy-D-glucose into the cerebral ventricles (Berthoud & Mogenson, 1977; Miselis & Epstein, 1975), suggesting the presence somewhere in the brain of glucoreceptors associated with feeding. The locus of these central glucoreceptors is uncertain, however, at the present time.

The possibility that there are glucoreceptors in the liver was first suggested by Russek (1963), who observed that the infusion of glucose into the hepatic portal system reduced food intake in dogs. He attributed this satiety effect to signals from hepatic glucoreceptors being transmitted to the CNS. A few years later, Niijima (1969) reported that the number of action potentials recorded from the vagus nerve of the guinea pig was reduced when the liver was perfused with a glucose solution and concluded that glucoreceptors in the liver sent signals to the CNS. More recently additional support for the view that there are hepatic glucoreceptors has come from experiments in which feeding was initiated by infusing 2-deoxy-D-glucose into the hepatic portal system, and food intake of food-deprived animals was suppressed by infusing glucose into the hepatic portal system (Novin et al., 1973). Because vagotomy significantly attentuated feeding initiated by hepatic portal infusions of 2-deoxy-D-glucose, it appears that signals from hepatic glucoreceptors are transmitted along the vagus nerve (Novin, 1976).

There is evidence that signals from glucoreceptors in the liver reach the lateral hypothalamus via the vagus nerve and possibly the splanchnic nerve. In acute electrophysiological recording experiments similar to those of Niijima, neurons in the lateral hypothalamus were shown to respond when the liver was perfused with glucose, and the responses were attributed to afferent signals from hepatic glucoreceptors (Schmitt, 1973). Whether these inputs were associated with the initiation or termination of feeding could not be determined from such acute experiments. However, the possible role of hepatic glucoreceptors in the initiation of feeding is suggested by experiments in which feeding, initiated by perfusing 2-deoxy-D-glucose into the hepatic portal system, was disrupted when the lateral hypothalamus was lesioned (Epstein, 1971; Novin et al., 1976). According to Novin (1976), glucostatic signals are transmitted from hepatic (and also duodenal) glucoreceptors via the vagal and splanchnic nerves to the lateral hypothalamus, and these signals contribute to feeding and satiety (Figure 4.7). An alternative interpretation of the lesion data is that following lesions of the LH, there is a greater sensitivity to thirst and metabolic challenges such as 2 DG (Stricker, 1976). In other words, LH lesions may produce a nonspecific deficit, not only a feeding deficit.

There is also evidence, as indicated previously, that the body monitors fat stores, and the VMH has been implicated in the integration of lipostatic signals for the control of food intake and the long-term regulation of body energy and body weight. The nature of the signals and the site of the receptors is not known, although several speculative proposals have been made. Hervey (1969) suggested that a signal of fat stores might be provided by dilution of steroid hormones, known to accumulate in fat tissue. Baile and co-workers (1971) suggested that prostaglandins might signal adiposity. Another more recent suggestion is that the VMH contains "transducer cells" (designated "privileged cells" by Nicolaidis, 1974), which monitor local lipogenesis and contribute to the long-term regulation of food intake and energy balance (Panksepp, 1975a).

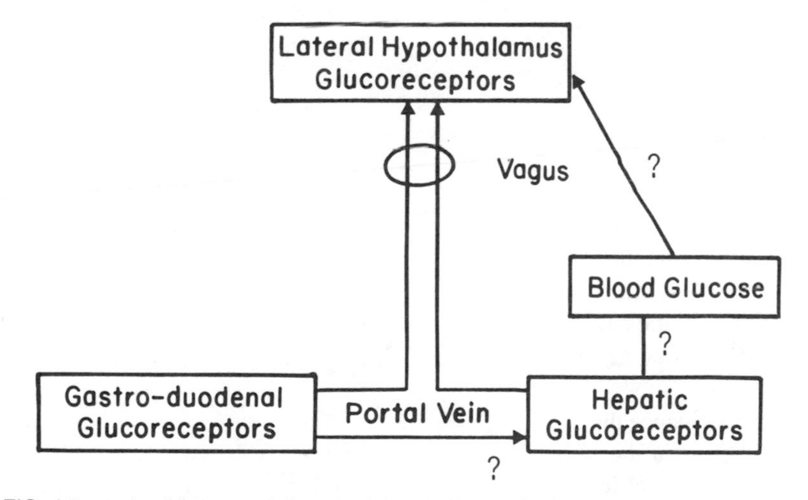

FIG. 4.7 A simplified control diagram of the possible mechanism of action of glucoreceptors. Hepatic and duodenal receptors require an intact vagus nerve to transmit glucostatic signals. It is still not known whether hypothalamic glucoreceptors modulate food intake directly or whether the LH simply "translates" and transmits messages from the peripheral glucoreceptors. (Modified from Mogenson, 1975.)

When animals have been food-deprived, the concentration of free fatty acids (FFA) in the blood is high due to lipolysis (conversion of fat stores to FFA). The intravenous injection of FFA in rats has been reported to increase food intake (Adair et al., 1968). Oomura (1976) has reported that glucosensitive neurons in the LH of the rat respond by increasing their rate of discharge when FFA are administered iontophoretically (Figure 4.8). On the other hand, FFA so administered to glucoreceptor neurons in the VMH inhibit their discharge. Oomura concludes that the hypothalamus is sensitive to FFA circulating in the blood and contributes to the control of feeding behavior. He suggests that FFA initiate feeding by activating neurons in the LH and inhibiting neurons in the VMH, which would (because of disinhibition) further increase the discharge of LH neurons. This interpretation is in accord with the dual-mechanism model, which assumes that the LH is the focus of a feeding system and the VMH of a satiety system (see Figure 1.8A).

Amino acids, important nutrients that have a major effect on food intake, might also provide a signal that contributes to the control of food intake (Harper et al., 1970). Panksepp (1975a) has reported that the direct administration of amino acids (e.g., glycine, alanine) into the lateral and ventromedial hypothalamus attenuates feeding, and he has suggested that there may be an "aminostatic mechanism" for the termination of feeding behavior. Panksepp points out that amino acids (e.g., GABA and glutamic acid) are formed during the

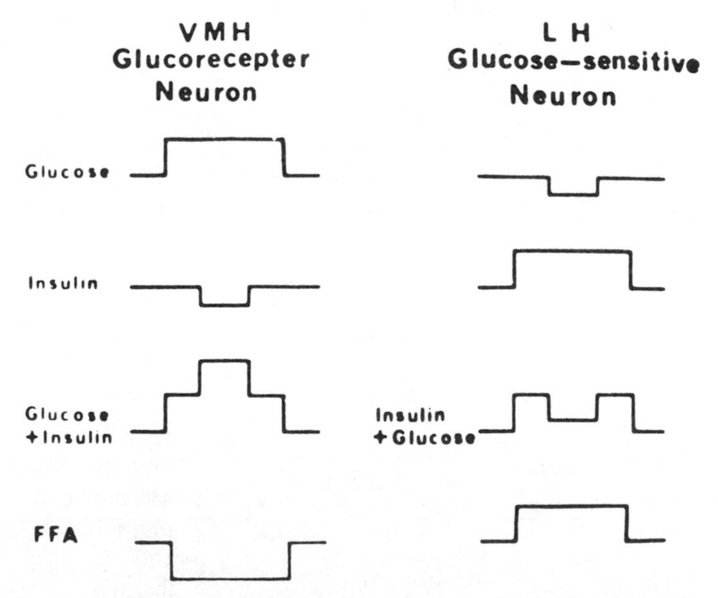

FIG. 4.8 Diagrammatic representation of hypothalamic neuronal discharge frequency after the microiontophoretic application of various solutions to single neurons in the rat. Upward deflection indicates an increase in discharge frequency. FFA = free fatty acids. (From Oomura, 1976.)

metabolism of glucose in the hypothalamus, and, therefore, ''aminostatic mechanisms'' could also be involved in mediating the effects of glucose on satiety. Again, although these are stimulating ideas, we do not yet know the neural substrates by which depletion and repletion of amino acids influence feeding behavior.

In summary, feeding behavior in the short term is related to the levels of glucose and free fatty acids in the blood and in the long term to body fat stores. It is widely assumed that the hypothalamus monitors blood levels of these nutrients directly and/or integrates signals from peripheral receptors. Lesions of the VMH in experimental animals have been interpreted as disrupting these integrative processes. It has also been proposed that obese humans have a deficit in responding to internal signals and rely heavily on external food-related stimuli. In this respect they have been compared by Schachter to rats with VMH lesions. However, much of the evidence concerning the central integrative mechanisms is indirect and inconclusive. The best evidence for glucoreceptors and fatty acid receptors in the hypothalamus comes from acute electrophysiological experiments, but these receptors have not been directly linked with feeding. It is possible that these receptors are concerned with autonomic and endocrine mechanisms for the regulation of blood glucose levels. Clearly, we know less about the neural mechanisms that subserve regulatory feeding than was assumed a few years ago when the classical dual-mechanism model was widely accepted.

Hormonal Effects on Food Intake: Neural Substrates

In this section we deal mainly with the effects of insulin and the estrogens on food intake. Other hormones, such as thyroid and adrenal cortical hormones, do influence food intake, but as indicated in an earlier section, the effects are secondary to changes in energy metabolism and expenditure.

The first evidence that insulin might act on hypothalamic feeding mechanisms came from experiments in which goldthioglucose, the compound shown to damage the VMH and produce hyperphagia and obesity (see p. 103), was observed to be ineffective in diabetic animals (Debons et al., 1968). Goldthioglucose damaged the VMH, and hyperphagia and obesity resulted only if the diabetic animals were administered insulin (Debons et al., 1969). From these observations and other evidence discussed in the previous section, it was concluded that insulin-sensitive glucoreceptors, present in the VMH, respond to the high plasma levels of insulin and glucose to produce satiety—the VMH being assumed to inhibit the LH to terminate feeding behavior (see Figure 1.8A). This proposal was recently supported by the results of electrophysiological experiments. Oomura (1976) showed that glucoreceptor neurons in the VMH increased their rate of discharge to the application of glucose when insulin was administered to the same neuron (Figure 4.8). Oomura suggested that insulin-sensitivity was due to the presence of insulin receptors on glucoreceptor neurons. On the other hand, when insulin was administered to glucosensitive neurons in the LH, their rate of discharge increased whereas their response to glucose was reduced (Figure 4.8).

These findings are consistent with the classical observation that the systemic administration of insulin initiates feeding. According to the dual-mechanism model for the control of food intake, this could be explained by insulin-inhibiting glucoreceptor neurons in the VMH and activating glucosensitive neurons in the LH. However, the exogeneous administration of insulin to induce feeding is not a physiological procedure, and it has been shown that plasma levels of insulin fall rather than rise just before a meal (Steffens, 1970). So the electrophysiological data of Oomura interpreted in relation to the dual-mechanism model does not seem to account for the occurrence of spontaneous feeding. Perhaps the glucoreceptor and glucosensitive neurons identified by Oomura are concerned with the central control of blood sugar levels and of metabolism rather than with the initiation of feeding behavior. Feeding behavior might be initiated by peripheral glucoreceptors, and, as discussed previously, the hypothalamus has been implicated in the integration of signals from these receptors (Schmitt, 1973).

Another possibility is that the glucoreceptors identified by Oomura may be concerned with the cessation rather than with the initiation of feeding. Steffens (1975) has demonstrated that the plasma levels of insulin and glucose are markedly elevated following a meal. As shown in Figure 4.8, this would activate glucoreceptor neurons in the VMH and inhibit LH glucosensitive neurons. Oomura's data fits this scheme well, and according to the dual-mechanism model, feeding would be shut off.

Estrogens inhibit food intake, and according to Wade and Zucker (1970b), this is due to a direct action of the hormones on the VMH, which inhibits the LH. They suggest further that food intake is not reduced by estrogen prior to puberty, because the effect of estrogens is inhibited by the high levels of growth hormone. According to Wade and Zucker (1970a), in rats, "the decline in growth hormone secretion which appears to occur at about 40 to 50 days of age coincides with the ability of estrogens to modulate food consumption [p. 218]." However, a more recent study shows that following lesions of the VMH, food intake is still reduced by the administration of estradiol, suggesting that other structures contribute to the inhibitory effect of estrogens on feeding (Reynolds & Bryson, 1974).

Recent evidence suggests that estradiol acts on the brain after being metabolized to catecholestrogen-2-hydroxyesterone (Fishman, 1976). It is of interest that young women with anorexia nervosa convert estradiol to catecholestrogen-2-hydroxyesterone to a greater extent than normal or obese patients. This metabolite of estradiol, which reduces food intake in rats, has been shown to inhibit the biosynthesis of catecholamines so that estrogens might exert their "appetite-suppressing" effects on catecholaminergic neurons.

Fishman (1976) has suggested that the suppression of food intake by estrogens may not be a direct effect of the hormones on appetite mechanisms but rather to the reduction of the set-point for body-weight regulation. A similar suggestion was made previously for the effects of insulin on food intake. According to Woods and co-workers (1974), the set-point for body-weight regulation depends on the ratio of insulin to growth hormones (I/GH ratio). To support this hypothesis, these authors review experimental evidence showing that plasma insulin levels are high and plasma growth hormone levels low in spontaneously obese and hyperphagic animals and that the administration of insulin or growth hormone to alter the I/GH ratio influences food intake and body weight. This hypothesis has been criticized and in particular the question raised as to whether plasma levels of insulin and growth hormone are "causes or consequences of feeding" (Panksepp, 1975b, p. 158).

Much work remains to be done to determine how insulin, estrogens, and other hormones exert their influences on feeding behavior.

Circadian Rhythms of Food Intake: Neural Substrates

Circadian rhythms of feeding behavior were demonstrated experimentally by Richter (1943) more than 30 years ago, but little has been learned since that time about the neural mechanisms that are responsible. The evidence is limited to a few studies of the effects of brain lesions on the normal day–night difference in food intake in rodents.

Most lesions have no or only small effects on the circadian rhythmicity of feeding. However, it has been reported that the ratio of nocturnal to diurnal feeding is increased by lesions of the lateral hypothalamus, although total food intake is reduced (Kakolewski et al., 1971; Rowland, 1976). Lesions of the ventromedial or dorsomedial hypothalamus, on the other hand, increase diurnal feeding more than nocturnal feeding, so that the day–night difference in food intake is reduced (Balagura & Devenport, 1970; Bernardis, 1973). Lesions of the suprachiasmatic nucleus, which receives retinal hypothalamic inputs (Moore, 1973) and is in a position to contribute to the entrainment of feeding and other circadian rhythms to the onset and offset of light, eliminates the circadian pattern of food and water intakes (Stephan & Zucker, 1972). When a reversible functional block of the suprachiasmatic nucleus is produced with colchicine (see p. 45), the circadian pattern of feeding is disrupted temporarily (Figure 4.9). It appears that the suprachiasmatic nucleus has an important role in the control of circadian rhythms, but whether it is the locus of the biological clock, an important component of the biological clock, or merely a stage of neural integration of visual signals, has not been established (Rusak & Zucker, 1975).

Of interest and possible relevance to understanding the neural control of the circadian rhythms of feeding behavior is the circadian fluctuation of the biogenic amines. In the rat the levels of noradrenaline in the hypothalamus and associated structures are low in the daytime and high at night (Manshardt & Wurtman, 1968; Reis et al., 1968). On the other hand, serotonin levels are high in the

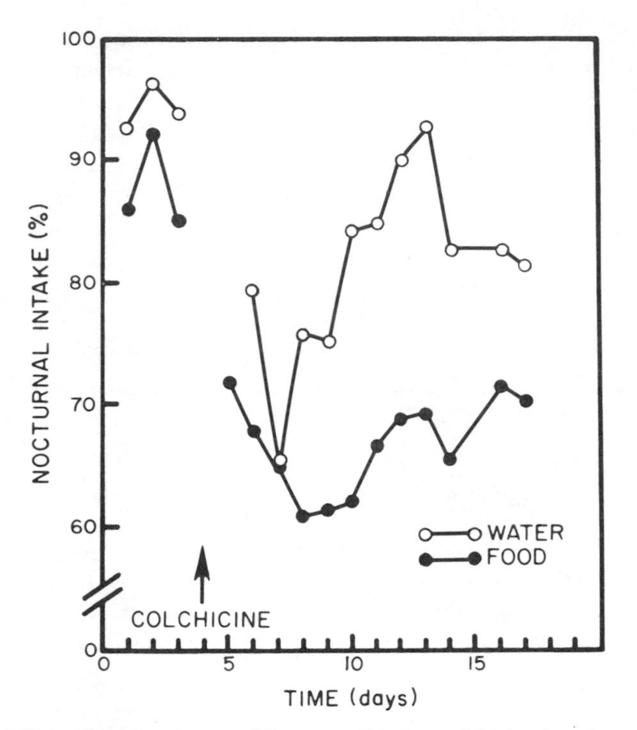

FIG. 4.9 Effect of injecting the reversible neuronal blocker, colchicine, into the suprachiasmatic nucleus of one rat. Before injection feeding and drinking both occur mostly during the night. After injection the percent of both activities that occurs during the night drops off markedly during the first-week postinjection and thereafter.

daytime and low at night. Whether these circadian patterns of noradrenaline and serotonin levels are controlled by the suprachiasmatic nucleus has not been investigated.

Biogenic amines have been implicated in the control of food intake: Feeding is initiated by the administration of noradrenaline to the hypothalamus (Grossman, 1962); feeding is disrupted by damage to dopaminergic and noradrenergic neurons (Oltmans & Harvey, 1972; Ungerstedt, 1971); and there is evidence that the anorectic drugs, amphetamine and fenfluramine, act via noradrenergic and serotoninergic systems respectively (Blundell et al., 1976; Garattini & Samanin, 1976). The ventral noradrenergic pathway is associated with the anorectic effect of amphetamine, as shown by the finding that amphetamine does not reduce food intake in rats with the ventral noradrenergic pathway lesioned (Ahlskog & Hoebel, 1973).

It is known that the sensitivity of animals to drugs varies in a circadian fashion (Scheving et al., 1974). There is a report that the effect on food intake of administering noradrenaline into the hypothalamus depends on the time of day and seems to be related to the circadian rhythm of endogeneous noradrenaline. When noradrenaline was administered to the hypothalamus in the daytime, a period when hypothalamic noradrenaline is low, it elicited feeding, but when it was administered to this region at night, a period when noradrenaline levels are high, it reduced food intake (Margules et al., 1972). Because noradrenaline and serotonin levels in the brain have peaks at night and in the daytime respectively, it might be expected that the anorectic effects of amphetamine and fenfluramine vary depending on when they are administered.

Nonhomeostatic Signals and Feeding: Neural Substrates

There have been two interpretations of the influence of taste, smell, and other nondeficit stimuli on food intake. The most popular view has been that odor, taste, and even the sight of food have modulatory effects on neural systems in the hypothalamus that subserve feeding to deficit or homeostatic signals (Stellar, 1960; Stevenson, 1967). An alternative view is that taste, smell, etc., do not merely modulate feeding initiated by deficit signals but are direct initiators of feeding behavior because of the rewarding or reinforcing properties of the taste or smell (Pfaffmann, 1960).

Although the neural mechanisms for the modulatory influence of stimuli on food intake have not been elucidated, there is indirect evidence implicating the amygdala and other limbic forebrain structures. Because the amygdala receives olfactory, taste, visual, and other sensory inputs and has strong fiber projections to the hypothalamus, via the stria terminalis and ventral amygdala–fugal pathway, this might be the route for mediating the modulatory influences of sensory stimuli on feeding behavior (Gloor et al., 1972). Consistent with this suggestion, it has been reported that electrical stimulation of the amygdala reduces food intake (Fonberg & Delgado, 1961; Fonberg, 1963; Grossman & Grossman, 1963) and produces ejection of food placed in the mouth (MacLean & Delgado, 1953; Robinson & Mishkin, 1968). The suppression of feeding from stimulation of the amygdala is eliminated by sectioning the stria terminalis (White & Fisher, 1969). The results of electrophysiological experiments suggest that the attenuating effects of amygdala stimulation on food intake mediated via the stria terminalis may involve the activation of inhibitory interneurons that encircle the VMH (Dreifuss et al., 1968; Renaud, 1976). A structure strongly implicated in feeding behavior (see Figure 1.8).

Another possibility suggested by Cabanac (1971) and Hoebel (1971) is that taste and other nondeficit stimuli are themselves modulated by homeostatic imbalance and that deficit signals rather than modulating food intake initiated by deficit signals. Cabanac arrived at this view from experiments in human subjects in which the palatability of glucose solutions depended on the energy state or degree of hunger or satiety (see p. 94). Hoebel observed that self-stimulation of the LH increased when the rat was food-deprived and decreased when it was supersatiated (for further details, see p. 218). According to Hoebel (1971), "the lateral hypothalamus generates a combination of reward and aversion which shifts along a continuum depending on the animal's homeostatic balance [p. 543]."

Electrophysiological evidence supporting the view that the incentive properties of taste and other stimuli for the initiation and maintenance of feeding behavior are modulated by the internal homeostatic state of the animal have been reported by Rolls and his co-workers. Recordings were made of action potentials from single neurons in the LH of monkeys (for procedure, see Figures 2.20 and 2.21). Some of the hypothalamic neurons changed their rate of discharge when the food-deprived monkey was presented with food. An example is shown in Figure 4.10A, in which a neuron decreased its firing rate when glucose solution was placed on the tongue but not when other solutions or foods (e.g., water, sodium chloride solution, or peanuts) were given to the monkey. This type of hypothalamic neuron did not merely respond to taste as a neuron in the taste pathway but appeared to respond to a particular rewarding taste, as shown by the following evidence. If the monkey was food-deprived, the neuron had a large response to glucose solution, but after becoming satiated to the glucose, the response of the neuron diminished toward zero. Rolls suggests that the neuron is signaling food taste reward and not hunger.

Some hypothalamic neurons changed their rate of discharge to the sight of food. An example is presented in Figure 4.10B, in which the discharge rate of a

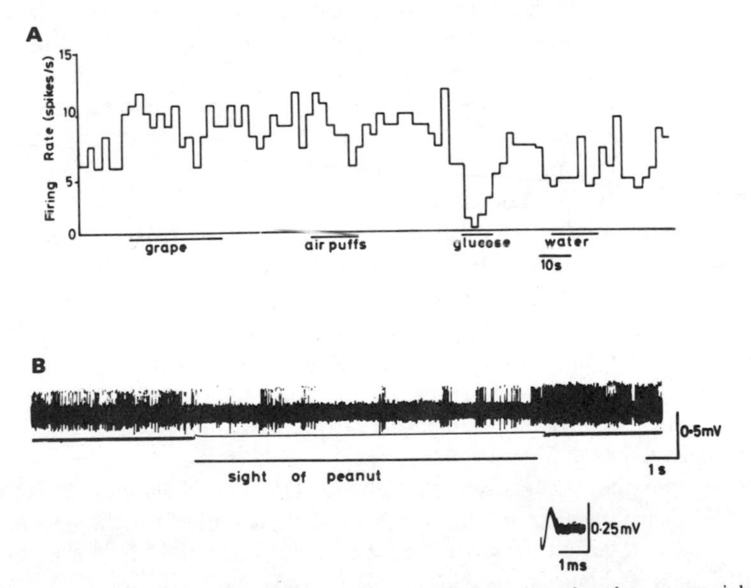

FIG. 4.10 Neurons of the hypothalamus of the monkey that respond to the taste or sight of food. A. The discharge of a neuron that was inhibited when glucose solution was administered to the tongue but which did not change its rate of discharge when other foods were placed in the mouth. B. A neuron inhibited by the sight of a peanut, a preferred food. (From Rolls, 1975.)

neuron was decreased when the monkey was shown a peanut. The neuron responded to foods that the monkey preferred but not to the sight of nonpreferred foods or to nonfood objects. Control tests showed that the neuron also did not respond in relation to motor movements, somatosensory stimulation, or swallowing. When the monkey became satieted to the initially preferred food, the hypothalamic neuron no longer responded.

A model formulated by Rolls (1975) to account for the electrophysiological data is shown in Figure 4.11. It assumes that the effects of food-related stimuli on lateral hypothalamic reward neurons is gated by neurons that signal "hunger."

The amygdala, which contributes to the influences of taste, visual, and other stimuli on food intake, is also implicated—along with other forebrain limbic structures and the cerebral cortex—in the influence of experiential and cognitive factors on feeding behavior. There is evidence to suggest that the amygdala is not only involved with mediating the influence of the incentive properties of these stimuli on feeding and other behaviors but that it makes, in addition, an important contribution to learning about stimuli that are rewarding or aversive to the animal. The deficits in feeding and other behaviors following lesions of the amygdala have been interpreted as a disruption of the association of stimuli with positive and negative rewards (Jones & Mishkin, 1972). The difficulty in distinguishing between edible and nonedible objects (for example, repeatedly taking

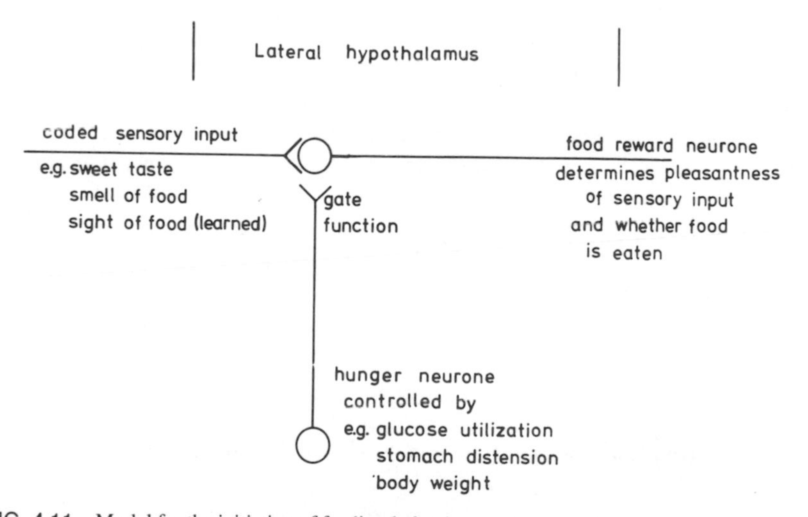

FIG. 4.11 Model for the initiation of feeding behavior consistent with the electrophysiological data shown in Figure 4.10 and described in the text. Sensory input corresponding to the sight and/or taste of food reaches the hypothalamic neurons providing the animal is hungry (e.g., has been food-deprived) and has a preference for the food. In other words, the coded sensory input is gated by hunger neurons to activate food-reward neurons. "Taken together with the evidence that lesions in this region disrupt food intake, the experiments provide strong support for the hypothesis that food intake occurs when hypothalamic reward neurons are gated by hunger to respond to sensory inputs such as the sight, smell, and taste of food" (Rolls, 1975, p. 29).

objects to the mouth, designated as "oral tendencies") seen after lesions of the amygdala may be due to the disruption of the associative processes by which the object acquires "motivational significance" for the animal. This behavioral deficit was first observed in monkeys following surgical ablation of the temporal lobes, which included removal of the amygdala (Klüver & Bucy, 1937).

Lesions of the prefrontal cortex have been reported to increase food intake (Kolb & Ninneman, 1975). A prominent change in humans and in higher mammals following frontal lobe damage is a tendency to perseverate behavioral responses. Whether this is the reason for overeating and the mild obesity after frontal lobe damage is not clear. Rolls (1975) has speculated that the perseverative tendency may be an indication that the orbitofrontal cortex is involved in disconnecting stimulus-reinforcement associations—the establishing of which the amygdala contributed to initially.

Catecholamines and Feeding Behaviors

The first evidence that implicated catecholaminergic neurons in feeding behavior was the demonstration that the application of norepinephrine to the LH initiated feeding behavior in rats (Grossman, 1962). It was reported later that lesions of the LH, which produce aphagia, reduce norepinephrine levels in the forebrain (Heller & Moore, 1965) and that the aphagia and loss of body weight following LH lesions is reversed when norepinephrine is infused into the cerebral ventricles (Berger et al., 1971). Subsequently, when monoamine-containing neurons were visualized with the histofluorescence technique (see Figure 2.24), it soon became apparent that dopaminergic (DA) as well as noradrenergic (NA) neurons projecting rostrally through the region of the hypothalamus are associated with aphagia and adipsia following LH lesions (Ungerstedt, 1971).

In an important pioneering study, referred to in an earlier section of this chapter, Ungerstedt (1971) demonstrated that aphagia and adipsia resulted from the destruction of the DA nigrostriatal pathway (projecting from the substantia nigra to the neostriatum), using the neurotoxic compound 6-hydroxy-dopamine (6-OH-DA). This study not only introduced a procedure, which when properly used permitted selective destruction of DA and/or NA neurons, but indicated for the first time the possible role of DA neurons in feeding behavior. Interestingly, the aphagia and adipsia reported by Ungerstedt following the selective damage of the DA nigrostriatal pathway were more severe than previously observed in the classical LH syndrome following electrolytic lesions. This difference was confirmed a few months later in a study by Oltmans and Harvey (1972), who compared bilateral electrolytic lesions of the nigrostriatal pathway with similar lesions of the LH, involving the medial forebrain bundle, which contains NA fibers as well as DA fibers.

Several investigators have produced damage to catecholaminergic neurons by administering 6-hydroxy-dopamine into the cerebral ventricles (Zigmond & Stricker, 1972; Breese et al., 1973; Fibiger et al., 1973). Two types of deficits, correlated with the amount of depletion of dopamine in the neostriatum, have been reported. When striatal DA levels are moderately reduced and NA levels are markedly depleted, the animals are indistinguishable from controls except in response to homeostatic challenges such as glucoprivation, hypovolemia, or exposure to cold. They do not make appropriate feeding or drinking responses to these challenges (Zigmond & Stricker, 1974). With large depletions of dopamine in the neostriatum, there are severe deficits in spontaneous feeding and drinking as well as ingestive responses to the challenges. The animals die unless given food and water by stomach tube, and if dopamine is depleted by 98% or more, they never recover from the aphagia and adipsia. If dopamine in the neostriatum is depleted by 90–95%, feeding and drinking gradually recover in a manner similar to that previously reported following electrolytic lesions of the LH, and the same stages of recovery are observed (e.g., Teitelbaum & Epstein, 1962). Zigmond and Stricker (1974) attribute the return of feeding and drinking to a functional recovery of damaged DA fibers and increased synthesis and release of DA from the remaining undamaged fibers.

It seems clear that aphagia and adipsia are associated with damage to DA neurons, especially those that constitute the nigrostriatal pathway (A9). However, it is not clear from the evidence whether damage to NA neurons or to DA neurons (A10) in the ventral tegmental area also contributes to the deficits in feeding and drinking. For example, Marshall et al. (1974) were reluctant to attribute the aphagia and adipsia following administration of 6-OH-DA to damage to DA nigrostriatal neurons exclusively. In studies involving the administration of 6-OH-DA into the cerebral ventricles (e.g., Zigmond & Stricker, 1974), the A10 dopamine system should be damaged as extensively as the A9 dopamine system. More work is needed to elucidate the role of the A9 and A10 dopamine systems and of the noradrenergic systems in feeding behavior.

Another important consideration is whether the deficits in food and water intake are specific to feeding and drinking behavior or are associated with general behavioral deficits. There is a good deal of evidence that following damage to the DA nigrostriatal pathway (which may include damage to other DA and NA pathways), the animals are hypoactive, have difficulty in initiating behavioral responses, and show sensory neglect and a disruption of behavioral arousal (Marshall & Teitelbaum, 1973; Myers & Martin, 1973; Ungerstedt, 1974). It appears that the deficits are not specific to feeding behavior and, furthermore, that there are severe deficits in the consummatory phase of ingestive behaviors that complicate the interpretation of possible effects on the appetitive phase (e.g., the initiation of ingestive behaviors).

The LH Syndrome is Associated with Multiple Deficits

The aphagia and adipsia resulting from LH lesions, originally reported by Anand and Brobeck (1951), was studied extensively by Teitelbaum and Epstein (1962), who provided a sophisticated analysis of the stages of recovery. They interpreted the deficit in feeding behavior as a "motivational deficit" rather than the result of "motor failure"; in a study with Rodgers, they showed that following LH lesions, a motivational deficit persisted after the initial motor deficit was over (Rodgers et al., 1965). Furthermore, because the LH animals did not recover feeding and drinking responses to metabolic and dehydrational challenges, Epstein (1971) concluded that the LH is the integrative site for "hunger" and "thirst" signals.

In recent years the results of a variety of experiments have cast doubt on this simple interpretation. Motor deficits, in particular when lesions impinge on the DA nigrostriatal pathway, were alluded to in the previous section. Sensory neglect has been observed following LH lesions due to damage to DA neurons and/or sensory input fibers (Marshall et al., 1974). Behavioral or motivational inertia, not specific to feeding or drinking behavior, may also occur. Animals with LH lesions may be more susceptible to a variety of stressors (Stricker & Zigmond, 1976). It is now generally accepted that the LH syndrome is associated with multiple deficits; the emphasis on deficits in feeding behavior for a number of years was apparently because feeding is easy to measure and because aphagia has disastrous consequences for the animal.

One of the sensory inputs that may be disrupted by LH lesions is trigeminal stimuli from the mouth, tongue, and snout. Zeigler and co-workers have reported striking deficits in food intake following peripheral trigeminal deafferentation in the pigeon and rat (Zeigler, 1976; Zeigler & Karten, 1974; Zeigler et al., 1975). Lesions of the rostral projections of the trigeminal nucleus at different rostral–caudal levels also produced aphagia in the rat and pigeon similar to the classical LH aphagia (Figure 4.12). These investigators suggest that disruption of the sensory–motor control of feeding may result from damage to the trigeminal projections when the lateral hypothalamus is lesioned and may contribute to the lateral hypothalamic syndrome. Because trigeminal projections are just dorsal to

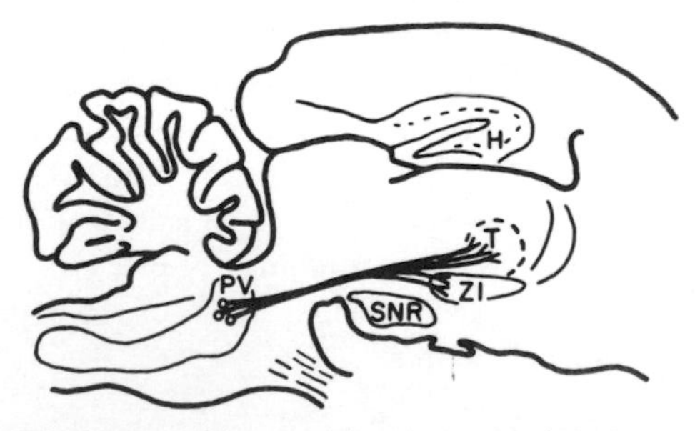

FIG. 4.12 A saggital diagram of the rat brain showing trigeminal nerve pathways. The cell bodies of the trigeminal nucleus are at PV, and the axons finally terminate in the thalamus at T. Branches (collaterals) are given off to the zona incerta (ZI) and posterior thalamus. (Modified from Zeigler & Karten, 1974.)

the substantia nigra, they may also be damaged and contribute as well to the aphagia following lesions of the substantia nigra.

Current Status of the Hypothalamus in Feeding Behavior

It is appropriate to ask, after briefly reviewing the recent developments in the two previous sections, whether the hypothalamus has any role at all in feeding behavior, and if it does what that role is.

Clearly, a variety of deficits result from damage to the LH, and the results of the classical lesion experiments have had to be reinterpreted in relation to the visualizing and mapping of monoamine pathways—the new chemical anatomy of the brain (see Figure 2.24). This reinterpretation also applies to the role of the VMH in food intake, because hyperphagia and obesity have been observed following lesions of the ventral noradrenergic bundle (Ahlskog & Hoebel, 1973) and following damage to serotonergic neurons (Saller & Stricker, 1976). However, it does not necessarily follow from these recent developments that the hypothalamus plays no role in feeding behavior. In fact, most of the studies that have led to the reinterpretation of earlier results and theoretical models have not provided any new insights in answering the fundamental questions (e.g., what initiates feeding, what terminates feeding). Rather, the major contribution of these studies has been to suggest that feeding deficits from LH lesions may be related to disrupting the consummatory phase of feeding behavior and to direct attention to the motor mechanisms of motivated behaviors. They have not contributed to understanding the neural substrates for the initiation of feeding behavior.

A major lesson from research on the neural substrates of feeding behavior during the last few years is the emphasizing of the necessity for caution in the interpretation of experimental findings. We have already dealt with the lack of specificity of the lesion technique and the difficulty in deciding whether lesions of the LH, the VMH, or the nigrostriatal pathway, produce a deficit in a particular behavior such as feeding, without producing general behavioral deficits. Electrical and chemical stimulation and other techniques used to investigate the neural substrates of feeding are also not always specific for the manipulation of a particular neural pathway or system or for eliciting a specific behavioral response.

There was a good deal of optimism when electrophysiological techniques were introduced for the study of hypothalamic mechanisms in the control of food intake. Glucosensitive neurons were identified in the LH and the VMH (Anand et al., 1962; Oomura et al., 1969), and by using the technique of iontophoresis, some of these neurons have been shown to be glucoreceptive (Oomura, 1976). However, definitive evidence that they are involved in behavior, specifically in the initiation or termination of feeding behavior, is lacking.

The best evidence implicating the hypothalamus in feeding behavior comes from chronic electrophysiological recording experiments demonstrating food-related neurons in the lateral hypothalamus (Rolls, 1975). As described previously, there are neurons in the LH that respond to the sight or taste of foods preferred by the animal but not to nonpreferred foods or to preferred foods following satiety (see Figure 4.10). These observations suggest that such LH neurons associated with feeding behavior are part of a complex neural system involving amygdala, prefrontal cortex, and other forebrain structures. In recent years there has been increasing interest in the role of the hypothalamus in blood sugar regulation and in the control of energy metabolism. It has been known for some time that energy expenditure increases during behavioral arousal and that the hypothalamus influences energy metabolism via the autonomic nervous system and its control over the anterior pituitary (Bray, 1976; Panksepp, 1975a). The hypothalamus is also involved in the preabsorptive release of insulin (Louis-Sylvestre, 1976) and in the hormonal regulation of blood glucose during and following absorption of food from the gastrointestinal tract. These functions are important, because the homeostasis of body energy depends on the control of both energy expenditure and energy intake (Barnwell, 1977). The hypothalamus by virtue of its strategic location in the brain—having functional relationships with the pituitary gland, the autonomic nervous system, limbic forebrain structures, and neural systems for motor control—is in a position to influence both energy expenditure (metabolism) and energy intake (feeding). The unique contribution of the hypothalamus may be in the "functional coupling" of the physiological regulations and behavioral responses that maintain the relative constancy of the internal environment and at the same time enable the animal to adapt to a complex and continuously changing external environment (see Figure 1.9).

SUMMARY

Food is the source of energy for various bodily functions, and feeding is a self-regulatory behavior in that energy intake tends to match energy expenditure, with body weight remaining relatively constant. The availability of nutrients in the body is monitored, and both the initiation and the termination of feeding behavior are influenced by glucostatic, lipostatic, aminostatic, and other signals.

However, feeding behavior cannot be accounted for in terms of a simple feedback mechanism that is directly related to current energy expenditures. An appreciable amount of energy is stored in the body so that, although energy expenditure is continuous, food intake can be periodic; most mammals are meal eaters. Another complication is that following the ingestion of food, there is a delay for digestion and absorption before the nutrients are available to body tissues. In order to avoid excessive caloric intake, feeding behavior must terminate before nutrient-deficit signals have disappeared. Signals from the oral cavity and gastrointestinal tract are utilized and contribute to preabsorptive satiety. Food intake is also influenced by taste and palatability and by experiential, social, and cultural factors. These factors are particularly important in man and may lead to a caloric intake that is in excess of energy expenditure; obesity is a serious problem in affluent societies.

For a number of years, the direction of research and interpretation of experimental findings were strongly influenced by the dual-mechanism model, which assumed that the ventromedial hypothalamus integrated ''satiety signals'' and that the lateral hypothalamus integrated ''hunger signals.'' In spite of the emphasis on homeostatic signals and mechanisms and on the contributions of the hypothalamus to feeding behavior, there is still much uncertainty about the locus of receptors, neural pathways, and integrative mechanisms. That this classic model is overly simplistic, especially for dealing with spontaneous feeding in the natural environment of the animal, has become increasingly apparent in recent years. Little attention was given for some time to the role of external stimuli, palatability, and experiential and cognitive factors in the initiation of feeding and in the control of food intake. As a consequence the neural substrates for these factors, which are particularly important in man and higher mammals, have not been investigated to any extent and are largely unknown. Another reason for the limited understanding of the neural mechanisms is that complex forebrain structures are involved.

An important development in recent years has been the vigorous investigation of the relationship of dopaminergic and other transmitter-specific pathways to feeding behavior. These important studies have been a major reason for a reassessment of the classical dual-mechanism model and a reinterpretation of the results of earlier lesion and stimulation studies. They have also directed attention to the contribution of motor and sensory–motor deficits to the disturbances in feeding behavior following lesions of the hypothalamus and associated structures. However, in spite of the significance of these investigations, they do not provide answers to the basic questions about the control of food intake; much remains to be learned about the neural mechanisms for the initiation and termination of feeding behavior and for determining the choice and relative amounts of various foods eaten. These questions, as well as the neural substrates of feeding behavior, can be fruitfully approached from the perspective of considering the various factors that initiate and contribute to food intake.

5

Drinking Behavior

The drinking of water and other fluids is a behavior that occurs at regular and, for many animals, frequent intervals. Farm animals drink water several times a day, and so does the dog kept as a house pet. Rats used in laboratory experiments and allowed free access to water drink also at frequent intervals, particularly at night. Man not only takes a glass of water from time to time through the day, especially when the weather is hot, but also coffee, tea, and other beverages with meals, as well as between meals. Water is essential for life, and because an appreciable amount of water is lost from the body each day—and in most species it cannot be stored—it must be replaced regularly by water intake.

Depletions of water are detected by the body, and deficit signals initiate both the oral intake of water and appropriate physiological responses for its conservation. Accordingly, drinking, like feeding, can be considered a self-regulatory behavior that contributes to body-water homeostasis. The relation of water intake to water balance is dealt with in the first section of this chapter.

Water intake does not only occur in response to water-deficit signals; the drinking of water and a variety of fluids is associated with meals and, in man, with various social activities. Typically, man as well as the rat and other animals take in more water than they need for body-water homeostasis. Drinking is not only a regulatory behavior occurring in response to water deficits; it is influenced by a number of other factors—palatability, circadian rhythms, and social, experiential, and cognitive factors—and is one of several motivated behaviors in which animals engage. In considering the neural substrates of drinking behavior, which is the major objective of this chapter, it is first appropriate to consider the characteristics and determinants of drinking as a motivated behavior.

CHARACTERISTICS OF DRINKING BEHAVIOR

Because our emphasis is on the factors that initiate and contribute to water intake, we are concerned mainly with the appetitive phase of drinking behavior, consisting of the locomotor and other behavioral responses involved in seeking water and in orienting and positioning the body so that it can be ingested. The consummatory phase that follows—the lapping and swallowing of water—is more stereotyped. In the rat, the favorite research subject for investigating the neurobiology of drinking behavior, water is ingested at a rather constant rate of approximately 7 laps per second (Stellar & Hill, 1952), and because this response is present at birth (Epstein, 1976a), it is thought to be controlled by preprogrammed circuits in the brainstem.

When a rat has been without water overnight, it goes promptly to the water spout when a water bottle is placed in its cage and begins to drink. If the rat has been trained previously to traverse a straight alley or a complex maze to obtain water, it will run promptly along the alley or through the maze to the water. Then it will lap at the spout for several minutes with occasional pauses that gradually become longer until drinking eventually ceases completely (Figure 5.1A). The behavior is goal-directed (or "purposive") and persistent, like a number of other motivated behaviors.

By attaching a drinkometer to the water bottle on the rat's home cage, one can record the temporal pattern of water intake. After drinking has terminated in our initially water-deprived rat, it may not drink again for an hour or two, or even longer, depending on the time of day, the availability of food, and the nature of the diet. The recordings of drinking responses of a rat living in its home cage

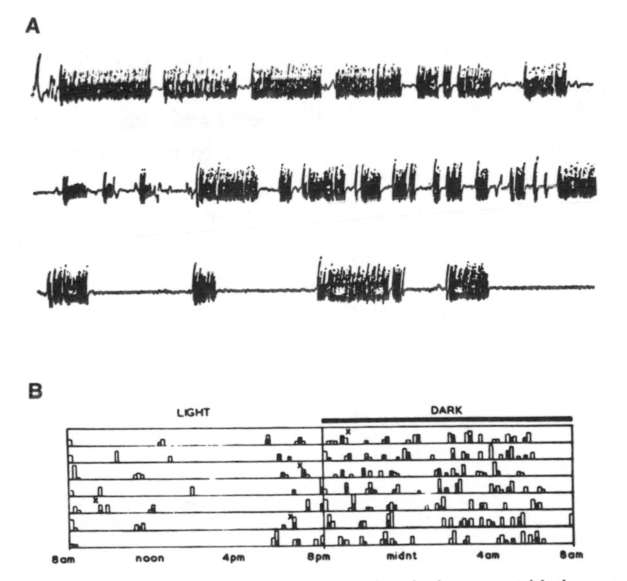

FIG. 5.1 A. Recordings on a smoked drum from an electrical contact with the water spout in a rat's cage, showing pattern of water intake after water deprivation. B. Record of drinking (open bars) and eating (solid bars) of a rat living in its home cage. Each row represents a single day with a 12-hour light–dark cycle. Note that most of the drinking and eating is done during the dark period. (After Oatley, 1973.)

with water and food continuously available are shown in Figure 5.1B. The record of laps on the water spout illustrates clearly that the drinking behavior is periodic, another prominent characteristic of motivated behaviors. The recordings also show that most of the drinking occurs at night and in association with feeding; about 80–90% of the total daily water intake in the rat occurs just before and after meals. This circadian rhythm of water intake is considered in more detail later.

In the rat and most other animals, water cannot be stored in the body in the way that energy is stored as fat tissue. There are physiological mechanisms for its conservation, as indicated earlier (see Figure 2.8), but normally the continuous loss of water from the body must be compensated by the relatively frequent intake of water. As a consequence drinking behavior has a relatively high priority in "motivational time-sharing."

WATER INTAKE AND WATER BALANCE

"Water is the single largest component of the body—life exists in an aqueous milieu" (Stevenson, 1965, p. 713). In the human about two-thirds of the total body weight is water, being highest during the first few months of life and decreasing with age as cellular mass increases. The proportion of body water varies considerably among individuals in relation to the amount of body fat. After puberty the relative amount of adipose tissue becomes increasingly higher with age in females as compared to males, thus body water as a percentage of total body weight decreases more in women than in men (Table 5.1). However, for any particular individual, body water is regulated within rather precise limits. As recognized by Claude Bernard a hundred years ago, the relative constancy of the content and composition of body fluids must be maintained.

There is a continuous exchange of water between the mammal and its external environment (Figure 5.2). In man this exchange of water with the environment

TABLE 5.1
Body Water at Various Ages
(as Percent of Body Weight)

					20–39 yrs.		40–59 yrs.	
	Newborn	6 mos.	2 yrs.	16 yrs.	M[a]	F	M	F
Total body water (TBW)	77	72	60	60	60	50	55	47
Extracellular fluid (ECF)	42	34	25	20	20	17	18	16
Intracellular fluid (ICF)	35	38	35	40	40	33	37	31

[a]M = male; F = female.

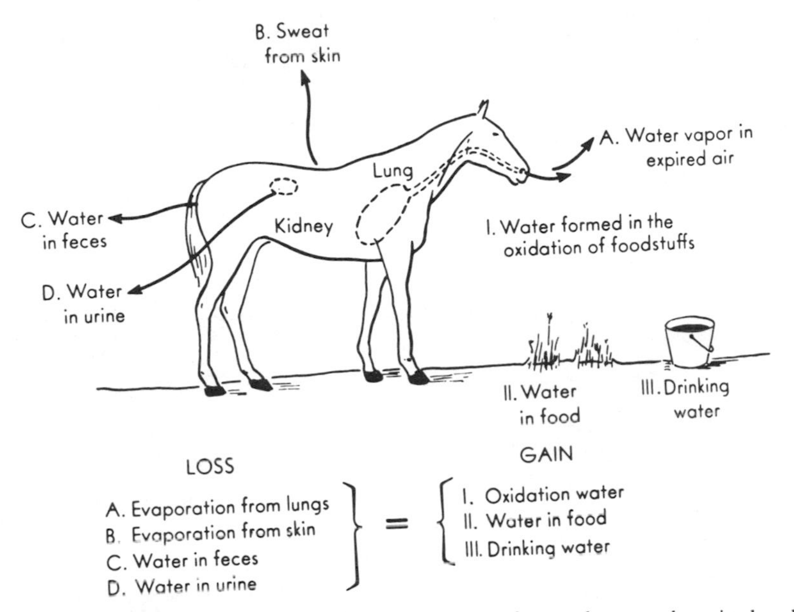

FIG. 5.2 In all mammals there is a continuous exchange of water between the animal and its external environment. Water is lost by evaporation from the lungs and the skin and also via urination and defecation. Water loss is compensated by the drinking of water so that body-water homeostasis is maintained. (From Schmidt-Nielsen, 1960.)

may be two or three liters a day and considerably higher in hot weather due to sweating. Some water is lost continuously from the skin and through respiration (insensible water loss) and a variable amount as sweat (sensible water loss) in the regulation of body temperature. A small amount of water is contained in the feces, but this can increase substantially during diarrhea. Normally, the largest amount of water lost is that from the kidney in the excretion of body wastes. The loss of water, which may vary considerably depending on the ambient temperature, the diet, and the state of health, is balanced by the drinking of water, so that the constancy of the internal environment is maintained.

Body water is distributed between two fluid compartments separated by the cell membranes. The water inside the cells (in the intracellular fluid compartment), which provides the aqueous medium for chemical processes in the cell, comprises about two-thirds of the total. The remaining one-third (in the extracellular fluid compartment), including water that is between the cells (the interstitial fluid) as well as the plasma of the blood, serves as a transport medium for energy nutrients, O_2, etc.

The body is responsive to depletions of the intracellular and extracellular fluid compartments (Figure 5.3). These compartments are monitored and signals integrated in the central nervous system to regulate the release of antidiuretic hor-

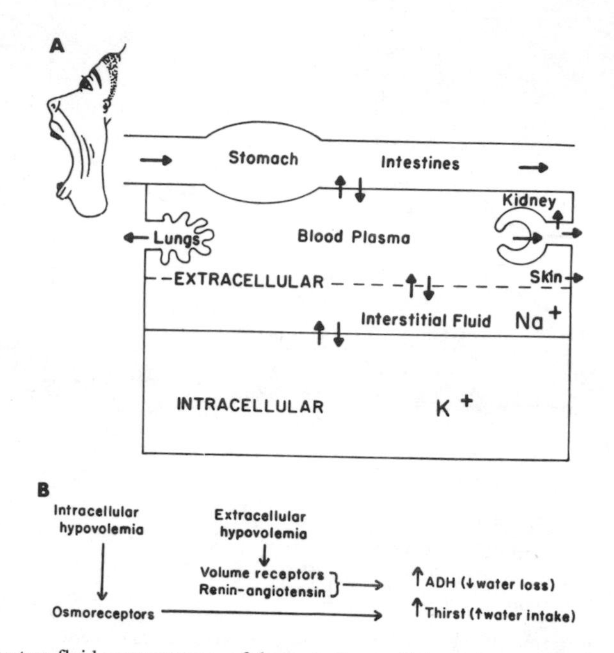

FIG. 5.3 A. The two fluid compartments of the body (intracellular and extracellular) are separated by cell membranes. The extracellular fluid compartment includes water that is between the cells as well as blood plasma. A continual exchange goes on in the body between these two compartments. B. The body is responsive to depletions of both fluid compartments. If dehydration occurs in the intracellular fluid compartment, osmoreceptors in the brain are activated that in turn cause release of antidiuretic hormone from the pituitary gland. This hormone decreases water loss from the kidney. Osmoreceptors also induce water intake. If hemorrhage occurs (extracellular hypovolemia), water conservation mechanisms and thirst are initiated via the renin-angiotensin system of the kidney.

mone (ADH) from the posterior pituitary gland, which controls the reabsorption of water by the kidney and thereby the loss of water as urine, and, as we see in the next section, to induce water intake. There is a good deal of evidence that there are osmoreceptors in the region of the supraoptic nucleus of the hypothalamus that signal changes in the osmolarity of the extracellular fluid. When activated these osmoreceptors send signals to the neurosecretory cells, which project to the posterior pituitary, to produce and release more ADH (see Figure 2.8). Depletions of the extracellular fluid compartment (extracellular dehydration or hypovolemia), following hemorrhage for example, also activate this ADH mechanism, resulting in reduced urine output. There is evidence that signals of extracellular dehydration come from volume (baro) receptors in the low-pressure side of the circulatory system (e.g., Gauer et al., 1970).

Renal mechanisms make it possible to conserve body water, and some desert mammals, such as the kangaroo rat, have kidneys with very high urine-concentrating capacities so that little water intake is required. This is an evolutionary adaptation made by such species to their environmental niche (Hudson, 1964). However, most animals drink water and other fluids on a regular basis, and for man and other mammals, the oral intake of water is quite substantial (see Figure 5.2). Drinking, a behavioral adaptation, and the variable excretion of water by the kidney, a physiological regulation controlled by ADH, are complementary in water balance.

Drinking: An Aspect of Body-Water Homeostasis

When a rat, dog, or other laboratory animal is water-deprived for a few hours, drinking is observed when water is made available. In a classic study, Adolph (1939) observed in the dog that the volume of water intake was proportional to the degree of dehydration (Figure 5.4). The two fluid compartments can be depleted selectively—the intracellular compartment by administering hypertonic saline and the extracellular compartment by administering polyethylene glycol as described in the next section—and when this is done, drinking occurs.

From the viewpoint of regulatory physiology, water intake contributes to body-water homeostasis, and drinking behavior is an example of a self-regulatory behavior (Richter, 1943). As an aid in understanding the mechanisms that subserve water intake in relation to body water balance, it is helpful to use system analysis and a control diagram of the sort shown in Figure 5.5. Signals that arise from intracellular dehydration and/or extracellular dehydration are assumed to be detected by osmoreceptors and volume receptors respectively.

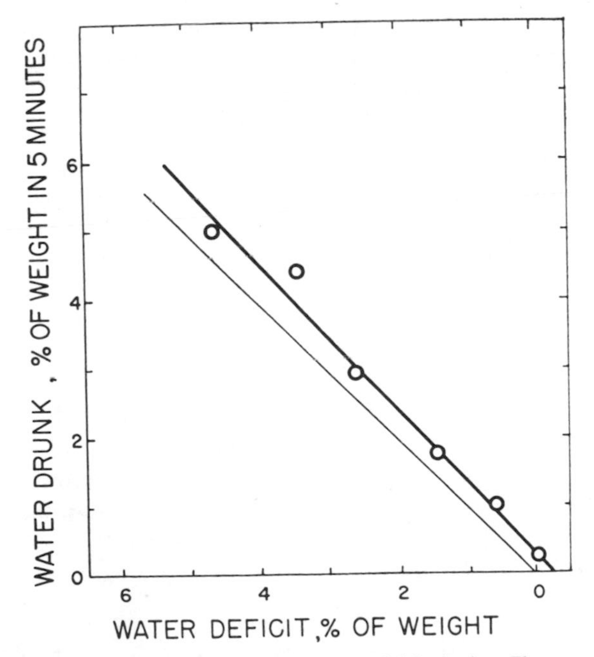

FIG. 5.4 Water drinking in dogs after varying degrees of dehydration. The amount of water drunk is directly proportional to deficits of water in the body. (From Adolph, 1939.)

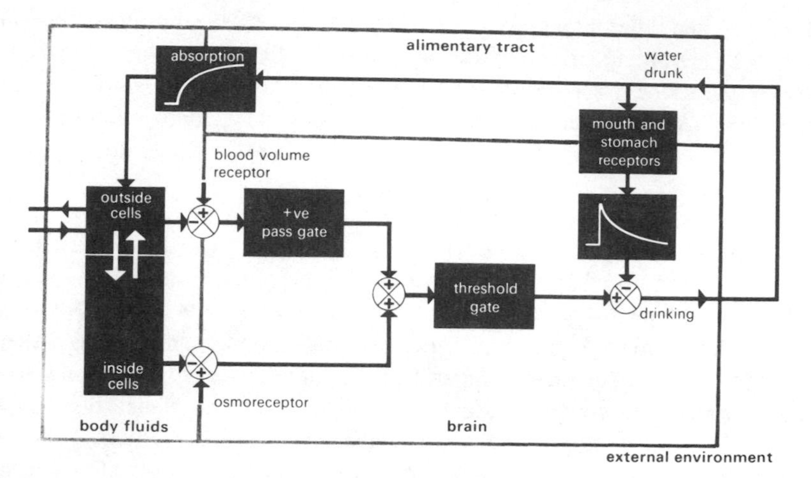

FIG. 5.5 A model for water intake as formulated by Oatley (1972). It is assumed that drinking is initiated by depletions of the two body-fluid compartments. Dehydration produces error signals from a set-point for body-fluid balance. Because of delay for absorption of water, it is assumed that metering of water intake in the mouth and stomach inhibits drinking prior to attenuation of the error signals.

These receptors send nerve impulses to integrative sites in the brain, and "command" signals are generated to initiate drinking behavior. The oral intake of water gradually replenishes the deficit to restore body-fluid balance. However, this is not instantaneous, because ingestion and absorption of water from the gastrointestinal tract takes time (see Figure 5.5).

The regulation of water intake depends on other signals besides those arising from intracellular dehydration and extracellular dehydration and, as suggested by Adolph (1964), is under multifactor control. Gastric distention has been shown to inhibit drinking in the dog (Adolph, 1950) and could serve as a signal for the termination of water intake, at least in certain species and under certain conditions. However, this is not the whole story; when an animal is prepared surgically with an esophageal fistula, so that water ingested via the mouth does not reach the stomach, the dog still terminates drinking, although the volume of water intake is greater (and drinking resumes after a shorter interval presumably because the water-deficit signals have not been eliminated as in the intact animal) (Towbin, 1949). This cessation of drinking has been considered evidence that the oral metering (of oropharyngeal sensations) of water intake provides a signal to "turn off" drinking before the water is absorbed and before the dehydration signals from the body-fluid compartments are eliminated.[1]

Signals from the mouth can enhance as well as attenuate water intake. As

[1]A study by Blass (1974b) showed that after being water-deprived for 24 hours, rats replenished only about 80% of the water deficit during the next several hours. The intracellular fluid compartment was restored, and somewhat overhydrated, but there was a substantial extracellular dehydration (for details, see Figure 11 of the article by Blass).

indicated in a later section, sweet and salty solutions of appropriate concentrations increase fluid intake (and bitter or other aversive tastes reduce intake). The temperature of the water may also influence intake, especially when the weather is very hot or cold. The classic oral signal is "dry mouth," but its role is not as crucial as initially proposed by Cannon (1918). Animals in which the salivary glands have been removed, drink more frequently, but unless they are given dry food, which increases their prandial drinking considerably, total water intake is not increased (Epstein et al., 1964). Dry mouth can influence and modulate drinking, but it is not the crucial signal for the control of water intake.

Oral and postingestional signals occur together (see Figure 5.5), but ingenious procedures have been devised to separate them experimentally. The chronic esophageal fistula, used in the middle of the nineteenth century by Claude Bernard (1856), makes it possible to investigate oral factors in the absence of gastrointestinal factors. With this procedure it has been observed that animals increase their intake of highly palatable solutions, the major difference being that total intakes are greater because gastric distention does not occur. On the other hand, the influence of postingestional factors can be studied in the absence of oral factors by infusing water or liquid food through a tube that is inserted into the nostrils, extending through the nasal passage, pharynx, and esophagus to the stomach (Figure 4.3). Animals fitted with this "electronic esophagus," as it has been designated, maintain normal water and calorie intakes, and when water intake requirements are increased or the diet is diluted or made more dense, they adjust intakes appropriately. Under these laboratory conditions, smell, taste, and other oral stimuli are not necessary for the regulation of intakes. However, this does not mean that such stimuli do not contribute when they are available. For the animal in its natural habitat, these stimuli may be very important, especially in discriminating and procuring food and water. This point is made by Epstein (1967) in comparing animals maintained in the apparatus shown in Figure 4.3 with those in a complex environment:

> They are not required to discriminate the edible from the inedible or poisonous, and they are performing a well-learned task that requires little motivation. As we have emphasized . . . , oral factors play their most important roles when choices must be made, when incentives are required to elicit ingestion, and when the performance of the behaviors leading to food and water must be aroused and sustained. The intragastric self-injection work shows that when the rat is not faced with the problems of detection and discrimination and when motivational demands are slight, the quantitative control of intake to meet nutritional needs is achieved in the absence of specific information from the head receptors [p. 211].

FACTORS THAT INITIATE AND CONTRIBUTE TO DRINKING BEHAVIOR

The several determinants of water intake are discussed in this section to provide an orientation and perspective for considering later the neural substrates of drink ing behavior.

Water-Deficit Signals

Following water deprivation, rats and other laboratory animals will reliably drink when water is available and will acquire and perform complex behavioral responses to reach water. It is assumed that the drinking behavior is initiated by signals that result from the depletion of body water. Because body water is distributed between two compartments, there are two kinds of signals: from depletion of the intracellular fluid volume and from depletion of the extracellular fluid volume. After a period of water deprivation, both kinds of signals occur, but as indicated previously, there are procedures for producing these signals separately.

When a hypertonic solution, such as a 2% NaCl solution or a 5% sucrose solution is injected, water passes from inside cells to the extracellular space because of the osmotic gradient produced by the increase in Na ions. In this way the volume of water inside the cells is reduced, and intracellular hypovolemia occurs (Figure 5.6, left side). In early experiments involving injections of hypertonic saline solutions, investigators were unable to demonstrate a direct relationship between degrees of cellular dehydration and quantity of water intake (e.g., Holmes & Gregersen, 1950). The volume of water intake was less than that required to restore body fluids to isotonicity presumably because of a reduction in water loss from the kidney in response to the osmotic stimulus. However, Fitzsimons (1961a) subsequently showed that in rats with kidneys removed, water intake was precisely that required to restore isotonicity (Figure 5.7). Even when water was withheld for 24 hours after the hypertonic saline injection, the nephrectomized rats drank until the isotonicity of body fluids was restored, indicating that adaptation to the osmotic stimulus did not occur.

In dog and man (Wolf, 1950) and in intact and nephrectomized rats (Fitzsimons, 1961a), it has been demonstrated that the threshold change in intracellular

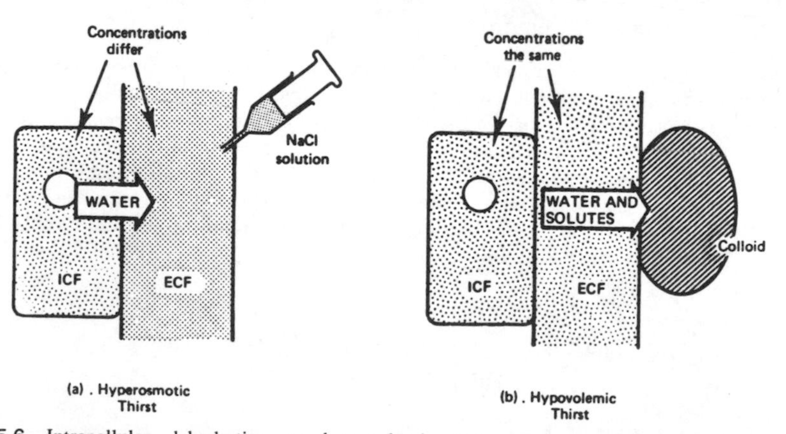

FIG. 5.6 Intracellular dehydration or hypovolemia (A) and extracellular dehydration or hypovolemia (B) can be produced artificially in the laboratory. (From Van Sommers, 1972.)

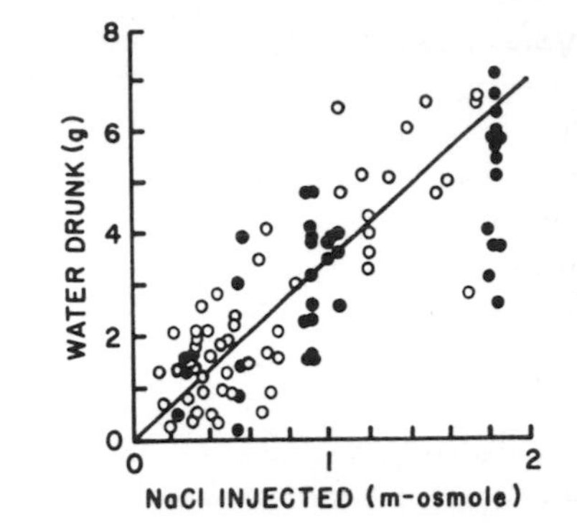

FIG. 5.7 Intracellular (osmotic) thirst in nephrectomized rats. The ordinate represents the osmotic pressure (milliosmoles/100 g initial body weight) of NaCl infused intravenously, rapidly (solid circles) or slowly (open circles). Water intake (measured as change in body weight/6 hours) was directly proportional to osmotic pressure regardless of infusion rate. (After Fitzsimons, 1963.)

fluid for the initiation of drinking is 1–2%. The threshold for the renal antidiuretic response is similar (Verney, 1947), which is evidence that behavioral (drinking) and physiological (ADH release-producing antidiuresis) responses are complementary in maintaining body-water homeostasis.

Water intake following hemorrhage in rats is shown in Figure 5.8. Intake increases in relation to degree of hemorrhage, but with severe hemorrhage, nonspecific effects or hemorrhagic shock disrupt drinking behavior. A reduction in extracellular fluid volume is also produced by injecting a hyperoncotic colloid, such as polyethylene glycol, dissolved in isotonic saline, either into the intraperitoneal cavity or subcutaneously (Figure 5.6, right side). The sequestering

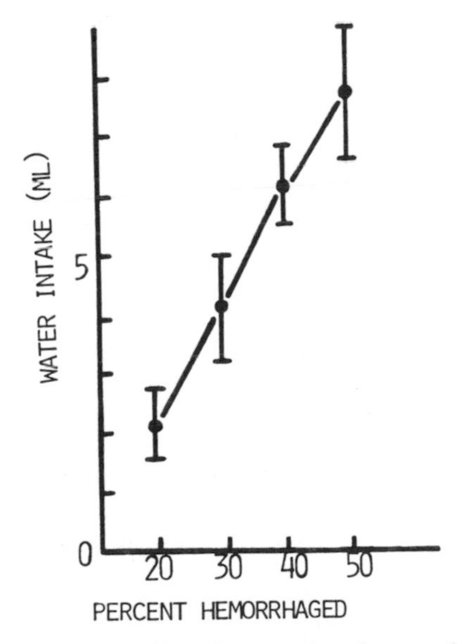

FIG. 5.8 Water intake of 10 rats subjected to varying degrees of hemorrhage (extracellular hypovolemia). (After Russell, Abdelaal, & Mogenson, 1975.)

of water at the injection site results in an extracellular hypovolemia without changing osmolality of body fluids. When Fitzsimons (1961b) treated rats in this way, they drank water during the next few hours. In subsequent experiments, Stricker (1966) administered polyethylene glycol by subcutaneous injection and showed that water intake was proportional to the concentration of the colloid administered.

Hormones

When an animal is hemorrhaged or is administered polyethylene glycol, there is an increase in the blood level of Angiotensin II, a hormone released by the kidney. These procedures, as well as ligation of the abdominal vena cava, reduce blood flow through the kidney and activate the renin-angiotensin system (Fitzsimons, 1970). Hemorrhage, the administration of polyethylene glycol, and caval ligation, all induce water intake in the rat. The rat also drinks when Angiotensin II is infused into the bloodstream (Fitzsimons & Simons, 1969). These and other observations have led Fitzsimons to propose that Angiotensin II mediates drinking to extracellular dehydration or hypovolemia (Figure 5.9). We see later that there is evidence that Angiotensin II acts directly on the brain to initiate drinking behavior.

Other hormones are indirectly related to water intake because of their effects on water and electrolyte balance. If the production and release of antidiuretic hormone is disrupted by lesioning the supraoptic nucleus or removing the post-pituitary gland (see Figure 2.8), animals drink large amounts of water. The drinking is secondary to the diuresis of water, which occurs in the absence of

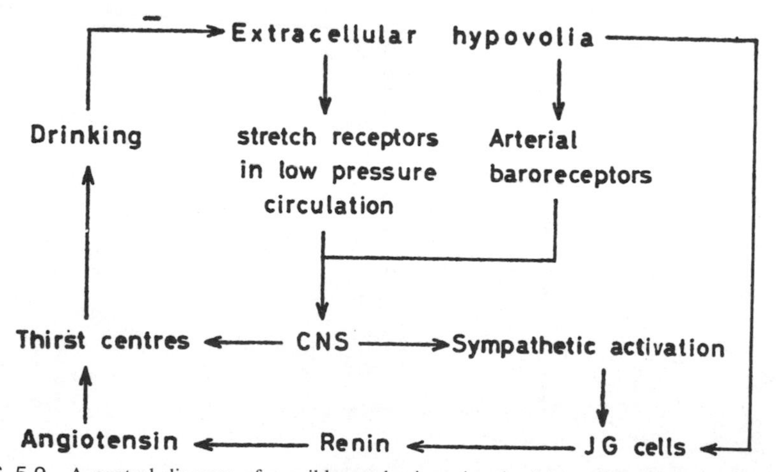

FIG. 5.9 A control diagram of possible mechanisms involved in angiotensin-induced drinking. (From Fitzsimons, 1970.) The renin-angiotensin is carried via the bloodstream to the brain to activate neural systems for the control of drinking behavior.

ADH, and is presumably initiated by water-deficit signals. Similarly, rats drink large volumes of NaCl solution when the adrenal gland is removed; in the absence of the hormone aldosterone, released from the adrenal cortex, which promotes the reabsorption of water by the kidney, rats make up for the excessive loss of sodium from the body by drinking water containing NaCl (Richter, 1943). As indicated previously, this is an example of a behavioral response compensating for the loss of a physiological regulation. Richter used to great advantage the strategy of disrupting a particular physiological regulation in order to show an exaggerated, compensatory behavioral adaptation or regulation.

Circadian Rhythm

Water intake is not uniform throughout the day and night but follows a circadian pattern. Because it is related to the 24-hour cycle of sleep and waking, drinking behavior occurs more frequently during the period when the animal is awake and active. In the rat, 80–85% of water intake occurs in the dark (Fitzsimons & Le Magnen, 1969). This is also the case in other nocturnal animals, whereas animals that are active in the day and sleep at night drink more water during the light period.

Most drinking behavior in the rat is associated with food intake, which is also circadian. About 80–90% of water intake is associated with meals. If rats are given food only in the daytime, or one or two 1-hour meals, water intake during the day increases from 10–15% to 50–55% of the total 24-hour intake (Fitzsimons & Le Magnen, 1969). However, drinking appears to have its own circadian rhythm and is not merely initiated by feeding; the two rhythms can be dissociated or decoupled. When food is not available, drinking continues to occur primarily at night, although the total intake is reduced (Morrison, 1968). This is illustrated in Figure 5.10, in which a rat that received half of its daily ration in the day and half at night still drank 65–70% of its total daily water intake at night.

Taste, Oropharyngeal, and External Stimuli

Drinking behavior is also influenced by factors not related to water-deficit signals or the status of body-fluid compartments. In this section we consider water intake associated with feeding (prandial drinking) and the influence of taste and oral factors. Nocturnal drinking in the rat and the circadian pattern of water intake observed in many species have already been considered.

In man, fluids are taken with meals. In the rat, drinking is associated with feeding. As shown in Figure 5.11, small draughts of water intake alternate with eating small amounts of food during a meal. It has been suggested that this drinking behavior is initiated by dehydration (e.g., water-deficit signals) due to salivary and gastrointestinal secretions and to oral cues from eating dry food (Kissileff, 1973). However, drinking frequently occurs just prior to a meal before

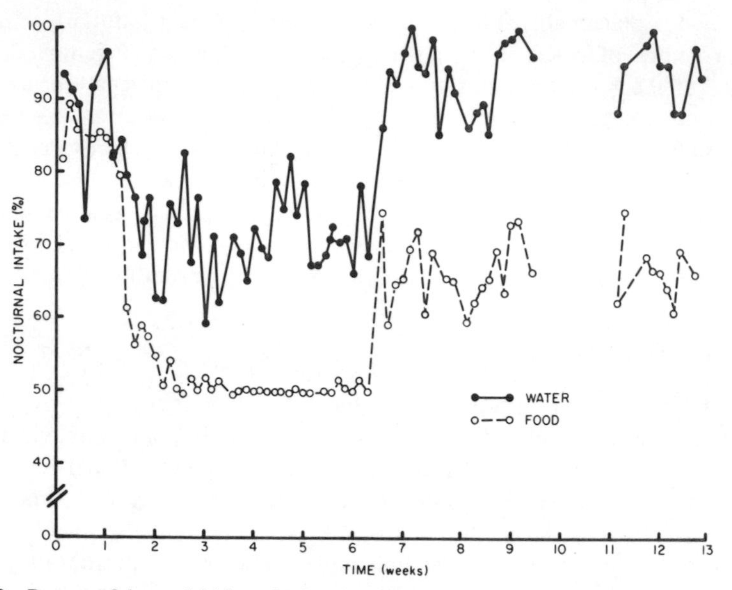

FIG. 5.10 Percent of the total 24-hour food and water intake occurring during the night in one rat. At the end of the first week, it was forced to eat more food during the day by being allowed only 50% of its normal 24-hour intake during the night. Water was available ad libitum. Note that although nocturnal drinking was reduced under these conditions, it was still well above food intake. Ad libitum feeding was continued again at the sixth week. (Unpublished data from Morris & Mogenson, 1977.)

digestive secretions begin. Laboratory studies have also shown that taste influences the choice and total intake of fluids. Rats show a preference for NaCl solution in the range of .1–1%, drinking more of these solutions than water (Weiner & Stellar, 1951). Rats also drink more water when saccharin is added and less when quinine is added; saccharin is particularly effective when added to a glucose solution, increasing intake five- or sixfold (Valenstein et al., 1967). Humans frequently prefer certain beverages to water: The palatability of fluids

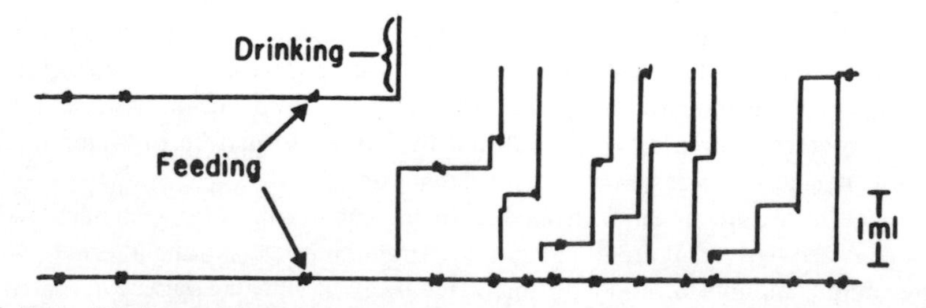

FIG. 5.11 Drinking and feeding patterns of a normal rat for 12 hours produced by the recorder pen of a drinkometer and a foodometer. The record is for a period from 6:00 p.m. to 6:00 a.m., when the rat does most of its eating and drinking. Note the prandial (meal-associated) drinking and on occasion the drinking prior to feeding. (After Kissileff, 1969.)

and total intake are increased by altering their taste, visual appearance, and temperature. Cold drinks are preferred in hot weather and hot beverages in cold weather. Laboratory studies in rats have shown that thermal oropharyngeal sensations are rewarding, and the licking at an air stream, which apparently cools the tongue as a consequence of evaporation, increases with water deprivation (Oatley & Dickinson, 1970).

The classical view has been that taste and oropharyngeal stimuli increase or decrease water intake that has been initiated by water-deficit signals (dehydration) but do not initiate drinking (see pp. 111–112 for further discussion of this view). In other words, these stimuli have only a modulatory role in the control of water intake in relation to body-water homeostasis (Adolph, 1967). This view has been succinctly summarized by Kissileff (1973) as follows: "In short, imbalance in body water arouses the urge to drink, previous experience and taste direct the selection of the drinking tube, and postingestional factors in combination with taste and water deficit determine how long drinking will continue [p. 179]."

An alternative view is that taste, oropharyngeal, and external stimuli can initiate drinking behavior in the absence of dehydration (Young, 1967; Pfaffmann, 1960): "Tastes can have hedonic value capable of arousing drinking in the need-free animal" (Kissileff, 1973, p. 176); "Rats undoubtedly drink to meet their water needs but they also drink fluids because they like their taste" (Epstein, 1967, p. 207). The second view has been gaining in popularity. After a comprehensive review of the literature on thirst, Fitzsimons (1972) concluded that animals normally do not drink in response to water-deficit signals; rather, water intake is influenced by taste and oropharyngeal stimuli, habit, and circadian rhythms, so that drinking usually occurs in the absence of and in anticipation of needs for water.[2] He designates this as secondary drinking, defined as water intake when the animal is in positive fluid balance, in contrast to primary drinking, defined as water intake in response to body-water deficits (e.g., intracellular and/or extracellular dehydration). Later, when discussing the neural substrates of drinking behavior, we see that memory and learning mechanisms are utilized if taste and other nonhomeostatic stimuli are to initiate and not merely modulate drinking behavior.

If, as proposed by Fitzsimons (1972), normal or spontaneous water intake is typically secondary drinking, then it follows that animals drink more water than they need to in order to maintain body-fluid homeostasis. In the laboratory rat, this has been demonstrated by providing a liquid diet from which water requirements are met. As shown by the open columns at the bottom of Figure 5.12, the

[2]Fitzsimons (1972) has emphasized the importance of secondary drinking as follows:

> When food and water are freely available, when climatic conditions are stable, and when the animal's activities remain the same from one day to the next, thirst is probably never experienced. In normal circumstances drinking is largely anticipatory of future needs for water and seems to be governed by oropharnygeal cues from the diet, by habit, and by an innate circadian rhythm; it is not dependent on a present need for water [p. 548].

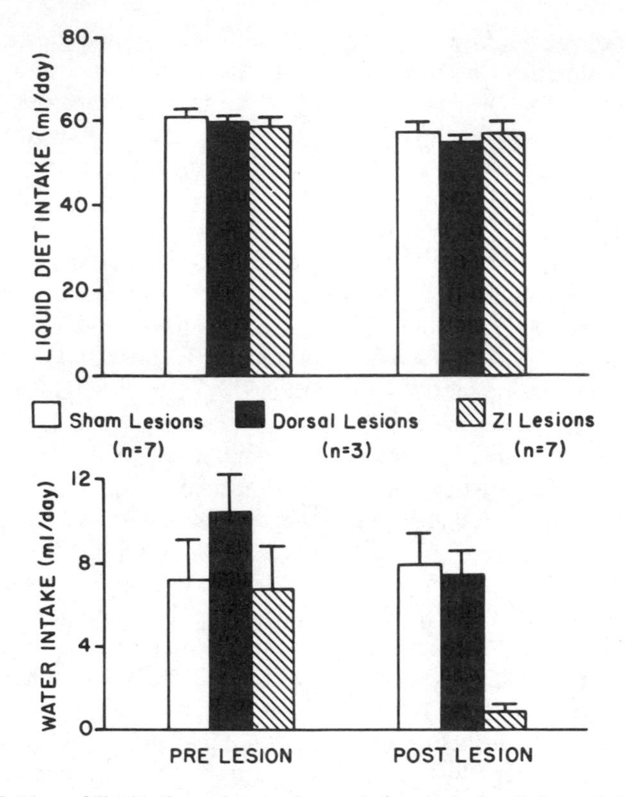

FIG. 5.12 Intakes of liquid diet and water in rats before and after lesions of the zona incerta (crosshatched columns). Note that only rats with lesions of the zona incerta reduced their water intake; those with sham lesions or lesions dorsal to the zona incerta did not. (From Evered & Mogenson, 1976.)

animals still drink an appreciable volume of water. The disruption of this water intake by lesions of the zona incerta is considered later. Another way to demonstrate that water intake exceeds need is to infuse water or other fluids by chronic intravenous or intragastric catheter. Results from a recent long-term study (Rowland & Nicolaidis, 1976) of this sort are shown in Table 5.2. A substantial volume of oral water intake occurred even when the amount of water infused was several times the normal daily intake of the animals. Rowland and Nicolaidis (1976) conclude that there is a ''nonhomeostatic contribution to normal drinking'' and suggest that ''there may be an urge to drink through the mouth independent of need condition [p. 7].''

Because an excess of water is readily excreted from the body (see Figure 2.8), it makes sense that neural mechanisms have evolved that enable an animal to ingest water in the absence of and prior to the occurrence of dehydrational signals (Mogenson & Phillips, 1976). We return to a consideration of these mechanisms later in this chapter.

TABLE 5.2
Effects of Continuous IG or IV
Infusions of Water on Oral Drinking[a]
(After Rowland & Nicolaidis, 1976)

Infusion ml/24 hr	Water drunk (ml/24 hr)	
	IG, 0%	IV, 0%
0	29.5±1.5	31.5±2.8
18	20.0±3.1	21.5±4.5
30	12.3±3.2	23.0±2.5
40	13.5±1.8	20.7±1.7
60		22.2±2.4
80	9.8±1.7	23.0±2.2
96	8.6±1.7	14.6±0.7
108		27.0±3.0
144		17 3.0
180	10.5±3.5	16.0±1.0
270		9.8±2.8

[a]Values are means ±SE. Route of water infusion is indicated IG = intragastric; IV = intravenous.

Experiential and Cognitive Factors

Fitzsimons has made a case for the importance of external stimuli, habit, and experience in the initiation of water intake and has proposed that secondary drinking occurs normally so that primary drinking is an emergency mechanism. These factors are particularly important in the control of water intake in man and higher mammals. What, when, and how much of beverages are ingested, are very much influenced by social and cultural factors (see p. 96).

For a number of years, experiential and cognitive factors were neglected in studies of the neural substrates of drinking behavior, which were carried out mainly in the rat using physiological challenges and deprivation procedures. As a consequence our theoretical views and models have been overly simplistic. However, in order to have an adequate understanding of how the brain controls drinking behavior, the role of experiential and cognitive factors must be included.

ONTOGENY OF DRINKING BEHAVIOR

The water requirements of the newborn mammal are met by suckling the mother's milk, and the drinking of water does not begin until some time later. The rat, for example, does not drink water until the third week of life. As long as the mother has an adequate supply of milk during the first 2 weeks or so, it is unlikely that the infant rat experiences body-water deficits. The suckling of milk is controlled, as indicated in Chapter 4, by the fullness or emptiness of the stomach. The newborn rat sleeps most of the time and is aroused by an empty stomach to root for the mother's teat and to suckle.

It has been shown that the absence of water drinking in the rat during the first 2 weeks or so of life is not because of a failure to respond to water-deficit signals (Figure 5.13). In tests performed under conditions not requiring appetitive responses, which the rat pup is incapable of performing, water was ingested in response to cellular dehydration (produced by the subcutaneous injections of hypertonic saline) at 2 days of age, to extracellular hypovolemia (produced by the subcutaneous injection of polyethylene glycol) at 4 days of age, and to the beta-adrenergic activation of the renin-angiotensin system at 6 days of age (Epstein, 1976a). The rat pup was able to make consummatory responses to these challenges during the first few days of life by licking from a syringe containing water, but it was not yet able to make the appetitive behavioral responses that precede the consummatory phase. By the third or fourth week, when weaning takes place and the neural and muscular systems have matured sufficiently, the rat pup is capable of the appropriate appetitive responses. The results of the careful studies of Epstein (1976a) on the ontogeny of drinking behavior in the rat have been summarized as follows:

> As revealed by consumatory responding, the physiological controls of feeding and drinking mature early and in sequence, the controls of drinking reaching competence weeks before they can be employed in behavior. The physiological controls await the maturation of capacities for sustained appetitive behavior. These appear in the last week of the suckling period, yielding animals that are prepared for adult-motivated ingestive behavior on the eve of weaning [p. 201].

After weaning, the nocturnal pattern of water intake begins in association with the nocturnal pattern of feeding. Experiential factors presumably begin to influ-

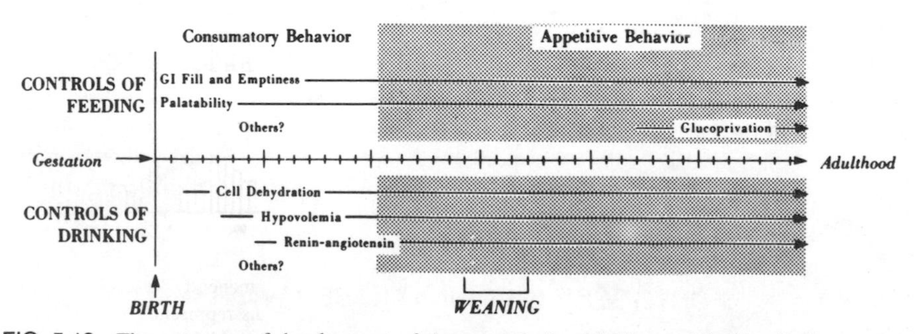

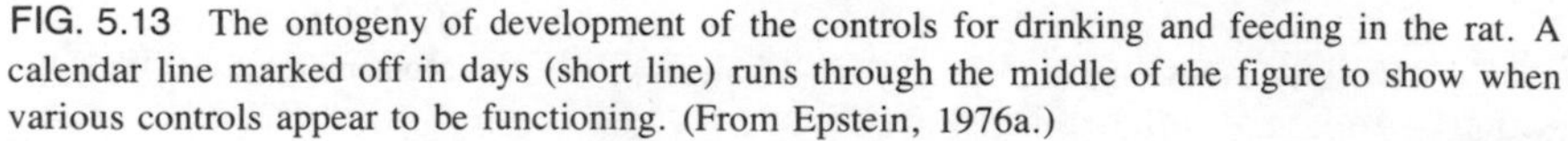
FIG. 5.13 The ontogeny of development of the controls for drinking and feeding in the rat. A calendar line marked off in days (short line) runs through the middle of the figure to show when various controls appear to be functioning. (From Epstein, 1976a.)

ence drinking behavior at a later date, but the course of development of this important determinant of water intake has not been investigated.

NEURAL SUBSTRATES OF DRINKING BEHAVIOR

In the 1950s and 1960s, drinking behavior was thought to be controlled by the hypothalamus. Lesions of the lateral hypothalamus were reported to cause adipsia in the rat (Anand & Brobeck, 1951; Montemurro & Stevenson, 1957), and electrical stimulation of this region elicited drinking behavior (Greer, 1955; Mogenson & Stevenson, 1966). The results of one of these classic experiments are shown in Figure 5.14. These observations were interpreted as evidence of a thirst center in the lateral hypothalamus, and the same model (see Figure 1.8a, which was widely used to account for the neural basis of feeding, was also used in dealing with the neural substrates of drinking behavior.

However, in recent years, as indicated in Chapter 4, there have been reservations expressed about the appropriateness of this model, and the results of lesion and stimulation experiments have been reinterpreted. It appears that the disruption of drinking when the LH is lesioned is not necessarily due to damage to an integrative system for thirst signals. Instead, the lesions damage dopaminergic fibers projecting rostrally through this region causing a general behavioral deficit, which accounts, at least in part, for the adipsia as well as aphagia. In the absence of an appropriate model to organize the experimental evidence, we begin—as with feeding behavior in the previous chapter—from the

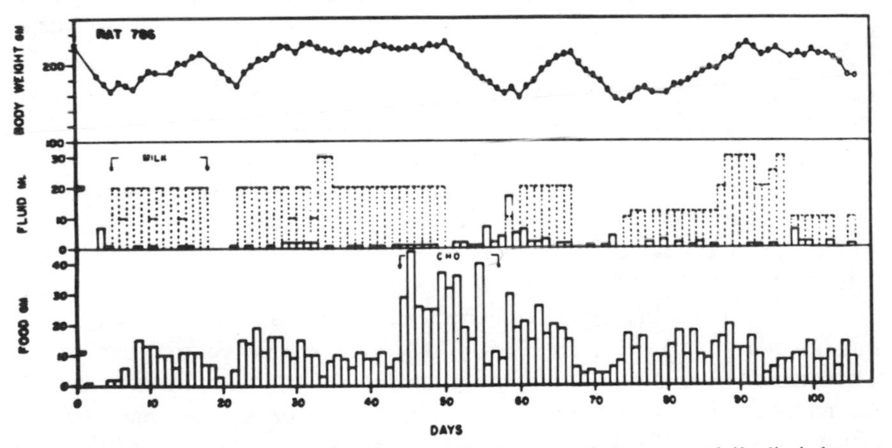

FIG. 5.14 Food and water intake and body weight of a rat made "permanently" adipsic but not aphagic by lesioning the lateral hypothalamus bilaterally. Dotted columns represent the volume of water (or milk as indicated) administered by stomach tube. The rat was fed a high-fat diet from Days 45 to 57, when it was also fed a high-carbohydrate diet (CHO). (After Montemurro & Stevenson, 1957, as modified by Stevenson.)

viewpoint of the factors that initiate and contribute to water intake. We return again at the end of this section to a brief consideration of the role of the hypothalamus in drinking behavior.

Neural Mechanisms for Primary Thirst

As indicated in an earlier section, the body can detect water deficits. Animals drink following water deprivation—after the intracellular fluid volume is depleted by subcutaneous or intraperitoneal injections of hypertonic saline and after the extracellular fluid volume is reduced by hemorrhage or the subcutaneous injection of polyethylene glycol (see Figure 5.6). Because drinking behavior initiated by these procedures is in response to water deficits, it is called *primary drinking* in contrast to drinking in the absence of and prior to water deficits, designated as *secondary drinking* (Fitzsimons, 1972). Most of the investigations of the neural mechanisms that subserve water intake have been concerned with primary drinking. Some of the relevant evidence is considered in the following.

Drinking to intracellular dehydration: Neural substrates. When an animal is injected with hypertonic saline, there is an increase in the osmolality of the extracellular fluid as the saline is absorbed from the gastrointestinal tract and distributed in the extracellular fluid compartment, and there is a concomitant decrease in the intracellular fluid volume (see Figure 5.6). It is now widely assumed that the biological transducer that detects these changes is an osmoreceptor.

Verney (1947) was the first investigator to postulate that osmoreceptors account for the diuresis that resulted from infusing hypertonic saline into the carotid artery. A few years later, Andersson (1953) demonstrated copious drinking in the goat when hypertonic saline was infused into the cerebral ventricles or hypothalamus using chronic cannulae. This was the first evidence for the presence in the brain of osmoreceptors that initiate drinking behavior.

The administration of hypertonic saline to the region of the supraoptic nucleus causes an increase in plasma ADH and causes antidiuresis (reduced urine production), and lesions of the supraoptic nucleus result in diuresis suggesting that osmoreceptors are located in this region of the hypothalamus (see Figure 2.8). This has been confirmed by recording from neurons in the supraoptic nucleus, which demonstrated that the neurons respond to injections of hypertonic saline into the carotid artery (see Figure 2.8). Although these osmoreceptors control the release of ADH in response to changes in osmolality of the blood, they are apparently not involved in the control of water intake, because administering hypertonic saline to the supraoptic nucleus does not initiate drinking behavior.

In recent years there has been a good deal of interest in trying to locate the site of osmoreceptors that initiate drinking behavior in response to intracellular dehydration. Using electrophysiological recording techniques, neurons that respond to injections of hypertonic saline into the carotid artery (Vincent et al.,

1972) have been identified in the LH, anterior hypothalamus, and preoptic region, and in some experiments, the electrophysiological responses have been associated with drinking behavior (see Figure 5.15). Such observations indicate that there are osmosensitive neurons in the brain, but it does not necessarily follow that they are osmoreceptors. They could be neurons that are part of a neural pathway subserving water intake in response to intracellular dehydration. By using the technique of microiontophoresis, in which hypertonic saline is administered by one barrel of a multibarrel pipette while recording from a single neuron (see Figure 2.23), it is possible to identify osmoreceptive cells. Oomura and co-workers (1969) performed such experiments in anesthetized rats and identified osmoreceptors in the LH. However, they did not perform behavioral experiments to see whether administering hypertonic saline to this same region

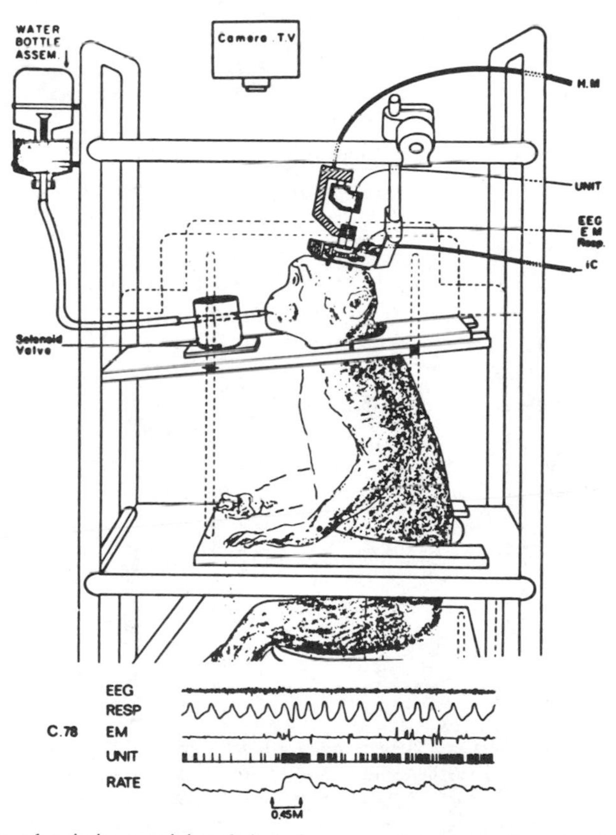

FIG. 5.15 A monkey in its restraining chair during a recording session. Note that the monkey has free access to the drinking spout while recordings are being made from the brain. When hypertonic saline is infused via a catheter into the intracarotid artery (IC), the discharge rate of the neuron in the hypothalamus (UNIT) is increased. The other recordings are of EEG, respiration, and electromyograph (EM). The bottom record is the integrated response of the hypothalamic neuron. (After Vincent, Arnauld, & Bioulac, 1972.)

elicits drinking behavior; but other investigators have reported that administering hypertonic saline to the lateral hypothalamus does not reliably elicit drinking behavior (e.g., Grossman, 1962). So present evidence does not permit any conclusion as to whether osmoreceptors for thirst are in the LH.

The administration of hypertonic saline or hypertonic sucrose to the preoptic region, one of the sites shown with electrophysiological recording techniques to contain osmosensitive neurons (Malmo & Mundl, 1975; Vincent et al., 1972), has been reported to initiate drinking behavior in the rat (Blass & Epstein, 1971; Blass, 1974a) and in the rabbit (Peck & Novin, 1971). However, there are reservations about concluding from these observations that there are osmoreceptors in the preoptic region, because the cannulae for administering the hypertonic solutions to the preoptic region passed through the lateral ventricles. It is possible that the hypertonic solutions diffused along the cannula shaft into the CSF and reached a receptor site in the cerebral ventricles (Johnson & Epstein, 1975; Johnson & Buggy, 1976).

Lesions of the preoptic region disrupt drinking behavior initiated by injecting hypertonic saline subcutaneously (Blass, 1974b), but this is not definitive evidence that osmoreceptors are located there. The lesions could have disrupted part of a neural circuit or integrative mechanism for drinking to intracellular dehydration, which has been the usual interpretation for the disruption of drinking to a hypertonic saline challenge following lesions of the LH (Epstein, 1971). More definitive evidence is obtained by infusing hypertonic saline or sucrose into the preoptic region using cannulae that are angled to bypass the lateral ventricles (Figure 5.16a). By using this procedure, Kucharczyk and Mogenson (1976) were

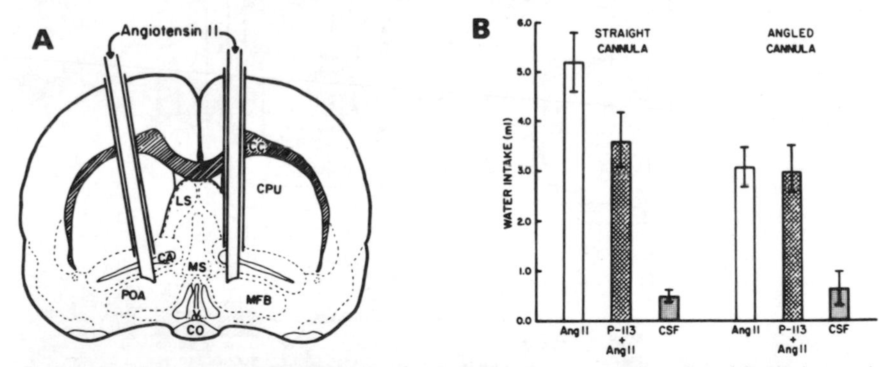

FIG. 5.16 Experimental evidence implicating the preoptic area as well as the subfornical organ in drinking induced by Angiotensin II (ANG II). A. A cross section of the rat brain showing the location of the cannulae in the preoptic area. The cannula on the left was angled to bypass the lateral ventricle (angled cannula). The cannula on the right passed through the lateral ventricle (straight cannula). B. Water intake during a 30-minute period when infusions were made either through the straight or angled cannulae. When ANG II (open columns) was infused via the angled cannula to bypass the ventricles and thus the subfornical organ, drinking was decreased, but it was still significantly greater than that observed after infusion of CSF (stippled column) by the same route. When an ANG II analogue (P113) was injected prior to ANG II as a blocker to prevent diffusion of ANG II to the subfornical organ, drinking still occurred (crosshatched columns). (From Assaf & Mogenson, 1977.)

able to elicit water intake in rats, suggesting that there are indeed osmoreceptors for drinking behavior in this region of the brain (Figure 5.17).

If osmoreceptors are in the preoptic region, where are the signals transmitted, and where do the neural pathways that carry these signals interface with the neural mechanisms for the motor control of drinking behavior? Kucharczyk and Mogenson (1975) demonstrated a disruption of drinking initiated by administering hypertonic saline into the preoptic region when lesions were made in the far lateral hypothalamus, suggesting the possibility that a neural pathway subserving drinking to intracellular hypovolemia projects through this region. However, Stricker (1976) has suggested that lesions of the lateral hypothalamus disrupt drinking after physiological challenges that initiate drinking and feeding not because this area contains specific neural integrative mechanisms or pathways for drinking and feeding behaviors but rather because the lesions make the animals susceptible to stress. This interpretation seems appropriate when peripheral challenges (e.g., hypertonic saline or 2-deoxy-D-glucose administered subcutaneously) are used, because they are quite stressful, but it seems unlikely that the administration of hypertonic saline or sucrose into the preoptic region is stressful. In any case, these same lesions do not disrupt drinking when Angiotensin II is administered to the preoptic region (Figure 5.18).

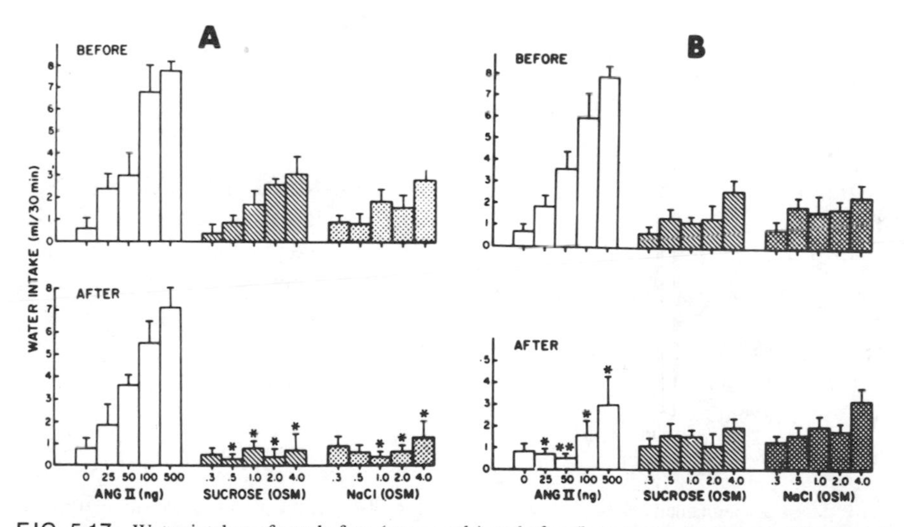

FIG. 5.17 Water intakes of rats before (top panels) and after (bottom panels) lesions of the lateral hypothalamus (LH) in response to intracranial infusions of varying concentrations of Angiotensin II (ANG II), sucrose, or NaCl into the preoptic area. Lesions of rats in A were in the far-lateral LH, and in B, they were in the midlateral LH. Vertical lines represent SEM, and asterisks represent statistical significance ($p < 0.05$). Note that rats with far-lateral lesions reduced their water intake after sucrose and NaCl infusions, which indicates the interruption of a neural pathway mediating intracellular thirst. Rats with midlateral lesions reduced their water intake after ANG II, which indicates the interruption of a neural pathway mediating extracellular thirst. (From Kucharczyk & Mogenson, 1976.)

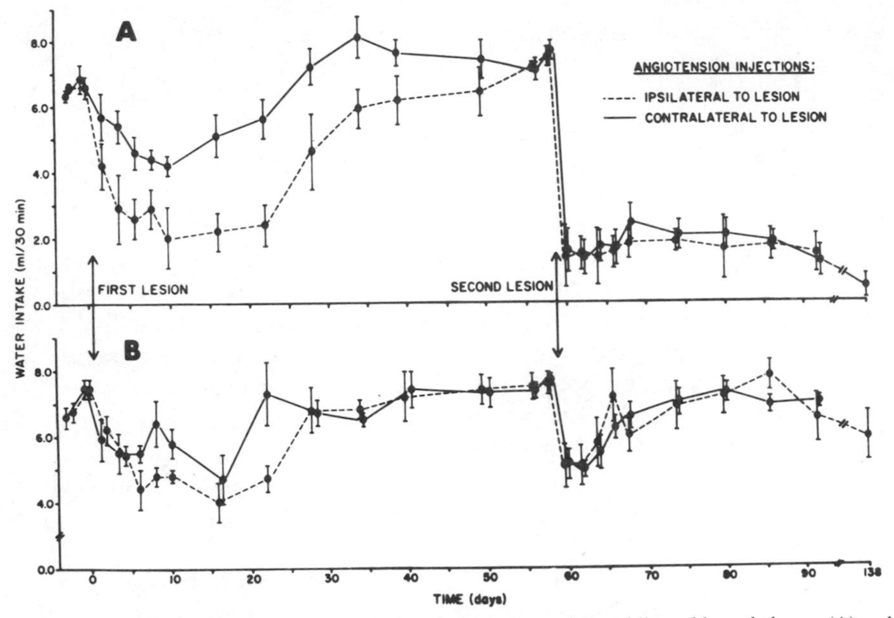

FIG. 5.18 Thirty-minute water intakes of rats with lesions of the midlateral hypothalamus (A) and far-lateral hypothalamus (B) after an intracranial injection of Angiotensin II (ANG II) into the preoptic area. A unilateral lesion was made on Day 0, and the second lesion on the opposite side of the brain in each rat was made on Day 58. The second lesions in the midlateral hypothalamus produced a severe and chronic decrease in drinking elicited by ANG II, implicating this area in extracellular thirst. (From Kucharczyk & Mogenson, 1975.)

It has been suggested that "command" signals for drinking behavior might go from the lateral hypothalamus to the midbrain (Mogenson & Huang, 1973). However, Kucharczyk and Mogenson (1975) have been unable to disrupt drinking initiated by administering hypertonic saline to the preoptic region, with lesions of the lateral and ventral midbrain. Recently Malmo (1976) reported that the discharge rate of neurons in the midbrain dorsal to the central gray responded when hypertonic saline was injected into the carotid artery. This region needs to be investigated as a possible projection site of the putative osmoreceptors in the preoptic region.

There has been a great deal of interest in the rostral third ventricle as a possible site of osmoreceptors. Johnson and Buggy (1976) have reported that lesions of the organ vasculosum in the optic recess of the third ventricle disrupted drinking after a hypertonic saline challenge. When deprived of water, the lesioned animals died after a shorter period than controls, suggesting that this structure contains osmoreceptors that contribute to the control of ADH as well as to thirst. The possibility that hypertonic saline administered to the preoptic region reaches the organ vasculosum must be considered.

Two other possibilities need to be considered when dealing with the neural mechanisms for drinking behavior initiated by intracellular hypovolemia. One is that osmoreceptors are in the periphery, perhaps in the liver as suggested by electrophysiological recording experiments (Niijima, 1969), and that signals from these peripheral osmoreceptors reach the preoptic region and/or lateral hypothalamus (Schmitt, 1973). Another suggestion is that there are sodium receptors, not osmoreceptors, and that they are on the walls of the third ventricle (Andersson, 1971). Both possibilities need further investigation.

Drinking to extracellular dehydration: Neural substrates. Extracellular dehydration or hypovolemia is an effective stimulus for drinking behavior as demonstrated with hemorrhage or the administration of polyethylene glycol (see Figures 5.6 and 5.8). It has been suggested that there are volume (or stretch) receptors in the low-pressure side of the circulatory system—in the low-pressure capacitance vessels or in the distensible veins near the heart. These postulated receptors have been implicated in the control of ADH and aldosterone secretion and in hemodynamic adjustments (e.g., blood pressure regulation) (Gauer et al., 1970; Fitzsimons, 1972). Because ligation of the abdominal vena cava in the rat, which reduces venous return to the heart, induces drinking (Fitzsimons, 1964), volume receptors may also contribute to water intake in response to extracellular hypovolemia. However, a role for these signals in the control of spontaneous drinking has not been demonstrated, and the neural mechanisms are not known.

Another possibility, which has received a good deal of attention in recent years, is that drinking after extracellular dehydration is mediated by the renin-angiotensin system and that Angiotensin II acts on the brain (see Figure 5.9). Experiments in which Angiotensin II has been administered directly to the brain using chronic cannula to initiate drinking have implicated the preoptic region (Epstein et al., 1970; Swanson & Sharpe, 1973) and the subfornical organ (Simpson & Routtenberg, 1973). However, when measures are taken to prevent Angiotensin II from reaching the subfornical organ—by placing cannulae into the preoptic region at an angle and using an analogue of Angiotensin II as a blocker—drinking is still elicited by the administration of Angiotensin II to the preoptic region (Figure 5.16B). Other evidence of an indirect nature also implicates the preoptic region as a receptive site for Angiotensin II. Lesions of the medial aspect of the lateral hypothalamus have been shown to disrupt (to a much greater degree) drinking induced by administering Angiotensin II to the preoptic region when the lesion is ipsilateral to the site of angiotensin administration as compared to a contralateral lesion (Figure 5.18). Furthermore, these lesions of the medial aspect of the lateral hypothalamus do not disrupt drinking induced by administering Angiotensin II to the subfornical organ (Kucharczyk et al., 1976). These observations cannot be accounted for by the hypothesis that Angiotensin II administered to the preoptic region must reach the subfornical organ for drinking to be initiated, and it appears that the preoptic region, as well as the subfornical organ and/or third ventricle, is a receptor site for Angiotensin II (Mogenson et al., 1977).

When evidence is available about the locus of receptors for Angiotensin II (as with osmoreceptors considered earlier), it is possible to begin investigating the neural pathways and integrative mechanisms that subserve drinking behavior to extracellular dehydration. The results of electrophysiological recording experiments suggest that a neural pathway from the preoptic region passes caudally through the medial aspect of the LH to the midbrain. Angiotensin II was administered to the preoptic region in doses that initiated drinking; the animals were later anesthetized, and electrophysiological records were made of action potentials from neurons in the LH and midbrain. The administration of Angiotensin II to the preoptic region was observed to change the rate of discharge of many of these neurons (Mogenson & Kucharczyk, 1975). In other experiments, Kucharczyk and Mogenson (1975) lesioned the medial LH, or the rostral aspect of the midbrain, and observed that drinking initiated by administering Angiotensin II to the preoptic region was disrupted. These lesions had no effect on drinking initiated by administering Angiotensin II to the SFO or by administering hypertonic saline to the preoptic region, suggesting that a neural pathway specific to angiotensin-induced drinking projects through the medial aspect of the LH to the midbrain.

These conclusions are quite tentative, and the role of central Angiotensin II receptors for drinking initiated by extracellular hypovolemia is uncertain and controversial. All workers in this field are not convinced by the evidence that Angiotensin II from the kidney acts on the brain to initiate drinking behavior. Angiotensin II is a relatively large molecule and may not cross the blood–brain barrier. Also, the doses of Angiotensin II administered centrally or into the bloodstream have typically been very high. When administered to the brain, for example, Angiotensin II could constrict blood vessels or have other nonspecific effects (Fitzsimons & Nicolaidis, 1976). Because the subfornical organ is outside the blood–brain barrier and is responsive to low doses of Angiotensin II, there has been a great deal of interest in this ventricular structure as a possible receptive site for Angiotensin II (Epstein, 1976b). Another possibility, because the components of the renin-angiotensin system are present in the brain (Ganten et al., 1971), is that Angiotensin II of peripheral origin acts on receptors in the subfornical organ or third ventricle and that Angiotensin II produced in the brain, perhaps in response to peripheral volume receptors, acts on the preoptic region (Mogenson et al., 1977).

Of great interest is a recent report suggesting that Angiotensin II may be a CNS neurotransmitter and that the preoptic region is a site where it is in high concentration (Fuxe et al., 1976). These observations, based on experiments in which the technique of immunohistochemistry was used, raise the possibility that Angiotensin II administered to the brain may act at synapses in the way that the central administration of noradrenaline initiates feeding by acting at noradrenergic synapses in the hypothalamus (Miller, 1965).

Clearly, much research is needed in this active and interesting field before we can have an understanding of the neural substrates for drinking behavior initiated by extracellular hypovolemia.

Circadian Rhythms and Water Intake: Neural Substrates

Lesions of the suprachiasmatic nucleus have been reported to disrupt the circadian rhythm of water intake in the rat, but it is not clear whether or not this is the result of damaging part of the "biological clock" for the control of drinking behavior (Stephan & Zucker, 1972). Lesions of this nucleus also disrupt the sleep–waking cycle, and it could be that drinking simply depends on the animal being awake. Another possibility, because water intake is food-associated, is that the disruption of the nocturnal pattern of drinking behavior is secondary to the disruption of the nocturnal pattern of feeding behavior. This second possibility is unlikely, however, in view of the report that when food is not available to rats, drinking behavior continues to be nocturnal although total water intake is reduced (Morrison, 1968).

Further research is needed to find out whether there is a separate biological clock that controls the circadian pattern of water intake and, if there is a separate clock, to determine its neural substrates and how it interfaces with neural systems that initiate drinking behavior (see pp. 280–281).

The Influence of Taste and Oropharyngeal Stimuli and Experiential and Cognitive Factors on Drinking Behavior: Neural Substrates

As indicated in an earlier section, drinking behavior is enhanced by dry mouth and taste, associated with meals, and influenced by habit and past experience. As we discuss the little that is known about the neural mechanisms that subserve these influences on water intake, we are concerned with an issue raised previously—whether these factors merely modulate drinking initiated by water-deficit signals or whether they also initiate water intake.

The classical view to account for the high percentage of water intake that occurs in association with feeding is that salivary and gastrointestinal secretions result in dehydration and water-deficit signals. Food may also cause a dry mouth, especially when dry food is ingested. Food-associated drinking is very prominent in rats recovered from adipsia and aphagia produced by lesions of the lateral hypothalamus. The recovered animal does not respond to dehydration challenges, its drinking is entirely prandial (e.g., occurs during meals in response to dry mouth), and the pattern of water intake is similar to that of the rat with salivary glands removed (Kissileff, 1973). Lateral hypothalamic lesions may disrupt the neural control of salivation, and it has been suggested that prandial drinking in the LH rat, as in the desalivate rat, is in response to dry mouth and difficulty in swallowing food. The importance of these oropharyngeal stimuli has been demonstrated by infusing small amounts of water via a cheek fistula in recovered lateral rats and showing that prandial drinking disappeared (Kissileff, 1973). Water infused into the stomach had no effect on the prandial drinking.

These observations suggest that oropharyngeal and water-deficit stimuli are important in food-associated drinking, but this is not the whole story. A bout of drinking often occurs before a meal (see Figure 5.11), and neither dehydration, secondary to salivary or gastric secretions, or dry mouth can account for this water intake. According to some authors, this is anticipatory drinking and evidence that the rat and other animals possess the neural machinery (e.g., representational processes) that enables the animal to drink prior to and in anticipation of water needs (Fitzsimons, 1972; Oatley, 1973).

That taste influences fluid intake is clearly demonstrated by adding saccharin or salt to the water. However, it is not clear according to some investigators whether this is merely because taste modulates water intake when drinking has been initiated by water-deficit signals or whether taste and other nondeficit signals can themselves initiate water intake. Inputs from taste and other sensory stimuli reach the amygdala and other forebrain structures (Gloor, 1960; Norgren, 1976; and see Figure 2.12). According to the classical view, these stimuli, by activating limbic forebrain structures, exert modulatory influences on a basic integrative system in the hypothalamus for the initiation of drinking behavior (Mogenson & Huang, 1973). Proponents of the alternative view, that such stimuli can initiate and not merely modulate drinking behavior, have suggested that inputs from these stimuli reach neural systems concerned with reinforcement (Pfaffmann, 1960; Hoebel, 1971; Rolls, 1975). Reinforcement, according to studies of brain-stimulation reward, is subserved by neural systems in the hypothalamus, amygdala, and associated forebrain structures (see Chapter 8). Recent experiments by Rolls and co-workers (1976) in which recordings were made from single neurons in the lateral hypothalamus of unanesthetized monkeys are relevant to this view. These investigators observed that neurons changed their rate of discharge when the monkeys were presented fruit juice that was highly preferred but that the neurons no longer responded when the animal was satiated by the juice (see Figure 4.10). They gave a reward interpretation to their findings and suggested that the activity of "water-reward neurons" or "food-reward neurons," involved in drinking and feeding behaviors, is influenced by taste. In other words, it is suggested that taste modulates the reward value of fluids and thus of fluid intake.

Although it is customary to assume that complex forebrain structures, and especially the cerebral cortex, are involved with memory, learning, and cognitive processes, there is not much definitive evidence concerning the neural mechanisms for these integrative activities. In fact, in studying the neural substrates of drinking behavior, most of the work has been on the mechanisms for primary drinking (e.g., in response to intracellular and extracellular dehydration), and the neural mechanisms for secondary drinking have been ignored. There was considerable excitement in my laboratory when it was observed that lesions of the zona incerta, an area just dorsal to the lateral hypothalamus, produced a 25–30% reduction in daily water intake (Figure 5.19). The water intake of animals with zona incerta lesions is less variable than for control rats, as indicated in the figure, and no deficit in drinking in response to hypertonic saline and polyethylene glycol challenges occurred; it appeared that the lesioned animals were relying on primary thirst signals and that secondary drinking was disrupted (Evered & Mogenson, 1976). This suggestion is supported by the observation that when rats were maintained on a liquid diet, lesions of the zona incerta disrupted the drinking of water, which is presumably in excess of need (see Figure 5.12).

At first it was concluded the zona incerta lesions had destroyed a neural pathway that contributes to secondary drinking. However, additional experiments showed that the deficit was probably not specific to secondary drinking; the results suggested that lesions of the zona incerta produce subtle motor or sensory–motor deficits and that secondary drinking is more susceptible to disruption. The lesioned animals had a reduced lap volume (Evered & Mogenson,

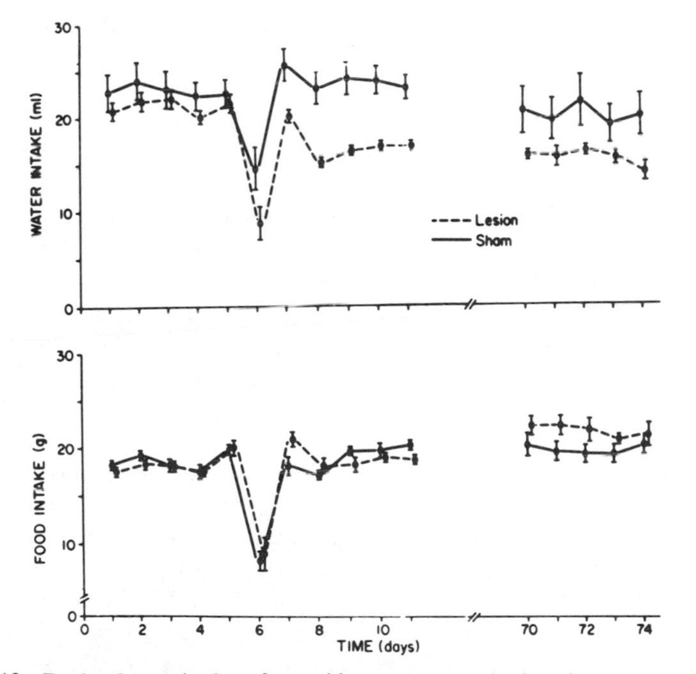

FIG. 5.19 Food and water intakes of rats with a permanent reduction of water intake (hypodipsia) produced by lesions of the zona incerta (ZI). Compensatory reductions in water losses maintained fluid balance. This finding illustrated an attenuation of what has been termed secondary drinking (drinking independent of water needs for fluid homeostasis). (From Evered & Mogenson, 1976.)

1976) and were not able to extend their tongues as far as control rats (Brimley & Mogenson, 1977). In spite of these deficits in the lap response, it appeared that they responded to water-deficit signals even though this involved making more laps of the tongue to obtain a given volume of water. It turns out that these experiments, like those involving damage to the DA nigrostriatal pathway (see p. 000), have more to do with the consummatory phase of water intake than with the initiation of drinking behavior (the appetitive phase). This does not necessarily mean that neurons of the motor system have been damaged. One possibility is that lesions of the zona incerta disrupt fibers of the central trigeminal system that transmit signals from the oropharyngeal region to the thalamus (Zeigler & Karten, 1974).

Although the results of these experiments are somewhat disappointing in relation to earlier optimism, they have drawn attention to the neural substrates of secondary drinking. There will not be a complete understanding of the neural mechanisms that control water intake until we can elucidate the neural mechanisms by which experiential and cognitive factors influence fluid intake. This is especially the case in man, because what, when, and how much of fluids are ingested, are frequently influenced by social and cultural considerations (Rozin, 1976). Elucidating these neural mechanisms will involve an understanding of memory, learning, and related cognitive processes, one of the most difficult tasks in the neurosciences, and it may be many years before we have an adequate neural model for these processes.

The Hypothalamus and Drinking Behavior

The hypothalamus was first implicated in the control of water intake by reports of adipsia following lesions of the LH (Anand & Brobeck, 1951) and of drinking elicited by administering hypertonic saline into the cerebral ventricles or hypothalamus (Andersson, 1953). Subsequently there have been many further studies that suggest that this small structure at the base of the brain is involved in drinking behavior: Drinking is elicited by electrical stimulation of the lateral hypothalamus (Greer, 1955; Mogenson & Stevenson, 1966) and by the administration of acetylcholine or carbachol to the hypothalamus (Fisher & Coury, 1962; Grossman, 1962); lesions of the LH disrupt drinking to dehydration challenges (Epstein, 1971); osmosensitive neurons have been identified in the hypothalamus in acute electrophysiological recording experiments (Oomura et al., 1969); and neurons that discharge in association with drinking have been observed in chronic recording experiments (Vincent et al., 1972). However, as a consequence of recent developments that have shifted attention away from the hypothalamus and promoted a reinterpretation of early experimental findings and theoretical ideas, these observations are no longer considered unequivocal support for the presence in the hypothalamus of an integrative system for drinking behavior. These developments and their impact on current views about the neural substrates of feeding behavior were considered at the end of Chapter 4.

It should be pointed out again, as in the discussion of the neural mechanisms that subserve feeding behavior, that an over-reaction is to be avoided. To exclude or ignore the hypothalamus in considerations of the neural control of water intake would be a serious mistake. This small but strategically located structure is in a position to exert important influences on limbic forebrain, midbrain, and extrapyramidal structures, on the pituitary gland, and on the autonomic nervous system. Its unique role may be to contribute to the overall coordination of drinking behavior with the endocrine and autonomic responses that maintain body-water homeostasis as the animal adapts to a continuously changing external environment. As indicated previously, this contribution to the ''functional coupling'' of endocrine, autonomic, and behavioral responses may be the distinctive contribution of the hypothalamus to body-water homeostasis, as well as to body-energy homeostasis and thermoregulation.

SUMMARY

Water, the single largest component of the body, is essential for life. Because water is lost continuously from the body, it must be replaced at relatively frequent intervals by the intake of water in order to maintain body-water homeostasis. Drinking behavior is periodic, and in many animal species, it is circadian, occurring during the part of the day when the animal is more active. In rats, 80–90% of the drinking behavior is at night, and most of its water intake is associated with food intake.

Drinking behavior is initiated by depletions of one or both of the body-water compartments—that is, by intracellular dehydration and by extracellular dehydration. These water-deficit or homeostatic signals can be very effective initiators of behavior; following a period of water deprivation, drinking behavior is especially persistent. Drinking behavior is also influenced by taste, by oropharnygeal stimuli, and by experiential and cognitive factors. These factors not only influence or exert modulatory effects on drinking, which is the result of deficit signals, but they can also initiate water intake in the absence of deficit signals. This has been called secondary drinking and is believed to represent the major source of water intake in a number of species. Accordingly, although drinking behavior may be initiated by water-deficit signals and contributes to body-water homeostasis, it is less of a ''homeostatic drive'' than thermoregulatory behavior or feeding behavior.

Drinking behavior is under the control of the central nervous system. Because a number of factors contribute to water intake, the neural mechanisms appear to be rather complex and are not well understood. The body can detect water deficits, but there is still uncertainty about the locus of the receptors for these signals. It has not been possible, therefore, to determine the neural pathways and integrative systems for primary thirst. Lesion and stimulation studies have implicated the hypothalamus and limbic structures, but the precise neural mechanisms for the control of water intake are poorly understood. There is even less known about the neural substrates of secondary drinking, but complex mediating processes represented in the cerebral cortex and associated limbic forebrain structures are thought to be involved.

6
Sexual Behavior

Boris Gorzalka and
Gordon Mogenson

Sexual behavior is normally included as an example of one of the more important motivated behaviors, but it is usually emphasized that sexual behavior differs from the other so-called biological drives in that it is not associated with homeostasis. In comparison to drinking and thermoregulatory behaviors, for example, there are no deficit signals that initiate the appetitive or courtship phase of sexual behavior, and copulation, the consummatory phase, does not restore a homeostatic imbalance. So sexual behavior does not contribute to the homeostasis or preservation of the individual animal, although it does have biological significance by contributing to the perpetuation of the species.

This does not mean that sexual behavior is any less important from the motivational point of view. In fact, as indicated in the following sections, sexual behavior shares the same motivational characteristics as the other behaviors considered in this book and is influenced by several of the same factors. Hormonal and experiential factors are particularly important, and their relative contribution depends on the species, age, and sex of the animal.

Sexual behavior in the broader sense includes courtship, mating (copulation), reproduction, and care of the offspring (parental behavior). In this chapter the discussion deals mainly with mating behavior.

CHARACTERISTICS OF SEXUAL BEHAVIOR

Sexual behavior differs from the behaviors considered in the three previous chapters in that it is nonhomeostatic, but, nevertheless, it is similar in terms of behavioral characteristics. It is goal-directed (or purposive), intense and persistent, and periodic, and under certain circumstances it can have a high priority in the "motivational time-sharing" of the animal.

There are numerous examples from a variety of species in which the sexually aroused male will seek and pursue a mate and make appropriate responses to arouse the female to become sexually receptive. You will no doubt have observed this behavior in dogs, cats, farm animals, and occasionally in animals in the wild. This appetitive phase, although highly variable from one species to another, is frequently quite stereotyped for a particular species, consisting of complex sequences of behavioral responses that appear goal-directed. Such sexual behaviors are typically very intense and persistent. For example, in dogs, cattle, and many other species, the male may pursue the female for hours or even days, and once she is fully receptive, they may copulate repeatedly. Receptivity, or the period of behavioral estrus, of the female rat lasts about 14 hours (Blandau et al., 1941), and during this time copulation may occur as frequently as 50 or 60 times. In laboratory studies the strength of the sex drive in male rats has been demonstrated in the obstruction apparatus; the male rat will cross a shock grid and tolerate intensely, painful foot shock to reach the estrus female (Warden, 1931).

Sexual behavior is periodic in many species, the periodicity typically being more prominent in the female. The female rat comes into natural behavioral estrus every 4 or 5 days. During this period the female will frequently respond to the presence of the male by characteristic hopping and darting movements, which are often accompanied by momentary body-freezing and so-called ear wiggling, the result of rapid vibrations of the head and ears. In response to mounting by the male, the sexually receptive female usually adopts the lordosis posture. This posture is characterized by a more or less pronounced concave arching of the back accompanied by an elevation of the hindquarters and deviation of the tail to expose the perineal region. As a quantitative index of female sexual receptivity, many investigators compute a lordosis quotient based on the probability of occurrence of the lordosis posture in response to mounts by the male (e.g., Whalen et al., 1971). The lordosis quotient is useful, because the appearance of lordosis is not an all-or-none phenomenon. Figure 6.1 illustrates copulatory responses in the hamster.

Seasonal breeders provide the most dramatic evidence of periodicity or cyclicity in male sexual behavior. In these annually breeding species, such as cat, sheep, and deer, reproductive behavior and physiology are linked to environmental changes.

Because of the intensity of the sex drive, some animals do not eat or sleep during the period of sexual receptivity, lose weight, and become dehydrated. It is well known that sexual behavior has a high priority in the motivational time-sharing of many species.

FACTORS THAT INFLUENCE AND DETERMINE SEXUAL BEHAVIOR

Homeostatic signals do not initiate sexual behavior, but other factors such as hormones and previous experience do contribute. Their role is discussed in the

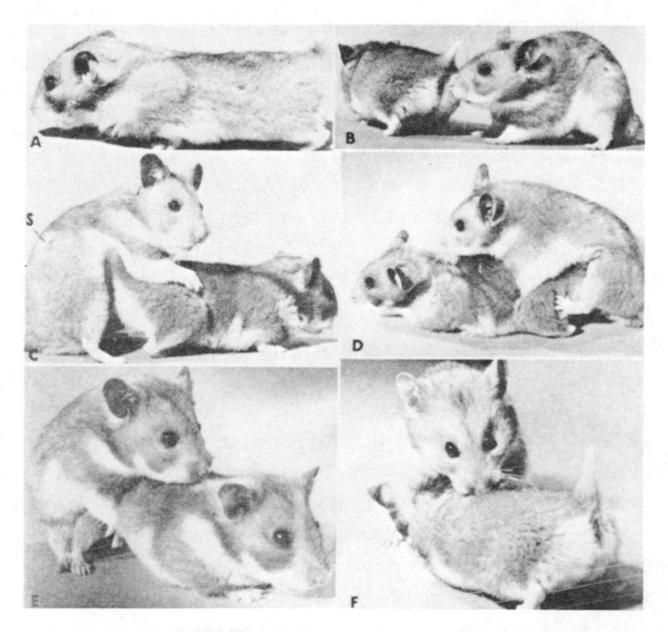

FIG. 6.1 Copulation in hamsters. A. Female in heat shows lordosis with rigid legs and erect ears and tail. B. Male investigates the perineum and cleans the vaginal orifice. C. Foreplay usually involves contact with female's pigment spot, which is located just under the male's right paw on the female's flank. Male's pigment spot is labeled *S*. D. Male, with right foreleg in contact with female pigment spot and left hind leg off ground, begins copulation. E. Male copulates with right foot on ground. F. Inexperienced male may mount incorrectly. (From Magalhoes, 1970.)

following, and we see that they contribute differentially depending on the sex, the age, and the species of the animal.

Hormones

Hormones contribute to sexual behavior in several ways: They influence the development of the CNS so that appropriate neural tissue becomes differentiated as a ''male brain'' or a ''female brain''; they influence the development of secondary sexual characteristics; and they activate neural circuits in the brain for sexual arousal. Ovulation—the release of an ovum from the ovary so that it is available for fertilization by the sperm—is also controlled by hormones in many species. In the next and subsequent sections, the details of these hormonal effects are considered.

Because our emphasis is on the initiation of motivated behaviors, it should be pointed out at this time that sexual behavior is much more dependent on hormones in some animals than in others. The female rat is an example in which copulatory behavior depends on elevated levels of ''sex'' hormones during estrus. On the other hand, hormones are less crucial in man and other higher mammals, especially when there has been previous sexual experience.

Biological Rhythms and Endocrine Cyclicity

Many female mammals show a cyclicity in ovarian steroid secretion. These ovarian fluctuations, as well as the concomitant changes in sexual receptivity, pituitary gonadotropin secretion, vaginal cornification, and uterine and ovarian morphology, are characteristics of the estrous cycle. In many primate species, these changes are accompanied by a cyclic uterine bleeding, or menses. Species in which menses occur are said to have a menstrual cycle rather than an estrus cycle.

In the female rat, there is a close relationship during the estrus cycle between sexual behavior and the production of estrogen and progesterone, pituitary gonadotropin production, and ovarian morphology, as shown in Figure 6.2. Analagous relationships during the menstrual cycle of the human female are illustrated in Figure 6.3. Although both species show marked hormonal cyclicity, the behavioral fluctuations observed in the human female appear rather trivial when compared with the dramatic changes of the rat estrus cycle. In humans, as well as in monkeys and apes—animals with a menstrual cycle—the relationship

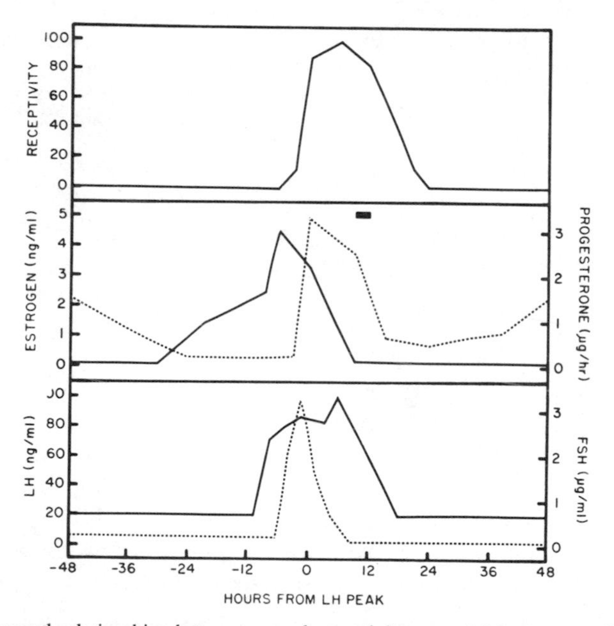

FIG. 6.2 Temporal relationships between sexual receptivity, gonadal hormone and gonadotropin levels in female rats. Levels of estrogen and the gonadotropin follicle stimulating hormone (FSH) are indicated by solid lines. Levels of progesterone and the gonadotropin luteinizing hormone (LH) are indicated by dotted lines. The solid bar indicates time of ovulation.

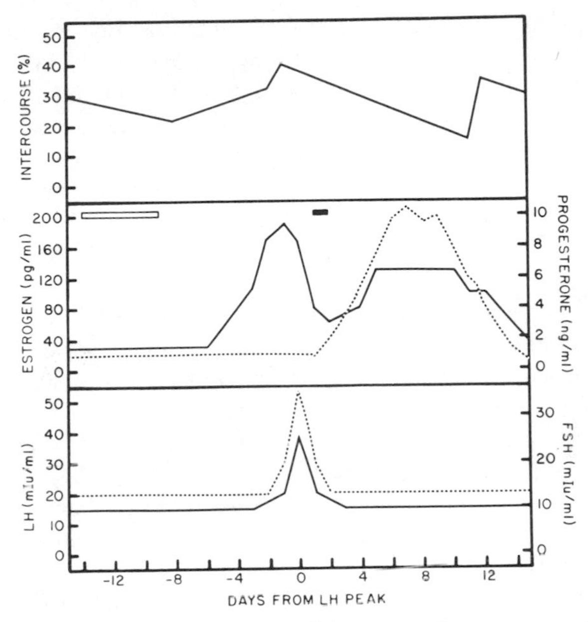

FIG. 6.3 Temporal relationships between sexual intercourse frequency, gonadal hormone and gonadotropin levels in human females. Levels of estrogen and the gonadotropin follicle stimulating hormone (FSH) are indicated by solid lines. Levels of progesterone and the gonadotropin luteinizing hormone (LH) are indicated by dotted lines. The solid and hollow bars indicate time of ovulation and menses, respectively.

between hormones and sexual behavior is much weaker than in the rat and lower mammals. The female rhesus monkey, like the human female, copulates to some extent during all phases of the cycle (Rowell, 1963).

Human females appear to show a significant cyclic fluctuation in frequency of sexual intercourse and orgasm during the menstrual cycle, but this behavioral fluctuation does not appear to be related to the secretion of estrogen, progesterone, or gonadotropins (Udry & Morris, 1968). Because human ovariectomy does not have any systematic effect on sexual behavior, as indicated in a later section, it seems more likely that other factors associated with the menstrual cycle [e.g., variation in adrenal androgens, aldosterone, norepinephrine or dopamine (Janowsky et al., 1971), or some psychological consequence of the physical symptoms of the menstrual cycle] account for these behavioral fluctuations.

There is some evidence of monthly changes in human birth rates (Cowgill, 1966), but these effects are minor and hardly permit classification of man as a seasonal breeder. Recent evidence does suggest, however, circannual rhythms in testosterone levels. In a study of young Parisian men, Reinberg and co-workers (1975) report that testosterone levels are lowest in April and rise progressively to maximal levels in October. Whether this circannual rhythm holds for other geographical locations remains to be seen. Sexual activity was also recorded for the males in this study. An annual variation occurred with a statistically significant peak in sexual intercourse and masturbation during October. Despite the correlation between testosterone levels and sexual activity, it is unknown whether the increase in sexual activity is secondary to the testosterone increase, whether the testosterone increase is secondary to the increase in sexual activity, whether both co-vary with some undetermined factor, or whether the correlation is fortuitous.

Reinberg and co-workers (1975) also report circadian rhythms in testosterone levels in the aforementioned subjects. The circadian peak seems to vary during the course of the year. For example, peak daily levels occur at about 8:30 a.m. in May and at about 3:00 p.m. in October. These circadian peaks probably do not correlate with sexual activity nor should they be taken to suggest an optimal time for sexual activity. Diurnal rhythms in sexual activity, if they exist, are more likely related to working and sleeping schedules than testosterone levels. Furthermore, among the numerous studies of circadian rhythms in testosterone levels that have been published in the last 10 years, there has been no general agreement concerning the circadian peak time. Perhaps these disagreements will be reconciled when circannual rhythms and geographical locale are taken into consideration.

External Stimuli

External stimuli serve two main functions; they contribute to sexual arousal, and they provide cues that guide the sexual behavior. It is well known, for example, that the colored plumage of some male birds may enhance sexual arousal and receptivity in the female. In some species the male bird engages in a "dance" or other ritualistic behavior to "interest" and/or arouse the female. Another widely quoted example is the stickleback, which engages in a highly stereotyped, species-specific behavior pattern prior to the female entering the nest to deposit her eggs (see Figure 1.2).

The elaborate behavioral responses of the appetitive or arousal phase may utilize visual, tactile, and other stimuli to initiate and guide the response components. In addition, the consummatory phase of copulation—including mounting, intromission, and ejaculation—also relies on sensory stimuli, particularly in the male of some species in which the motor aspects of copulatory behavior are more elaborate in comparison to those of the more passive female partner.

The profound influence of external stimuli on sexual functioning is well illustrated by the Coolidge effect in domestic and other animals. After achieving several consecutive ejaculations with a single female, a male animal will appear fatigued, and mating will cease. However, with the successive introduction of novel females, the apparently fatigued male will again resume copulatory activity. Thus a bull that normally might show up to 10 successive ejaculations with a single cow could show 50 consecutive ejaculations if presented with 5 successive cows. According to anecdotal evidence, this phenomenon has been termed the Coolidge effect in tribute to the observations of a form U.S. president at a poultry farm. Although the Coolidge effect has been reported in a variety of species, it is relatively weak in rodent species and quite powerful in bovine and ovine species.

Experiential Factors

Experiential factors are critical for the development of the sexual behavior repertoire. A variety of approaches have been employed for assessing the relative importance of these factors. These have included: (1) measuring the improvement in sexual performance as a function of consecutive tests; (2) determining the effects of prepubertal social isolation on subsequent sexual performance; and (3) examining the role of sensory input in sexually experienced and inexperienced animals.

Male mammals typically benefit from sexual experience. Appropriate copulatory responses increase and inappropriate responses decrease with each subsequent sexual encounter until a maximum level of efficiency is reached. This effect is apparent in most species including the rat. For example, when male rats are given a series of time-limited tests with receptive females, the percentage of males copulating shows a consistent increase whereas the latency to copulate shows a consistent decrease with increasing experience (Rabedeau & Whalen, 1959). This improvement in sexual performance with consecutive tests has not been demonstrated in female rodents, because initial female performance is indistinguishable from that of females that have mated many times. However, in many more highly developed species, females clearly do profit from experience. For example, sexual receptivity in female cats shows a progressive increase with successive mating tests (Whalen, 1963).

The effects of prepubertal social isolation on adult sexual behavior vary with the sex and species. In virtually all male mammals studied, early social isolation has proved detrimental to adult sexual performance. Impairments often include inappropriate mounting positions as well as a general reduction in sexual interest. These effects have now been demonstrated in rodents, cats, dogs, rhesus monkeys, and humans. Figures 6.4 and 6.5 illustrate male copulatory responses in normal monkeys and monkeys raised in isolation. Early experience seems to be of minimal importance for the development of adequate copulatory responses in lower mammalian females. However, social isolation is as detrimental in female primates as it is in males (Harlow, 1965). This would suggest that the neocortex is probably as important for the development of female as for male primate sexual behavior. Similarly, the effects of social isolation in nonprimates are consistent with the greater importance of neocortical tissue for male copulatory responses in these species.

FIG. 6.4 Typical copulatory behavior in rhesus monkeys. The male postures in a dorsoventral position while the female elevates her buttocks and tail while lowering her head and shoulders. (From Harlow, 1965.)

Postpubertal sensory deprivation also impairs sexual performance in a manner that varies with the sexual experience of the animal. Beach (1942a) has performed an analysis of this type in which visual, olfactory, or cutaneous stimuli were reduced either singly or in combination in male rats. In sexually inexperienced rats, combined reduction of any two senses prevented subsequent sexual activity. By contrast, only impairment of all three sensory inputs eliminated sexual activity in experienced rats. Similar studies of the female have not been reported. However, at least for responses to olfactory stimulation, sexual experience has a differential effect on males and females. Sexually experienced male rats prefer the odor of receptive females over nonreceptive female odor (Carr, et al., 1965). Inexperienced males and castrates show no preference. By contrast, receptive females prefer the odor of intact males over castrate male odor, and this occurs whether the females are experienced or not.

ONTOGENY OF SEXUAL BEHAVIOR

Sexual behavior is unlike other motivated behaviors in that it requires a relatively longer period of time before being manifested. Typically, complete patterns of sexual behavior are not evident before puberty. Components of the sexual pattern—particularly reflexive components—may be present at a very early age; for example, human male infants less than one year of age are capable of complete penile erection and pelvic thrusting in response to genital manipulation (Kinsey et al., 1948). Nonetheless, the complete repertoire of copulatory activity is dependent upon a variety of interacting factors. These include experiential influences as well as requisite development of neural, muscular, and hormonal systems.

The direction of development of these biological systems is largely determined by the genome. Within the genome are the sex chromosomes that determine the fate of the undifferentiated gonad. An XX combination results in the gonad becoming an ovary and an XY combination results in the gonad becoming a testis. From there on, the sex chromosomes have no known direct influence on subsequent sexual and psychosexual differentiation (Money & Ehrhardt, 1972). Secretions from the differentiated gonad then produce a variety of sexually

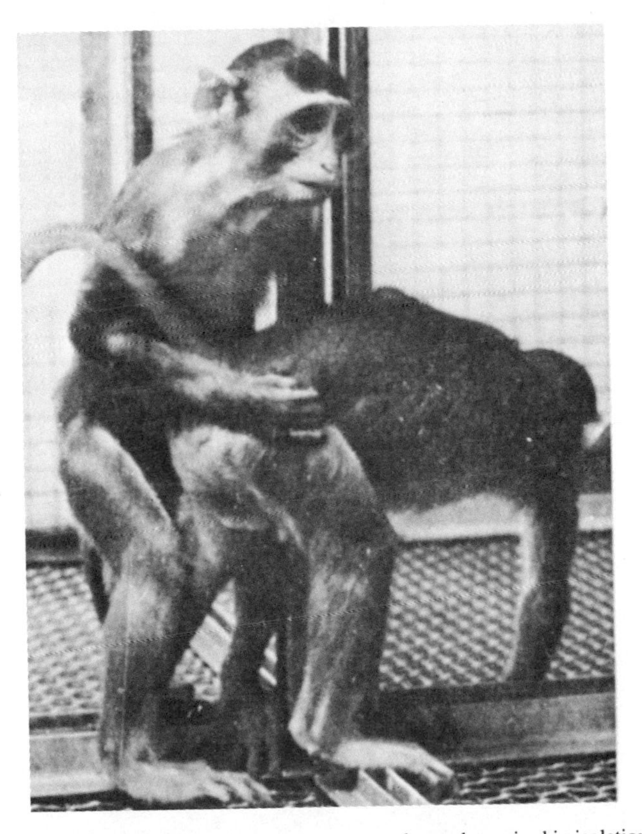

FIG. 6.5 Abnormal attempt at copulatory behavior by a male monkey raised in isolation. Although the female is receptive, the male's lateral mounting posture is awkward and ineffective. (From Harlow, 1964.)

dimorphic effects. As far as sexual behavior is concerned, the major effects are on the undifferentiated brain and genital system. In the presence of testicular secretions, continued differentiation produces the characteristic neural and reproductive structures of a male. In the absence of testicular secretions, female structures are produced. Ovarian secretions during this stage appear irrelevant for the development of female structures. Shortly before puberty, the gonads show a very marked increase in hormone secretion. In many species this correlates with the onset of sexual activity and, at least in humans, with the development of secondary sex characteristics.

Although sex chromosomes and gonadal secretions during development are a good predictor of adult sexual behavior in most species, these factors are not sufficient to account for human psychosexual differentiation. The behavior of parents toward their children may be an even more powerful predictor of adult gender identity than a child's sex chromosomes and fetal hormones. This point is dramatized in a case history reported by Money and Ehrhardt (1972). Two identical twin infants were being circumcised. During surgery a mishap occurred that resulted in the removal of the penis in one twin. After consultation with physicians, the parents chose to rear this child as a girl. Female external genitalia were constructed with plastic surgery and estrogen therapy was recommended. Over the years the parents treated one twin as a boy and reared the other as a girl. According to the parents, the boy has developed a male gender identity, and the girls has developed a female gender identity. The latter child developed a female gender identity despite the presence of XY chromosomes and fetal secretions of testosterone.

This case history should not be taken as evidence that biological factors play no role in human psychosexual differentiation. Hormone secretions are critical in determining whether the newborn infant will possess male or female genitalia. This in turn influences the parents' rearing practices as well as the child's later self-image. In the human, one can conclude that hormones influence physical differentiation directly and behavioral differentiation indirectly. In most other species, both processes are directly controlled by hormones.

Most experimental studies of the effects of hormones on the differentiation of sexual behavior have been done in the rat; the conclusions from these studies probably apply to many but certainly not all species. Table 6.1 summarizes the effects on adult sexual behavior of testosterone propionate (TP) treatment or gonadectomy at birth in male and female rats. When tested in adulthood for female sexual behavior (lordosis responses), all animals received estrogen and progesterone. When tested for male sexual behavior (mounts, intromissions, and ejaculations), all animals received testosterone propionate. Several conclusions are justified from the data. First, the complete sequence of male sexual behavior only occurs in animals exposed to androgens at birth. Second, female sexual behavior only occurs following the absence of androgens at birth. Third, mount-

TABLE 6.1
Relationships Between Hormonal Condition During Development and Sexual Responding in Adulthood
(From Whalen et al. 1971)

	Hormone condition during adulthood			
	Androgen		Estrogen and progesterone	
Hormone condition during critical period	Mounts	Intromission responses	Ejaculation responses	Lordosis responses
Male				
1. Testes intact	+++	+++	+++	–
2. Castrated at birth	+++	+	–	+++
3. Castration at birth + TP[a]	+++	+++	++	–
Female				
4. Ovaries intact	+++	+	–	+++
5. Ovariectomized at birth	+++	+	–	+++
6. TP at birth	+++	++	+	–
7. TP pre- and postrostrally	+++	+++	+++	–

[a]TP = Testosterone Propionate; +++ = a high degree; ++ = a moderate degree; + = a low degree; – = none.

ing responses in adulthood are independent of sex or early hormonal stimulation. Fourth, androgen-treated females are capable of showing the intromission and ejaculation reflex even though they lack a penis.

These results strongly implicate testicular secretions in the differentiation of sexual behavior. The mechanism by which this differentiation occurs is by no means resolved. Most investigators accept the view that the brain has been altered in some manner, but few agree on the nature of the change. One recent approach has involved a searching for male–female differences in brain anatomy using electron microscopy. Another approach has been to compare hormone metabolism in male and female animals, the assumption being that hormone metabolism is essential for the behavioral actions of hormones. Other investigators are currently comparing hormone receptors and measuring hormone-induced protein synthesis in male and female animals. In many instances, significant sex differences are being reported. It would be premature to suggest that any of the recently reported sex differences in brain anatomy or chemistry are crucial for behavioral differentiation. One might just as reasonably suggest that the differences account for male—female differences in neural control of gonadotropin secretion. Indeed, the differences may be irrelevant to both sexual behavior and gonadotropin secretion. Resolving the precise neural mechanisms of sexual differentiation remains a very challenging problem for the future.

A discussion of ontogeny would be incomplete without some statement about the effects of aging on sexual behavior. With a few exceptions, most studies of aging have been done in humans. The reader might already be aware that sexual arousal and arousability, as measured by frequency of orgasm, etc., decline in human males soon after puberty. Peak levels of sexual activity occur at 16–20 years of age and decline steadily thereafter. This decreased activity clearly cannot be attributed directly to a change in testosterone levels. In a recent sampling of healthy men between 20 and 80 years of age, there were no significant differences in testosterone levels (Comhaire & Vermeulen, 1975). Nonetheless, androgen treatment is occasionally effective in restoring sexual interest in aging men. This restorative effect is probably not specific to sexual activity. Clinical studies suggest that it is a secondary result of exogenous androgens maintaining positive protein balance, total body economy, and a general sense of well-being in the aging male (Masters & Johnson, 1966). These same researchers conclude that the decline in male sexual response can be attributed to one or more of these six factors: "(1) monotony of a repetitious sexual relationship (usually translated into boredom with partner); (2) preoccupation with career or economic pursuits; (3) mental or physical fatigue; (4) overindulgence in food or drink; (5) physical and mental infirmities of either the individual or his spouse; and (6) fear of performance associated with or resulting from any of the former categories [p. 264]."

Male sexual decline can also be partially attributed to a natural decrease in sensory perception. Age-dependent increases in sensory capacity have been well documented for vision, hearing, taste, olfaction, and touch (Weiss, 1959). Penile sensitivity decreases dramatically with age. Newman (1970) has reported penile thresholds of 4.9 volts in men aged 35–44 years, 8.4 volts in men aged 45–53 years, and 39.4 volts in men aged 65–74 years. These men were all sexually active, although the frequency of coitus was considerably lower in the older subjects. In 65–74-year-old men who were no longer sexually active, penile threshold rose to 176.2 volts. These results should be interpreted with caution, because it is unclear whether decreased penile sensitivity is a cause or a result of decreased sexual activity.

It is a common belief that human males show peak levels of sexual activity during late adolescence and that human females show peak levels of sexual activity during middle age. This notion probably arose from a superficial examination of the reports of Kinsey and co-workers. As shown in Figure 6.6, the sexual activity of human females is related as much to marital status as to age. In single women there is indeed a slight increase in sexual activity during middle age. However, married women show an age-dependent gradual decline in sexual activity, which approximately parallels the aging pattern of single and married

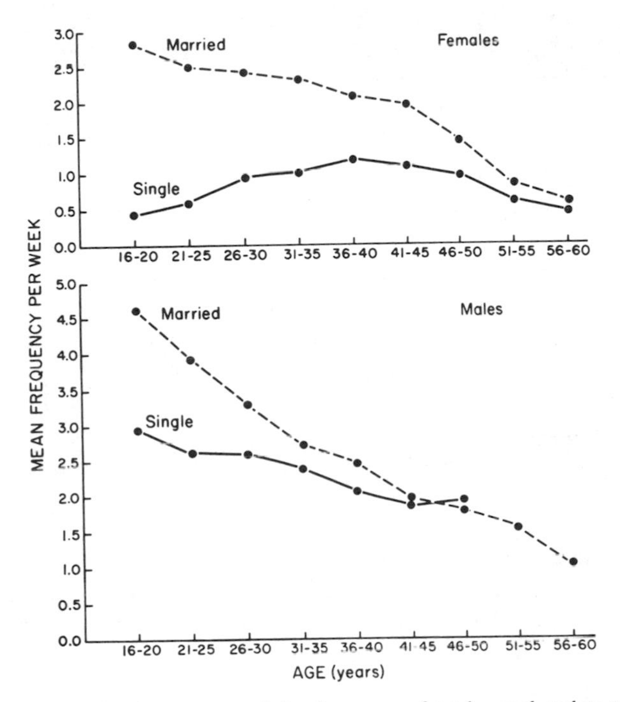

FIG. 6.6 Comparison of aging patterns of the frequency of total sexual outlets to the point of orgasm in men and women as a function of marital status. (After Kinsey, et al., 1948, and Kinsey, et al., 1953.)

men. The marked influence of marital status on female rather than male sexual activity might very well reflect social and cultural factors prevalent at the time of the Kinsey studies. It would be interesting to know whether marital status has as profound an effect on sexual behavior in the mid-1970s.

HORMONAL DETERMINANTS OF SEXUAL BEHAVIOR

Sex hormones have an important role and often a profound effect on sexual behavior. In this section we discuss sexual behavior as a function of hormone levels and consider evidence from studies involving: (1) removal of endogenous hormones; (2) replacement with homotypic exogenous hormones; and (3) replacement with heterotypic exogenous hormones.

Understanding a few basic concepts about steroid hormones should assist further exploration of hormone-behavior relationships. Steroid hormones are the major hormones produced by the testis, the ovary, and an endocrine gland present in both sexes, the adrenal cortex.

The major hormones of the testes are known as *androgens*. For sexual behavior, the most potent and therefore most important androgen appears to be testosterone. The major groups of hormones produced by the ovaries are the *estrogens* and the *progestins*. There are four estrogenic steroids of which the most potent is 17β-estradiol. The other three, estrone, estriol, and 17α-estradiol, are successively less potent. Although the ovary also produces a large number of progestins, only one, progesterone, appears to be particularly potent.

Often, androgens are termed male sex hormones, and estrogens and progestins are termed female sex hormones. These terms may be misleading, because both gonads actually produce all the sex steroids to some degree. For example, the horse testis produces more estrogens than any other known endocrine gland, and stallion urine is the richest known source of 17β-estradiol! Generally speaking, however, the testis produces more androgens than estrogens and progestins, and the ovary produces more estrogens and progestins than androgens. The terms male and female sex hormones may be misnomers, because the differences are largely quantitative rather than qualitative.

The adrenal cortex, in both males and females, also produces relatively small quantities of androgens, estrogens, and in some species, progestins. Although the adrenal cortex contains at least 30 different steroids, the predominant steroids in man appear to be cortisol, corticosterone, and aldosterone, and in the rat, mainly corticosterone.

All steroid hormones are linked through their biosynthetic pathways, and these pathways are identical in the testis, ovary, and adrenal cortex. Differences in steroid metabolism in the various endocrine glands appear to be more quantitative than qualitative. It is for this reason that steroid content differences in the gonads are also more quantitative than qualitative. Figure 6.7 illustrates some of the pathways of synthesis and metabolism as they occur in the steroid-producing glands.

Removal of Hormones in the Adult

Females. Ovariectomy eventually abolishes sexual receptivity in the female of all species, except in the human. In nonprimate species, the disappearance of sexual receptivity is virtually complete in a matter of hours or at most days after ovariectomy. However, in primates such as the rhesus monkey, it may take up to 3 months for all sexual activity to be eliminated (Michael et al., 1966). The hormonal emancipation of the female reaches its evolutionary peak in the human whose sexual receptivity is only rarely altered by ovariectomy (Filler & Drezner, 1944).

In contrast to the ovarian control of sexual receptivity by estrogen and proges-

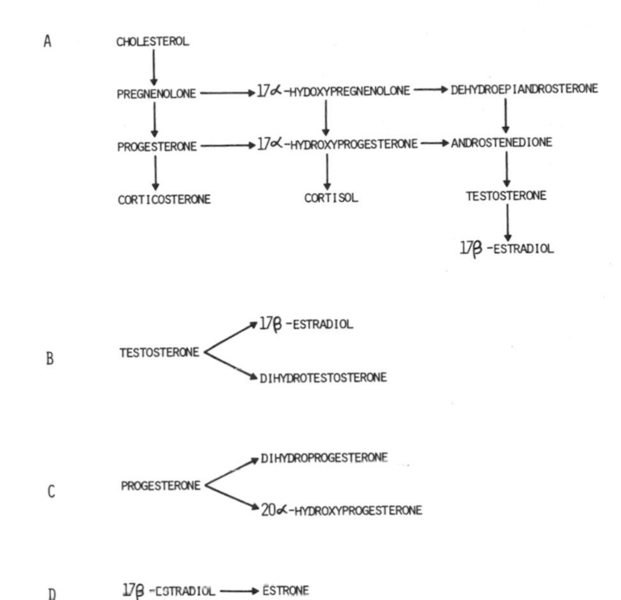

FIG. 6.7 Schematic representation of the biosynthesis and metabolism of steroids. A. Major pathways of steroid biosynthesis in adrenals ovaries and testes. B. Major pathways of testosterone metabolism in the brain. C. Major known pathways of progesterone metabolism in the brain. D. Major known pathway of 17β-estradiol metabolism in the brain.

terone in rodents and many other nonprimates, a more complex hormonal system seems to exist in nonhuman primates. Everitt et al (1972) have concluded, from studies of ovariectomy, adrenalectomy, and hormone replacement in the female rhesus monkey, that adrenal androgens regulate sexual receptivity, and that ovarian estrogen and progesterone modulate attractiveness of the female to the male.

Although ovariectomy does not significantly diminish human female sexual response, a dramatic reduction follows adrenalectomy (Waxenberg et al., 1959) or adrenal suppression following hypophysectomy (Schon & Sutherland, 1960). Of course it can be argued that because adrenalectomy and hypophysectomy are radical surgical techniques often performed in patients with cancer, the subsequent loss of libido is not particularly surprising. Although this is a possibility, studies of androgen replacement therapy are consistent with the view that adrenal androgens are involved in female sexual response.

Males. Sexual behavior in the male disappears following castration, with the time course varying as a function of the species. Not all components of the sexual repertoire are lost simultaneously; for example, castration of the rat results in the loss of the ejaculatory response followed successively by the loss of the intromission response and the cessation of all mount attempts (Whalen, et al., 1961). Although ejaculatory and intromission responses are eliminated in most rats 4 weeks after castration, a small proportion may continue to ejaculate for several months (Davidson, 1966b).

The duration of retention of copulatory responses following castration is greater in carnivores than in rodents. In some dogs, the entire pattern of sexual behavior may be retained for up to 5 years after castration (Beach, 1970). Sexually experienced cats may continue to mate for several months after castration, whereas in inexperienced cats, penile insertions are not observed beyond the first postcastration week (Rosenblatt & Aronson, 1958).

Postcastration copulatory responses are retained even longer in primates than in carnivores. Three years after castration in male rhesus monkeys, Phoenix (1974) observed mounting behavior in all of the animals, intromission in 60%, and ejaculation in 30%. Feinier and Rothman (1939) reported on the heroic efforts of one man, castrated at the age of 23, who was still having weekly coitus with his wife 30 years after the surgery. However, such a long period of retention is not usual. In an extensive investigation of several hundred men who had been legally castrated for sexual offenses, Bremer (1959) reported that all sexual interest and activity was eliminated within a year in two-thirds of the men. In the remaining subjects, sexual activity was present for 1–16 years. Similar results have been obtained in normal men following accidental castration. These highly variable effects of human castration might be partly related to the degree of precastration sexual experience. Studies of the cat (Rosenblatt & Aronson, 1958) and dog (Beach, 1970) indicate that postcastration sexual activity is much more likely to be present in sexually experienced rather than in inexperienced animals. However, experience per se does not entirely account for the dramatic individual differences in postcastration sexual activity; there is considerable variance in sexual performance both prior to and following castration (Rosenblatt & Aronson, 1958)

Beach (1947) suggested that there is an inverse relationship between the extent to which sexual behavior in male mammals is controlled by gonadal hormones and the degree of neocortical development. The postcastration differences in sexual activity of rodents, carnivores, and primates are consistent with this hypothesis, but when the different life spans of these animals are considered, it is not surprising that sexual activity is retained longest in primates and longer in carnivores than in rodents. Furthermore, there have been reports of intraspecies differences in postcastration retention of mating behavior greater than interspecies differences. It seems, therefore, that there may be no close relationship between the evolution of the neocortex and the degree of hormonal control of male sexual behavior.

It is not clear why sexual behavior in the male persists for months or years after castration whereas in the female—except for the human—sexual behavior is abolished almost immediately after ovariectomy. Adrenal androgens are apparently not responsible, because adrenalectomy (at least in non-primates) has been reported not to influence sexual behavior in male animals following castration.

Homotypic Systemic Hormone Replacement

Females. In most ovariectomized rodents, estrogen and progesterone restore full sexual receptivity; estrogen and progesterone act in a synergistic rather than in an additive manner (Boling & Blandau, 1939). Estrogen alone is sufficient to restore full receptivity in some species such as the cat and dog (Young, 1961). In the female rhesus monkey, adrenalectomy abolishes sexual receptivity, and administration of androstenedione, an adrenal androgen, restores receptivity (Everitt et al., 1972). Although testosterone injections also increase sexual receptivity in female rhesus monkeys (Trimble & Herbert, 1968) and women (Salmon & Geist, 1943), only very small quantities of this steroid are actually produced by the female.

Although ovarian steroids are not necessarily involved in primate receptivity, they do appear to regulate the attractiveness of the female to the male. Ovarian steroids presumably regulate attractiveness by means of olfactory cues to the male. Michael and Saayman (1968) observed that intravaginal application of estrogen was effective in increasing the frequency of male mount attempts. Progesterone, on the other hand, appeared to reduce the attractiveness of female primates. This inhibitory effect of progesterone on the attractiveness of females to males has been suggested for both rhesus monkeys (Michael et al., 1967) and humans (Udry et al., 1973).

Males. In castrated males of all species, exogenous testosterone restores the pattern of sexual behavior (Young, 1961). However, in contrast to the almost immediate restoration of female sexual receptivity following exogenous estrogen and progesterone, male sexual behavior returns quite gradually, often requiring several weeks of androgen treatment.

Male sexual behavior does not appear to be related in any simple, direct, or quantitative manner to circulating levels of androgens. First, as just indicated, in androgen replacement studies, mating behavior is not restored until weeks after the circulating steroid levels have returned to precastration levels. Second, as indicated by castration studies discussed in the previous section, mating persists long after gonadal androgens have been removed from the circulation. Furthermore, individual differences in male sexual performance are not necessarily related to levels of circulating androgens; in male guinea pigs (Grunt & Young, 1953) and rats (Beach & Fowler, 1959), individual differences in sexual performance prior to castration remain the same following castration and a constant treatment dose of androgen. Whalen et al. (1961) have suggested that the failure of some intact male rats to copulate is not simply a function of androgen deficiency, because noncopulators fail to mate even in the presence of massive androgen doses. Rather, noncopulators may be relatively insensitive to androgens. Recent theories have claimed that many hormones must interact with neural tissue by means of hormone-receptor proteins. It is interesting to speculate that noncopulators may have a hormone-receptor deficiency rather than a hormone deficiency.

Heterotypic Systemic Hormone Replacement

Females. Androgen treatment induces some degree of malelike sexual behavior and lower but significant levels of female sexual receptivity in female rats (Beach, 1942c) and other species. However, androgens do not induce the full degree of normal male or female sexual behavior. It has been suggested that androgens may induce female sexual receptivity by metabolic conversion in the brain to estrogens, a process known as *aromatization* (see Figure 6.7). Although the degree of aromatization of androgens is relatively small (Naftolin, et al., 1972), it should be noted that the amount of estrogen necessary to induce sexual receptivity is only about one-twentieth of the amount of androgen required (Whalen & Hardy, 1970). There is other evidence to suggest that androgen-induced sexual receptivity in females is mediated by aromatization; the administration of CI-628, an antiestrogenic drug that blocks sexual receptivity induced by 17β-estradiol (Whalen & Gorzalka, 1973), also blocks sexual receptivity induced by androgens (Whalen et al., 1972).

Males. Although early reports suggested that estrogen was only partially effective in restoring male sexual behavior in male rats (Ball, 1937; Beach, 1942b), recent reports indicate full restoration of male sexual behavior in castrates by exogenous estrogen (Södersten, 1973; Gorzalka et al., 1975). Although this result is consistent with the suggestion that estrogen is normally involved in the control of male sexual behavior in the rat (McDonald et al., 1970)—aromatization to estrogens being a prerequisite for androgen action—a more likely interpretation is that estrogen treatment may stimulate the release of sufficient adrenal androgens to maintain male sexual behavior in castrates. In support of the latter interpretation, Gorzalka et al. (1975) have reported that estrogen maintained ejaculatory responses in castrated rats but not in adrenalectomized, castrated rats, whereas androgen was equally effective in castrated and in adrenalectomized, castrated rats (Figure 6.8).

Castrated male rats administered estrogen initially will not show female sexual receptivity but may show low levels following several weeks of estrogen stimulation. Progesterone does not appear to facilitate receptivity in male rats beyond the levels seen with estrogen only (Whalen et al., 1971). Therefore, with respect to female sexual receptivity in males, animals appear only slightly responsive to estrogen and quite unresponsive to progesterone.

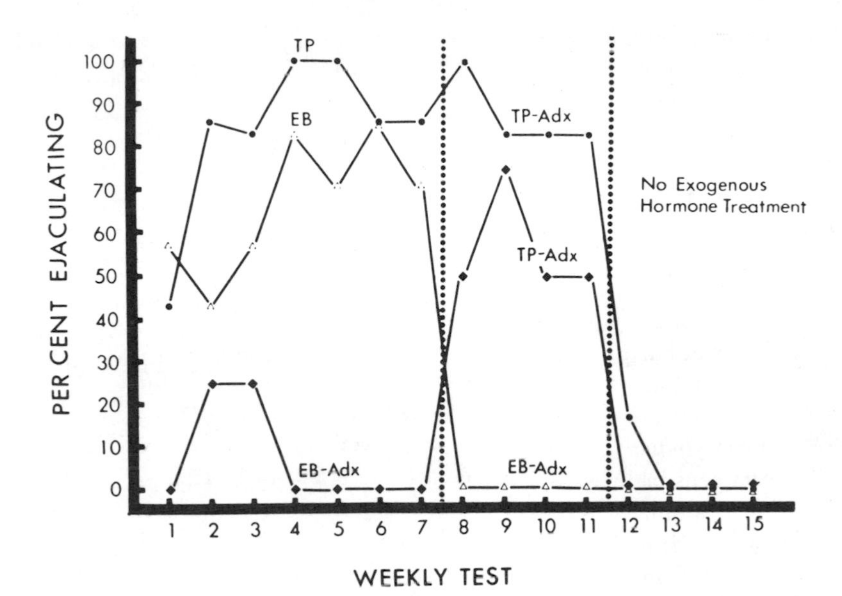

FIG. 6.8 Ejaculatory responses in adrenalectomized (ADX) male rats and castrated male rats treated with testosterone (TP) or estradiol (EB). Note that estradiol- and testosterone-treated rats ejaculated, but when they were adrenalectomized, only the testosterone was effective in maintaining ejaculation. (From Gorzalka et al., 1975.)

NEURAL DETERMINANTS OF SEXUAL BEHAVIOR

Early studies of mating behavior were preoccupied with the concept of a "sex center" in the central nervous system. It was once assumed that this brain center was endowed with all the features critical for mating behavior—the target for gonadal hormones and external stimuli, the integrator of sexual inputs, and the initiator of "command signals" to the effector mechanism for mating behavior. The notion of a center that integrates signals of the multideterminants of copulatory behavior is no longer accepted.

Copulatory behavior is the result of integrative activities at several levels of the CNS extending from the spinal cord to the cerebral cortex. Beach (1967) has proposed that copulatory behavior depends on spinal reflexes, which are normally inhibited by supraspinal neural structures. Sexual reflexes occur when these inhibitory structures are themselves inhibited, sexual reflexes being the result of disinhibition. In order for disinhibition to occur, appropriate inputs—either hormones or external stimuli (e.g., an appropriate mate)—must be present. In lower animals both types of stimuli seem critical for sexual activity to occur, but in many higher animals, as indicated in the previous section, sexual reflexes may persist for relatively long periods of time following gonadectomy.

Methodologies

Neural and endocrine systems interact to control sexual behavior in a manner unique among motivated behaviors, and somewhat different strategies have been used in the study of these neural systems. Furthermore, even when similar methods have been used, unique problems in data interpretation have become apparent. In the present section, we discuss some of the more widely used techniques for investigating neural systems that subserve sexual behavior.

Destruction of neural tissues. Lesions of a particular brain site may increase or decrease the frequency of sexual activity. If the effect is decreased sexual activity, this could be attributed to a variety of mechanisms. First, the impairment could be nonspecific if it is observed that the decreased sexual activity is accompanied by a general behavioral debilitation. Second, the impairment could be the result of the decreased production of gonadal hormones; atrophy of the gonads may occur—a "functional castration." This possibility can be tested experimentally by administration of gonadal hormones; if sexual activity is restored, then the lesion probably destroyed a part of the brain that controls hormone production. Third, the impairment could be caused by destruction of a region that is a target site for gonadal hormones; the animal produces normal amounts of gonadal hormones, but the brain target for the hormones has been destroyed. Fourth, the impairment could be caused by destruction of an area or pathway critical for sexual behavior that is not a target for hormones. The administration of gonadal hormones does not distinguish between the third and fourth possibilities; hormone treatment would not restore sexual activity in either case.

When lesions increase the frequency of sexual activity, a variety of mechanisms might also be involved. First, the higher frequency may be due to a general increase in arousal, activity, or some other nonspecific process. Surprisingly, most studies that report increased sexual activity following brain lesions fail to control for this possibility, which can be done by measuring other types of behavioral activity. Second, the increase in mating responses may be due to increased hormone sensitivity of a region normally controlled by the lesioned structures. This can be tested by gonadectomizing animals and demonstrating that lesioned animals require less hormone than nonlesioned animals for restoration of sexual activity. Third, the increase may be due to destruction of a region that normally acts to inhibit sexual activity in a manner independent of hormones. This possibility could be tested by comparing the effects of the brain lesion with the effects of massive doses of hormones. If the lesion increases sexual frequency and the hormone treatment does not, it is likely that one is dealing with a hormone-independent effect.

Cerebral implantation of hormones. Gonadectomy eliminates mating behavior, whereas systemic administration or cerebral implantation of the hormones restores copulatory activity. The implantation technique is particularly useful in determining the neural site at which a hormone acts. When a lesion of the CNS impairs mating behavior and systemic hormone replacement fails to restore the behavior, the hormone implant method provides a logical follow-up technique to determine whether the neural site is hormone-sensitive.

The method involves direct implantation of a crystalline steroid, usually testosterone, estrogen, or progesterone, into the brains of gonadectomized animals. Testosterone and estrogen implants are usually tested in the absence of any systemic hormone administration. However, in order to test the effects of progesterone implants, concurrent systemic administration of estrogen is required, because progesterone is behaviorally ineffective in the absence of estrogen. In testing the relative effectiveness of a particular hormone at a particular neural site, the investigator must be satisfied that the hormone has not diffused to an adjacent region; the hormone must not leak into the bloodstream, where it could reenter the brain and act at sites remote from the hormone implant; and implantation of inert steroids as a control procedure should produce no behavioral effects.

Electrical stimulation of the brain. In males but not females, direct electrical stimulation of certain hypothalamic and limbic structures increases the frequency of sexual responses in many species. This phenomenon is sometimes termed *stimulus-bound copulation,* because the males mate at above normal rates during electrical stimulation and at below normal rates during the intervals between successive stimulation periods.

Neurophysiological activity. Electrodes implanted in the brain can be used to monitor electrical activity in response to external stimuli (e.g., olfactory stimuli) and administered hormones. This method is particularly useful in demonstrating the interaction between external stimuli and gonadal hormones in the neural systems that subserve copulatory behavior. At least four kinds of observations have been made. First, there are regions that are hormone-dependent but stimulus-independent. Such an area shows electrical changes in response only to specific gonadal steroids, not to specific types of external stimuli. Second, there are areas that are stimulus-dependent and hormone-independent. An example is an area that responds only to tactile stimulation or perhaps only to genital stimulation. Third, there are areas that are both stimulus-dependent and hormone-dependent. For example, it has been reported in some female mammals that genital stimulation, but not other tactile stimulation, has specific effects in the presence of estrogen. The specific effects of genital stimulation do not occur in the absence of estrogen. Fourth, there are regions that are both stimulus-independent and hormone-independent. Areas that fail to show a specific electrical response to hormones or external stimuli may still be involved in the execution of copulatory response.

Radioactive tracers. Administration of gonadal hormones labeled with radioactive tracers provides a convenient technique for following retention patterns of the hormones in the brain. Typically, the labeled hormone is administered systemically, the brain is excised, and the radioactivity is measured. Hormones with quite similar behavioral effects such as estrogen and progesterone may show quite disparate patterns of brain retention. The retention patterns, although providing no behavioral information per se, can act as a guide to subsequent studies. If the concentration of radioactivity in a particular brain site is found to be high, this region is a candidate for investigation using hormone implantation and other techniques.

In the following section, we review experimental evidence concerning the role in copulatory behavior of the spinal cord, brain stem, hypothalamus, limbic system, and cerebral cortex. Evidence that a particular site exerts a hormone-dependent or hormone-independent role is derived largely from the five methodologies just discussed.

Spinal Cord

Certain spinal reflex responses are normally inhibited by higher brain structures. An example is the observation that genital stimulation usually does not elicit ejaculatory responses in the intact dog (Hart, 1967), but in the presence of a receptive bitch, genital stimulation does elicit ejaculatory reflexes. Presumably, the visual and olfactory stimuli associated with the female are sufficient to disinhibit the higher brain structures that normally inhibit spinal reflexes.

Furthermore, genital stimulation per se is sufficient to produce penile erection and ejaculation following spinal transection in dogs (Hart, 1967), rats (Hart, 1969), and a variety of other species. Following genital stimulation, most human male paraplegics are capable of erection and many ejaculate (Money, 1961). Plasma testosterone levels in paraplegics are similar to those in normal men (Mizutani, et al., 1972), and it is possible that the sexual reflexes referred to previously are hormone-regulated, with androgens acting directly on the spinal cord—a hypothesis that is supported by animal studies. Castration reduces sexual reflexes in spinal male rats (Hart & Haugen, 1968) and dogs (Hart, 1968), and the reflexes are restored when androgens are implanted directly into the spinal cord. Hormones may also regulate spinal sexual reflexes in the females of some species. Estrogen administration has been reported to facilitate sexual reflexes in spinal female cats (Hart, 1971) and dogs (Hart, 1970) just as androgens do in the male.

Brainstem

The brainstem appears to regulate copulatory activity in one of two possible ways. The first involves a brainstem system having an inhibitory function in the hierarchical organization of sexual reflexes. Under natural conditions, afferents from higher neural structures apparently disinhibit the brainstem, because large lesions at the level of the mesencephalic-diencephalic junction facilitate copulatory performance in male rats (Heimer & Larsson, 1964). These lesions produce a striking increase in ejaculatory frequency and a decrease in the postejaculatory interval—the refractory period between ejaculation and the resumption of mating behavior. Recent studies have demonstrated similar behavioral effects following rostral midbrain lesions in male rats (Barfield, et al., 1975; Clark, et al., 1975). Because these midbrain lesions disrupt noradrenaline and dopamine pathways, it is possible that one or both are normally involved in the brainstem regulation of sexual behavior.

There is evidence for a second type of brainstem system in the female of some species in which the hormone progesterone appears to act directly on the brainstem. Whether progesterone facilitates sexual receptivity by activating a brainstem mechanism or by disinhibiting an inhibitory brainstem mechanism remains to be determined. Two kinds of evidence implicate the brainstem in progesterone action. First, implants of progesterone into the mesencephalic reticular formation are much more effective than control implants at the same site or progesterone implants at other neural sites in facilitating sexual receptivity in rats (Ross, et al., 1971; Gorzalka, 1974). Similar effects have been produced with mesencephalic implants of 5α-dihydroprogesterone, a biologically active metabolite of progesterone. Second, in radioactive tracer studies, the greatest concentrations of radioactivity are detected in the mesencephalon following the administration of labeled progesterone or 5α-dihydroprogesterone (Gorzalka, 1974).

Hypothalamus

The hypothalamus influences sexual behavior by regulating the production of gonadal hormones, by acting as a target site for the gonadal hormones, or by acting as a target site for external stimuli involved in mating. In the control of hormone production, the arcuate nucleus and adjacent hypothalamic structures regulate gonadotropin release from the anterior pituitary (see Figures 2.7 and 2.9). The gonadotropins in turn regulate gonadal hormone secretion. Although sexual behavior is inhibited by arcuate lesions, the behavior can be restored by gonadal hormone administration.

Much of the current research on the role of the hypothalamus concerns areas such as the anterior hypothalamus and medial preoptic area. Lesions of these areas eliminate sexual activity, but the administration of hormones is not effective in restoring the behavior. The importance of the anterior hypothalamus for sexual behavior was first demonstrated by lesion studies. Ranson and his colleagues (Fisher, et al., 1938; Brookhart, et al., 1940) showed in female cats and guinea pigs that anterior hypothalamic lesions attenuate copulatory behavior, and estrogen administration failed to restore the behavior. Subsequently, lesions of the anterior hypothalamus were reported to abolish copulatory responses in male rats, and testosterone was not effective in restoring the copulatory behavior (Heimer & Larsson, 1966/1967). Other studies suggest a partial separation of the mechanisms for male and female sexual behavior in the hypothalamus. Singer (1968) found that anterior hypothalamic lesions but not medial preoptic lesions abolish female sexual behavior in female rats. On the other hand, medial preoptic lesions but not anterior hypothalamic lesions abolish male sexual behavior in androgen-treated female rats. This finding of functional localization has since been confirmed for male and female sexual behavior in male rats (Gorzalka, 1976, unpublished observations).

Although hypothalamic lesions typically attenuate sexual activity presumably by destroying the sites at which hormones act, some hypothalamic lesions actually increase sexual activity. Indeed, it has been suggested that the hypothalamus includes regions that inhibit sexual behavior in a manner independent of hormone levels. Law and Meagher (1958) reported that posterior hypothalamic lesions in the region of the mammillary bodies potentiate the display of female mating responses during diestrus in intact rats. Goy and Phoenix (1963) noted that hypothalamic lesions occasionally facilitate sexual receptivity in ovariectomized guinea pigs in the complete absence of exogenous hormones. These results are consistent with the concept of inhibitory regions for sexual behavior in the hypothalamus. Male sexual behavior is also facilitated by posterior hypothalamic lesions at the level of mammillary body. In particular, the number of ejaculations per mating test increases following damage to this area in male rats (Lisk, 1969). Lisk (1969) also suggests that mammillary lesions facilitate sexual behavior in sexually inactive male rats. By contrast, androgen treatment is ineffective in facilitating mating behavior in sexually inactive male rats (Whalen, et al., 1961), and so it seems unlikely that the lesion effect can be attributed to an alteration in hormone levels.

The concept of inhibitory hypothalamic systems disinhibiting lower neural structures, an earlier popular view, has been less prominent in recent years. Instead, there has been a great deal of interest in the possibility of facilitatory hypothalamic systems, the presumed sites of action of androgens and estrogens. Evidence of facilitatory hypothalamic systems from studies of the effects of lesions, hormone implantation, and electrical stimulation, and from studies using neurophysiology and radioactive tracer techniques, are in general agreement.

Studies involving cerebral hormone implantation, like the lesion studies, have implicated the anterior hypothalamus and medial preoptic area in copulatory behavior. Estrogen implants in these regions restore sexual behavior in ovariectomized rats (Lisk, 1962) and cats (Harris & Michael, 1964). Androgen implants have similar effects in castrated rats (Davidson, 1966a). The effective sites of androgen implants and behavioral results in castrated rats are shown in Figure 6.9. The site of androgen action in female primates appears to be similar; in adrenalectomized female rhesus monkeys, androgen implants in the region of the anterior hypothalamus and medial preoptic area restore sexual receptivity (Everitt & Herbert, 1975). Thus, although the hormonal control of receptivity in female primates differs from that of nonprimates, the major site of hormone action may be similar.

Electrical stimulation studies have provided evidence consistent with hormone implantation studies. There is an increased frequency of male sexual responses following stimulation of the anterior hypothalamus (Vaughan & Fisher, 1962) and medial preoptic area in rats (Malsbury, 1971; van Dis & Larsson, 1971).

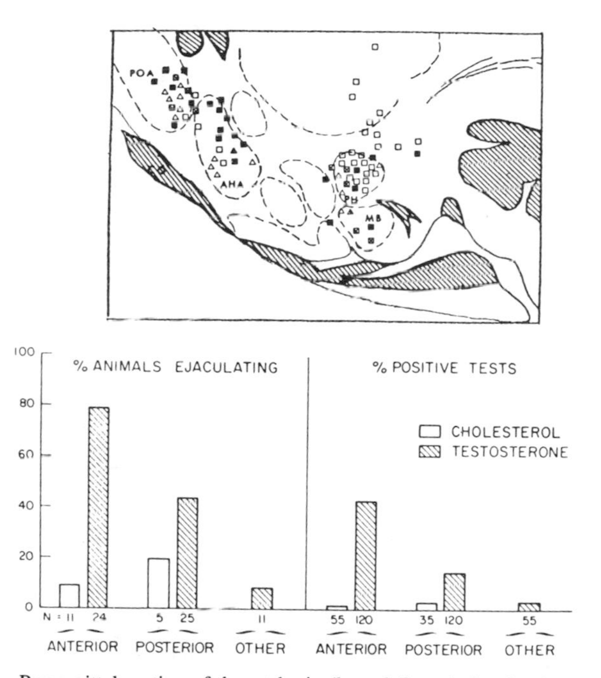

FIG. 6.9 *Top*—Parasagittal section of the rat brain (lateral .2 mm) showing location of centers for implantation of testosterone propionate (squares) or cholesterol (triangles). Open squares or triangles = no ejaculation; solid squares = more than one ejaculation, and crossed squares or triangles = one ejaculation. AHA, anterior hypothalamus; POA, preoptic area; CO, optic chiasm; PH, posterior hypothalamus; MB, mammillary bodies. *Bottom*—Comparison of the effects of anterior and posterior hypothalamic implants of testosterone on the sexual behavior of castrated male rats. (From Johnston & Davidson, 1972.)

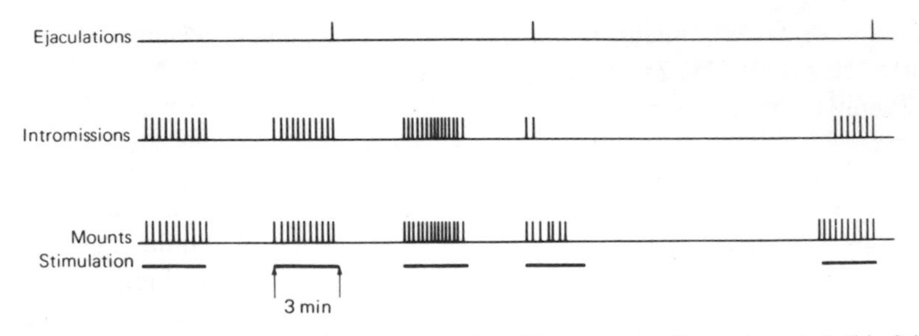

FIG. 6.10 Copulatory behavior of a male rat produced by successive 3-minute periods (black bars) of brain stimulation to the posterior hypothalamus. (After Caggiula, 1970.)

Stimulation-induced copulatory responses seem to be androgen-dependent, because this effect has not been reported in castrated animals. Furthermore, there is some evidence that neural pathways projecting caudally from the preoptic area through the medial forebrain bundle to the brainstem may transmit signals for copulatory behavior. Electrical stimulation of the medial forebrain bundle increases (Caggiula, 1970; see Figure 6.10) and lesions decrease the frequency of copulatory activity in male rats (Paxinos, 1974).

Electrophysiological studies have provided indirect evidence that gonadal hormones act on the anterior hypothalamus—preoptic region. Pfaff and Pfaffmann (1969) have reported that the response of single neurons of the preoptic region in male rats is modified by androgen administration; a control injection did not produce this effect.

Radioactive tracer studies provide further confirmation of the hypothalamic site of action of androgens and estrogens. Following systemic administration of labeled testosterone to rats, radioactivity was relatively concentrated in the anterior hypothalamus (Pfaff, 1968). In a comparison of the brain retention pattern of ovarian hormones, estrogen radioactivity was concentrated in the anterior hypothalamus, whereas progesterone radioactivity was concentrated in the mesencephalon (Whalen & Gorzalka, 1974).

Limbic System

The limbic system influences sexual behavior by altering sensitivity to gonadal hormones, by regulating sexual arousability, and by control of general processes such as attention and motor performance not specific to sexual activity. These three effects can be separated experimentally. Alterations in hormone sensitivity can be demonstrated when a limbic lesion increases or decreases the quantity of hormone required to maintain sexual activity in gonadectomized animals. The other two effects can be dissociated by determining whether a limbic lesion affects nonsexual as well as sexual activities.

The habenula and septum are two limbic structures that appear to influence sensitivity to hormones. Lesions of the habenular nucleus have been reported to significantly impair sexual receptivity in female rats treated with estrogen and progesterone (Modianos, et al., 1975). However, there are no differences between habenular-lesioned and sham-operated females in receptivity tests conducted after the administration of estrogen alone. Thus the habenula may be either a direct site of progesterone action or part of a system modulating the effect of progesterone on remote sites such as the midbrain. Interestingly, there are extensive connections between the habenula and both the preoptic area and the midbrain (Mok & Mogenson, 1974).

The septum appears to regulate female sexual behavior by an inhibitory mechanism; septal lesions in female rats increase sexual receptivity by increasing sensitivity to estrogen, but not to progesterone (Nance, et al., 1975). Septal lesions contrast with habenula lesions that appear to decrease sensitivity to progesterone but not to estrogen. The septum connects, directly or indirectly, to the habenula, preoptic area, medial forebrain bundle, and midbrain. All of these sites have now been implicated in the hormonal control of sexual receptivity.

Other studies have suggested direct involvement of limbic structures in sexual function and arousability. For example, electrical stimulation of the amygdala in the cat has been reported to produce erection of the penis, mounting, and ejaculation (Shealy & Peele, 1957). Penile erection has also been produced occasionally by stimulation of the septum in the squirrel monkey (MacLean & Ploog, 1962) and in man (Heath, 1964).

There are numerous reports of changes in the frequency of sexual activity following lesions of limbic structures. Because nonsexual activities are rarely monitored, it is difficult to conclude from most of these studies whether limbic lesions produce sex-specific or nonspecific effects. For example, it is claimed that hippocampal lesions reduce copulatory activity in cats and primates but produce no such effect in rodents. It is not known whether this species difference reflects a decrement in sexual arousability or more general behavioral decrements.

There has been a great deal of interest in the role of the amygdala in sexual behavior—in part because hypersexuality was a prominent component of the Klüver–Bucy syndrome (1937) in monkeys with extensive surgical damage to the temporal lobes, including the amygdala and hippocampus. Bilateral temporal lobectomy in a human male also resulted in hypersexuality (Terzian & Ore, 1955). These reports have been complicated by evidence that lesions restricted to the amygdala and adjacent neural tissue produce hypersexuality in male cats (Schreiner & Kling, 1953) and male monkeys (Kling, 1968). The effects of amygdalectomy on sexual behavior in cats have been variable, and there is evidence that the adjoining pyriform cortex rather than the amygdala per se may be critical (Green, et al., 1957). Damage to the pyriform cortex produces quite remarkable changes in the choice of sex object in cats. In particular, these animals mount other species, mount inanimate objects, and have been observed to engage in tandem copulations with several other male cats. Furthermore, there is a possibility that hypersexual behavior may develop spontaneously in unoperated cats exposed to a testing room at regular intervals (Hagamen, et al., 1963). In the rat there are several studies that demonstrate a decrement in sexual activity following amygdalectomy (Bermant, et al., 1968; Harris & Sachs, 1975; Michal, 1973). When comparing effects of amygdala lesions across species, it is important to remember that the amygdala is a collection of nuclei, not a unitary structure; apparent species differences following amygdala lesions may be partly related to the particular nuclei that are damaged.

The stimulation and lesion studies, some of which have been cited here, clearly implicate limbic forebrain structures in sexual behavior and suggest that these structures represent another level of neural integration in the hierarchical organization of the control of sexual behavior. In general, the effects of lesions of limbic structures are less dramatic than the effects of hypothalamic lesions on sexual behavior. Furthermore, the effects are not necessarily specific to mating behavior; the changes that occur following lesions to limbic forebrain structures may be secondary to a more general alteration in arousal, attention, motor performance, etc. Clearly, limbic structures exert both excitatory and inhibitory influences on copulatory activity; for the amygdala, and possibly other structures, whether the effects are excitatory or inhibitory seems to depend on the species.

Cerebral Cortex

Much of the evidence implicating the cerebral cortex in sexual behavior comes from lesion studies. Deficits in sexual performance following cortical lesions are difficult to interpret, because they could be attributed to disruptions of sensory functions, motor functions, complex sensory–motor integrations, general arousability, or sexual arousability. The relative importance of these and other cortical processes for sexual activity seem to depend on the sex and species of animal. The relevant evidence is reviewed briefly in the following.

In most male mammals, cortical lesions impair sexual performance. Beach (1940) reports that in male rats, lesions of more than 60% of the cortex completely abolish sexual activity, lesions of less than 20% of the cortex have little effect on sexual activity, and lesions of 20–60% of the cortex produce decrements that vary with the amount of tissue removed. Androgen treatment fails to restore sexual activity in these animals. Beach (1940) concludes that the size of the lesion, not the locus, is a crucial factor in determining the extent of the deficit in copulatory activity. This conclusion is challenged by Larsson (1964), however, who claims that frontal lesions are more effective than temporal, parietal, or occipital lesions. Although frontal cortex is perhaps more important than other cortical regions for sexual activity, no single area of the cerebral cortex is critical for this behavior.

Similar effects have been reported for the male cat (Beach, et al., 1955; 1956): Complete decortication eliminates sexual activity; and frontal lesions are somewhat more effective than other cortical lesions in impairing sexual activity. As with the rat, androgen treatment fails to restore sexual activity in decorticate cats.

The rabbit is one species in which cortical involvement is not critical for male sexual behavior. Cortical lesions fail to inhibit either sexual arousal or the motor responses involved in male sexual performance in the rabbit (Brooks, 1937; Stone, 1925).

There have been few systematic studies of cortical involvement in human sexual behavior. The majority of lesion studies have been concerned with the effects of temporal lobe lesions in patients with temporal lobe epilepsy. Numerous clinical reports suggest that temporal lobe damage may alter the intensity of sexual activity and the orientation of sexual interest (Epstein, 1973). However, appropriate control groups and sampling methods are often lacking in these reports.

Available evidence suggests that the cerebral cortex is of lesser importance in the female than in the male, at least in those species that have been studied. Surgical removal of the entire cerebral cortex fails to inhibit female sexual activity in female cats, guinea pigs, rabbits, or rats (Beach, 1944). However, cortical destruction does eliminate masculine responses such as mounting behavior in the female rat (Beach, 1943). This differential effect on sexual behavior may reflect the greater degree of complex sensory-motor integration in male sexual behavior in lower mammals. The role of the cerebral cortex in sexual behavior in female primates has not been studied, but from the detrimental effects of social isolation, one would predict a crucial role for the cerebral cortex in both male and female primate sexual behavior.

Beach (1967) has suggested that sexual receptivity in the female rat may be controlled by an inhibitory mechanism in the cerebral cortex, and Clemens, et al. (1967) have postulated that progesterone might facilitate sexual receptivity by inhibiting this mechanism (that is, by disinhibition). This suggestion came from the observation that both the administration of potassium chloride (KCl) to the cerebral cortex and the systemic administration of progesterone—both in estrogen-primed, ovariectomized rats—facilitated sexual receptivity. Because KCl causes spreading cortical depression and thus might be expected to disinhibit the inhibitory mechanism, these investigators proposed that progesterone acts in a similar manner to disinhibit the cerebral cortex. However, Whalen, et al. (1975b) showed that administering KCl to the cerebral cortex of adrenalectomized rats or to rats treated with dexamethasone—a drug that inhibits the pituitary-adrenal system—failed to facilitate sexual receptivity. Thus an alternative interpretation is that the application of KCl to the cerebral cortex has remote effects on the hypothalamus, which results in the increased pituitary release of ACTH and in turn of adrenal hormones such as progesterone; adrenal progesterone rather than cortical depression may be responsible for the facilitation of sexual receptivity observed by Clemens and co-workers.

FUTURE DIRECTIONS OF RESEARCH

Until recently, investigations of the physiology of sexual behavior were limited to the classical experimental techniques (see Chapter 2). For a number of years, most studies utilized the lesioning technique, and later it was complemented by brain stimulation and electrophysiological techniques. In recent years there has been an interest in the role of putative central neurotransmitters in sexual behavior. The visualizing of monoaminergic neurons, the mapping of their projections (see Figure 2.24), and the elucidation of the biosynthesis and synaptic action of several neurotransmitters has stimulated interest in using neurochemical and neuropharmacological techniques to investigate the neural substrates of sexual behavior. There has also been a great deal of work on hormone metabolism and hormone-genome interactions, the molecular mechanisms by which gonadal hormones interact with the CNS.

Neuropharmacological Studies of Sexual Behavior

The most extensively used experimental strategy has been to administer drugs that either deplete or enhance a particular neurotransmitter, such as acetylcholine or serotonin. For example, using pilocarpine, a drug that increases cholinergic activity, and atropine, an anticholinergic drug, cholinergic mechanisms have been implicated in sexual behavior in the rat (e.g., Soulairac & Soulairac, 1975). The drug, p-chlorophenylalanine (PCPA), a depletor of brain serotonin, has been used to investigate the role of central serotoninergic neurons in sexual behavior. The results have been contradictory, but a recent review of many studies suggests that PCPA facilitates male copulatory behavior in inexperienced but not in experienced animals (Whalen, et al., 1975a).

There has been a great deal of interest in the possible role of the catecholamines, and especially of dopamine, in sexual behavior. This was prompted by the observation that the administration of L-dopa, a precursor of dopamine, to patients with Parkinson's disease had an aphrodisiac effect (Barbeau, 1969). Subsequent studies of rats have shown that dopaminergic drugs increase sexual behavior and dopaminergic blockers reduce sexual behavior. For example, the administration of apomorphine, a dopamine agonist, increased copulatory behavior, and the administration of haloperidol, a dopamine antagonist or blocker, reduced copulatory behavior (Tagliamonte et al., 1974).

It is too early to draw definitive conclusions from the pharmacological studies. The effects of drugs often vary with dose, species, and sexual experience of the animal. Also, the effects of drugs may not be specific to sexual behavior. The purported aphrodisiac effects of L-dopa in Parkinsonian patients may be due to an improved psychological outlook or sense of well-being of the patient as his condition improves. This possibility must be considered seriously, because L-dopa is relatively ineffective in treating impotence in otherwise healthy men (Benkert, 1973). Neuropharmacological studies have implicated cholinergic, serotoninergic, and catecholaminergic systems. For adequate sexual behavior in the male rat, it appears that the ideal condition includes moderately high cholinergic and catecholaminergic activity and moderately low serotoninergic activity (Soulairac & Soulairac, 1975). As a tentative working hypothesis, it may be suggested that reduced or impaired sexual performance in the male rat is the result of a deviation of one or more of these neurotransmitters from an optimal level.

Hormone Metabolism and Sexual Behavior

A current hypothesis is that certain steroid hormones must be metabolized by target cells to a more active compound in order to exert their physiological effects. This view is based on the peripheral action of testosterone; the observation that the metabolite of testosterone, 5α-dihydrotestosterone (DHT), is more effective than exogenous testosterone in inducing the growth of the ventral prostate (Anderson & Liao, 1968). Testosterone and other androgens are metabolized in the rat brain to DHT as well as to estrogen (the latter process is called aromatization). A number of investigators are now interested in the possibility that aromatization and/or metabolism to DHT is necessary for testosterone to activate neural systems that control sexual behavior.

There is evidence that the central aromatization of androgens to estrogens is not a sufficient requirement. Although estrogens are as effective as androgens in maintaining sexual behavior in castrated male rats, estrogens do not maintain the ejaculatory response in the castrated male rat that has been adrenalectomized (Gorzalka, et al., 1975). Another possibility is that metabolism to DHT as well as to estrogen (aromatization) is necessary for the effects of testosterone on sexual behavior. In support of this hypothesis, Baum and Vreeburg (1973) have reported that a combination of exogenous estrogen and DHT is as effective in maintaining copulatory behavior as exogenous testosterone in the castrated male rat. However, this hypothesis must be viewed with caution, because it has been reported that cyprotetrone acetate, a drug with antiandrogenic and antiestrogenic properties, inhibits copulatory behavior induced by testosterone (Luttge et al., 1975). Furthermore, combined estrogen and DHT are ineffective in maintaining ejaculatory behavior in male castrates that have been adrenalectomized (Gorzalka, 1976, unpublished observations). Further research is needed to clarify the role of aromatization and metabolic reduction of testosterone in the behavioral effects of testosterone.

There is experimental evidence, which we are not able to review here, that suggests that female sexual behavior in rodents does not require the metabolism of estrogens (Beyer, et al., 1971) or progesterone (Whalen & Gorzalka, 1972).

Brain-Hormone-Genome Interactions and Sexual Behavior

According to a current hypothesis, gonadal hormone action is mediated by the following sequence of events; hormone → DNA → RNA → protein. As was true with the hormone metabolism concept discussed previously, this hypothesis is based primarily on studies of the effects of gonadal steroids on peripheral reproductive tissues. Following is a brief summary of the steps through which a hormone interacts with a target tissue cell. First, the hormone becomes bound to a specific cytoplasmic receptor protein in the cell membrane. Second, the hormone-receptor complex is transported to the nucleus. Third, the nuclear hormone-complex binds to specific sites on the DNA-containing genome. Fourth, DNA transcription results in the production of specific messenger RNA. Fifth, this hormone-induced RNA is transported back to the cytoplasm resulting in the synthesis of new proteins. Sixth, the new protein initiates the functional response characteristic of the particular target tissue. Guided by these conclusions based on studies of peripheral tissues, it might be possible to eventually understand hormone—brain-behavior interactions at the intracellular neuronal level.

Two approaches have been recently used to implicate protein synthetic events in the induction of lordosis behavior by estrogen and progesterone. One approach has been to place inhibitors of RNA or protein synthesis in the regions of the brain that regulate receptivity. A second approach has been to correlate fluctuations in brain RNA or protein synthesis with changes in endogenous or exogenous gonadal hormones. Both approaches have produced results consistent with the hypothesis that hormones regulate sexual receptivity by means of neuronal protein synthesis.

Implants of either actinomycin D, an RNA synthesis inhibitor, or cycloheximide, a protein synthesis inhibitor, into the preoptic area of rats block the effects of estrogen on sexual receptivity (Quadagno and Ho, 1975). Implants of these compounds elsewhere in the brain fail to block receptivity. However, studies of RNA and protein synthesis inhibitors might justifiably be criticized on the basis of the toxic effects of these drugs. Indeed, it is not surprising that a toxic drug implanted in the brain interfered with behavior. Perhaps the drug interferes by debilitating the animal in a nonspecific manner. In an attempt to control for this, Whalen et al. (1974) varied the time interval between implanting actinomycin D and behavioral testing and found that the critical interval for inhibiting receptivity was the length of time between the actinomycin D implant and the estrogen injection, not the length of time in which actinomycin D had been in the brain. This suggests that actinomycin D acts to block receptivity by interfering with some specific effect of estrogen rather than by interfering with the behavior of the animal in some nonspecific manner.

There appears to be some evidence of cyclic patterns of brain RNA synthesis (Salaman, 1970) and protein synthesis (MacKinnon, ter Haar, & Burton, 1972). These cyclic patterns are correlated to some extent with the changes in pituitary gonadotropin and ovarian steroid secretion during the estrus cycle. However, whether the cyclic neural RNA and protein synthetic events are related directly to behavioral receptivity remains an open question. It may very well be that these events regulate pituitary gonadotropin secretion. Brain protein synthesis may of course control behavioral receptivity indirectly via pituitary gonadotropin and ovarian steroid secretion. Nevertheless, research with synthesis inhibitors in estrogen-injected animals has suggested that protein synthesis is directly related to sexual receptivity (Quadagno and Ho, 1975). It is possible that brain protein synthesis is a necessary step for both the production of ovarian steroids and the steroid induction of behavioral receptivity.

Drugs that block protein synthesis have also been used to examine the neural mechanisms for the action of progesterone. Wallen and co-workers (1972) used the intracranial administration of cycloheximide (an inhibitor of protein synthesis) and observed that the inhibitory effect or refractory period—that follows the initial facilitation of estrous behavior—after administering progesterone was blocked. That is, female guinea pigs with cycloheximide-implants displayed lordosis behavior, whereas control animals did not. The possibility of nonspecific debilitation and toxicity effects were ruled out by this use of a positive end-point—the presence rather than the absence of a behavior. Because cycloheximide did not alter the facilitatory effects of progesterone in this study, it was suggested that only the inhibitory actions of progesterone might be mediated by protein synthetic events.

Research on sexual receptivity in the mouse suggests, though by no means proves, that progesterone might exert facilitatory effects on behavior through protein synthesis (Gorzalka & Whalen, 1974). The ovariectomized mouse, in contrast with the ovariectomized rat, which displays full sexual receptivity when estrogen and progesterone are administered for the first time, is completely unreceptive to male mount attempts following administration of estrogen and progesterone (Thompson & Edwards, 1971). It is only after several weeks of combined estrogen and progesterone treatment that female mice gradually become more responsive to males. It has been proposed that this long-term effect is due to a progesterone-dependent process, possibly involving protein synthesis (Gorzalka & Whalen, 1974).

In recent studies of progesterone and DHP effects in mice, Gorzalka (1974) has suggested that progestational compounds may act behaviorally by increasing the amount of functional receptor proteins. In the CD-1 strain of mouse, both progesterone and DHP are effective in facilitating estrogen-induced receptivity, whereas in the Swiss–Webster strain of mouse, progesterone is effective and DHP is ineffective (Gorzalka & Whalen, 1974), suggesting that steroid effects on receptivity are gene-dependent. Furthermore, the reciprocal crossbred offspring of CD-1 and Swiss–Webster parents are responsive to both progesterone and DHP (Gorzalka & Whalen, 1976). Perhaps by establishing the fundamental differences in brain chemistry between DHP-sensitive and DHP-insensitive groups, and by determining the deficits, we may have a powerful tool for understanding neural mechanisms of hormone action in general and of progestational action in particular.

In studies involving administration of radiolabeled progesterone and DHP to CD-1 and Swiss–Webster female mice, significant differences in brain retention of radioactivity were observed (Gorzalka, 1974). The CD-1 mice, which respond behaviorally to DHP, retained more radioactivity than Swiss–Webster mice following administration of radiolabeled DHP. Furthermore, this strain difference was observed only in mesencephalic tissue, not in diencephalic, telencephalic, or pituitary tissue. This mesencephalic specificity is interesting when one considers that this is the same region of the brain where progesterone and DHP implants act to facilitate receptivity in estrogen-primed animals (Gorzalka, 1974). Perhaps the strain difference in radioactivity retained in the mesencephalon indicates a greater amount of DHP receptor proteins in the CD-1 mouse. However, this is only a hypothesis based on an indirect experimental approach, and considerably more research is required before unequivocal statements can be made about the role of brain-receptor proteins in hormone action.

The present discussion has not dealt with male sexual behavior primarily because there have been few if any attempts to implicate protein synthesis in the behavioral effects of testosterone. It has been demonstrated that castration in the male rat decreases total protein content in the anterior hypothalamus, whereas androgen treatment restores it (Moguilevsky et al., 1971). Whether this is related directly to both gonadotropin secretion and sexual performance remains to be determined.

SUMMARY

Sexual behavior is largely controlled by hormonal, neural, and cognitive–experiential factors. The relative contributions of these factors to sexual intensity and direction depend in turn upon ontogeny, phylogeny, and chromosomal sex.

During prenatal and early postnatal development, hormones play an organizing role that ultimately regulates the intensity and direction of sexually motivated behavior. During postpubertal development, hormones assume an activational role that primarily allows regulation of sexual intensity. Furthermore, hormonal factors interact differentially with phylogeny and chromosomal sex. For genetic females, hormones, albeit different ones, appear essential for sexual motivation.

In most nonprimates, estrogen and usually progesterone serve this function. In many nonhuman primates, female sexual behavior is controlled by estrogen, progesterone, and androgens. Finally, in the human female, androgens seem to be the primary hormonal regulator of sexual motivation. By contrast, androgens regulate sexual activity in males of all species, and the relative importance of androgens for male sexual behavior varies with the species. Generally speaking, sexual activity is maintained for longer periods of time following castration in higher mammals than in lower mammals. Perhaps a greater reliance on neural and cognitive–experiential factors compensates for the apparent hormonal independence of highly developed males.

The relative importance of neural factors also varies with age, phylogeny, and chromosomal sex. Prior to puberty, even massive hormone administration typically fails to elicit the complete repertoire of sexual behavior. This partially reflects insufficient development of hormone-receptor systems in the brain. Furthermore, as with hormonal factors, neural factors interact differentially with phylogeny and chromosomal sex. For example, a portion of the neocortex seems crucial for the execution of sexual responses in lower mammalian males but is of lesser importance for sexual responses in lower mammalian females. In primates, an even greater portion of the neocortex is required for male sexual responses. Unfortunately, insufficient data are available for equivalent studies in female primates. The hypothalamus appears to be the major site of action of estrogens and androgens in both sexes for all species including primates. Furthermore, the mesencephalic reticular formation appears to be a major site of action of progesterone. More caudal regions of the central nervous system such as the myelencephalon and spinal cord directly control the reflexive mechanisms of sexual behavior. These reflexive mechanisms are directly facilitated by gonadal hormones in many species. In females, species differences exist. For example, estrogens facilitate spinal sexual reflexes in female cats and dogs but not in female rats.

Cognitive–experiential factors are critical for the development of the sexual behavior repertoire. Social isolation of male or female primates during prepubertal development severely impairs adult sexual behavior. This effect is consistent with the relative importance of the neocortex for primate sexual responses. In nonprimates the effects of prepubertal social isolation vary with the sex of the organism. Typically, isolation impairs sexual responses in male but not female nonprimates. It is important to distinguish experiential factors (obtained from social isolation studies) from cognitive factors. Consider, for example, that when given a choice, female rats will prefer one male to another. This would suggest that cognitive factors play a role in lower mammalian females as they do in other animals. Nonetheless, the relative importance of cognitive factors clearly varies with phylogeny. Males and females of higher species generally show more marked sexual preferences than males and females of lower species.

Sexual behavior is purposive, persistent, and periodic, and it occurs in accordance with priorities, characteristics it shares with the other motivated behaviors.

However, sexual behavior differs in several respects. First, it is important for the survival of the species rather than for the survival of the individual. Second, its initiation does not involve homeostatic or deficit signals. Third, the intensity of sexual behavior is not related in any systematic manner to deprivation conditions. Fourth, sexual behavior can be aroused by a large variety of conceivable, external stimuli. Fifth, it is not clear what constitutes the terminal goal response; the orgasm is absent in females of most species and the behavior passive and thus not obviously goal-directed. Indeed, when given a choice, female rats prefer a mounting, nonintromitting male to an intromitting male. In many males, arousal of the sex drive is as actively sought as its reduction. For example, male rats can be trained to perform a task in which intromission without ejaculation is consistently the reward. Finally, cognitive and experiential factors assume much greater importance for the intensity and direction of sexual behavior than for the motivated behaviors considered in the three previous chapters.

7
Sleep and Waking

Like most humans, you spend about one-third of your life asleep. Some animals, such as the cat, sleep 75–80% of the time, and certain animals such as the bear, sleep most of the time—for weeks or months during a period of hibernation. Sleep alternates with waking in a regular manner, and sleep–waking is one of the most important and prominent biological rhythms. Sleep has biological significance in the sense that we neither feel well nor perform well when lacking sleep and also in the sense that it competes with ingestive, agressive, and other behaviors. Although many behaviors are absent or occur only infrequently during sleep, sleep is not the absence of behavior; rather, as suggested by several authors in this field, sleep is a distinctive kind of behavior. Considered from the ethological point of view, for example, sleep is the final consummatory phase of a complex instinctive behavioral sequence (Moruzzi, 1969). The drowsiness or appetitive phase that precedes sleep frequently includes instinctive or species-specific patterns of behavior (e.g., returning to the home territory; building a nest or shelter). Moruzzi (1969) points out that sleep may occur in a specific territory (''the area of maximum security where the young are born and cared for''), and in some species ''a typical social pattern is closely associated with sleep [p. 182].''

Our main interest in this chapter is with the neural substrates of sleep and waking, but we begin, as in the other chapters, by first considering the characteristics of sleep and the factors that initiate and influence sleep. There has been a great deal of activity in the field of sleep research in recent years that has led to important empirical findings and to changes in our views about the neural mechanisms. One of the major developments discussed in later sections is a change from the view that sleep is a passive process—the absence of waking—to the view that sleep is an active process; in other words, there is a neural mechanism in the brain that produces sleep.

CHARACTERISTICS OF SLEEP

Sleep is "a readily reversible loss of reactivity to environmental events" (Mountcastle, 1974, p. 260) characterized in relation to the waking state by changes in the behavior of the animal, by physiological changes, and by changes in the patterns of electrical activity of the brain. In this section we examine the behavioral, the physiological, and the electroencephalographic changes associated with sleep.

Behavioral Characteristics

If sleep is the consummatory phase of a species-specific behavioral sequence, preceded by an appetitive or drowsiness phase, to what degree does it share the prominent motivational characteristics (purposiveness, persistence, periodicity, and priorities) of feeding, drinking, and other behaviors dealt with in this book?

Sleep is purposive in the Darwinian sense, because it has survival value and apparently fulfills a biological need. Following sleep deprivation, one's performance tends to be more erratic, more effort is needed to perform well at complex motor and mental tasks, and there is a feeling of drowsiness that becomes more intense as the period of deprivation is extended. Furthermore, the appetitive phase of sleep is purposive in the sense that an animal returns to its home territory or to its nest and, in some species, may actually construct a nest or shelter prior to going to sleep. In man, "getting ready for bed" not only involves a shift from other behaviors but frequently includes routine activities such as a "bedtime snack" and personal grooming (e.g., bathing, brushing teeth, etc.). For many species the responses that constitute the appetitive phase may be quite elaborate and persistent, and the consummatory phase or actual period of sleeping may last for several hours.

Periodicity is a very obvious characteristic of sleep, as is the periodicity of feeding and some of the other behaviors considered in this book. Sleep–waking is one of the prominent circadian rhythms in the human adult and in many other species. In human infants, sleep is polycyclic, with several sleep periods during the 24 hours (see Figure 7.1).

Sleep is an essential biological process, and because it has a relatively high priority, it is a very important factor in motivational time-sharing. Exploratory behavior and most goal-directed behaviors such as feeding and drinking are absent during sleep, although, as indicated previously, behaviors are not completely absent.

Physiological Changes

Physiological responses during sleep have been investigated extensively by Kleitman (1939) and other investigators. The major responses as compared to the

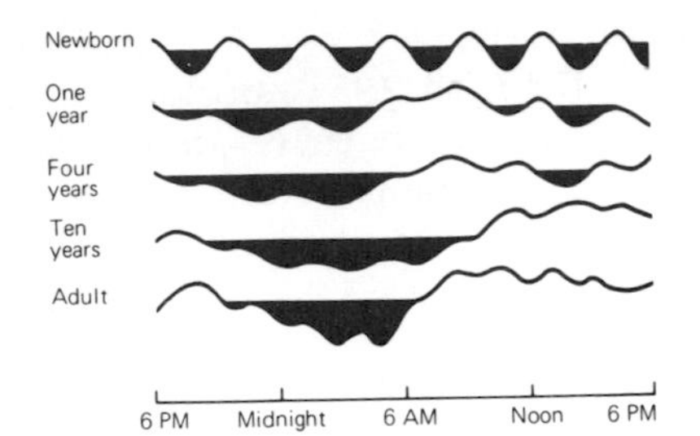

FIG. 7.1 A graphical representation of the polycyclic nature of sleep in the newborn human and the change in sleep–waking patterns with age. The black troughs represent periods of sleep. (After Kleitman, 1939.)

awake animal are summarized in Table 7.1. Accompanying the reduced skeletomotor activity are reductions in heart rate, blood pressure, body temperature, metabolic rate, etc., so that the general picture is one of energy conservation by the body. However, sleep is not a uniform state, and there are periods of "activated sleep" during which blood pressure, heart rate, and other physiological measures, including the electrical activity of the brain, are similar to those

TABLE 7.1
Physiological Responses During Sleep[a]
(Based on Kleitman, 1939)

Increase	Decrease	No change
CO_2 tension in blood	Plasma Ca and K	Plasma Na and Mg
	Heart rate	pH of blood
Blood volume	Blood pressure	Intestinal motility
	Respiratory rate	
	O_2 consumption	
	Metabolic rate	
	Body temperature	
	Reflex excitability	
	Secretion of bile	
	Urine volume	

[a]Subsequent studies have demonstrated other physiological changes, and it is of interest to note that the release of certain hormones is circadian. For example, the release of growth hormone increases during sleep, and the release of antidiuretic hormone also increases, accounting in large part for the reduced urine volume during sleep observed by Kleitman. On the other hand, the release of adrenal cortical hormones is reduced during sleep, reflecting a circadian fluctuation in hypothalamic-pituitary function.

recorded during wakefulness. These changes were first observed during sleep in human infants (Aserinsky & Kleitman, 1955). Because these responses occurred in association with rapid movements of the eyes, it was called rapid eye movement sleep (REM). Similar observations were made in adults by Dement and Kleitman (1955), who recorded the EMG from eye muscles during sleep. When the subjects were aroused during REM sleep, they reported the occurrence of dreaming.[1]

Electrical Activity of the Brain During Sleep

Following the discovery of electroencephalography (Berger, 1929)—a landmark in the modern era of sleep research—it was demonstrated that a high-frequency, low-amplitude (desynchronized) pattern of brain waves occurs during wakefulness and that this pattern changes to low-frequency, high-amplitude (synchronized) during sleep (Figure 7.2). For nearly three decades after these important observations, it was assumed that there was one kind of sleep, which varied in depth, characterized by low-frequency, high-amplitude brain waves. Then Dement and Kleitman (1957) observed that during periods of sleep when rapid eye movements occurred, the EEG of humans was desynchronized (comprised of high-frequency, low-amplitude waves). Similar observations were made in the cat (Dement, 1958). Because this desynchronized brain wave pattern was similar to that recorded during the waking state, this stage of sleep was designated fast wave or paradoxical sleep (Figure 7.3). Rapid eye movements also occur during this stage and, therefore, it is sometimes labeled REM sleep.

A comparison of REM or fast-wave sleep with slow-wave sleep is made in Table 7.2, and it is clear that from the physiological and behavioral point of view, they are quite different states.

[1]According to Teyler (1975):

Apparently we all dream, although some of us are better at recalling dreams than others. Experiments have shown that even persons who claim they "never dream" actually do, although not as often as other people. They apparently forget the dreams they do have. The "average dreamer" has five to seven per night with each dream lasting from 10 to 40 minutes. Dreams get longer as the night progresses. . . . Contrary to popular belief, we dream in "real time," not in super-fast or super-slow time. Many people dream in color and stereophonic sound! Most of us are unaware of the subject matter of most of our dreams. We generally remember a dream best if we are awakened in the midst of it. In detailed studies of dream content hundreds of people were queried about the contents of thousands of dreams. The picture that emerges is that novelty has a predominant role in dreams. A third of dream time is spent merely in going away from or toward something. A high proportion of dream time is taken up by active sports and a low proportion by dull routine. People report that the dreams following a rather dull day are sometimes spectacular and exciting, whereas the dreams following a day full of invigorating activities tend to be bland and tame. It is almost as if a dream were compensating for daily activity [pp. 109–110].

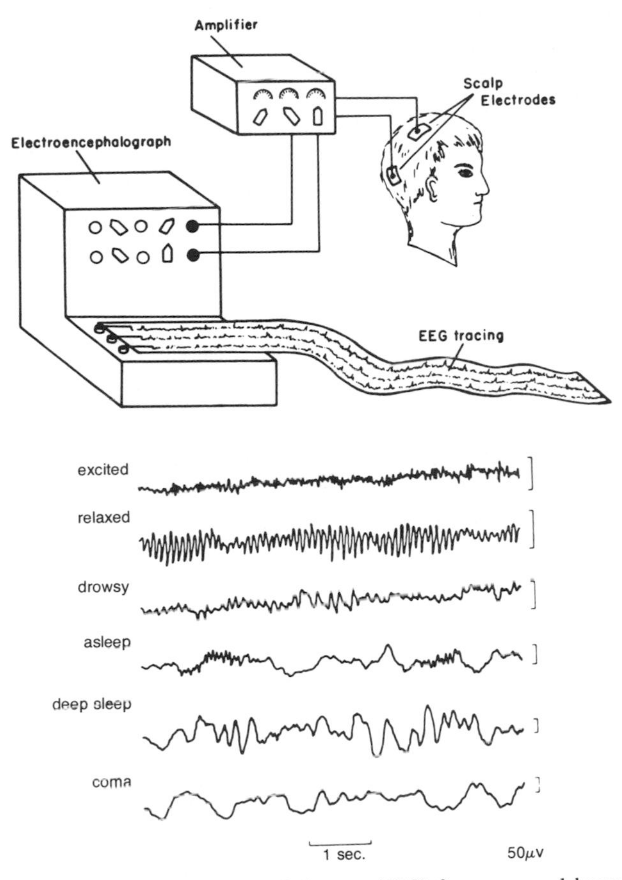

FIG. 7.2 A. Recordings of the electroencephalogram (EEG) from a normal human in different states of consciousness. B. The EEG during alert wakefulness is characterized by low amplitude (voltage) and high frequency, designated desynchronized, and during sleep the EEG is high amplitude, low frequency, designated synchronized. (From Penfield & Jasper, 1954.)

Periods of REM sleep occur about every 90 minutes, lasting 5–30 minutes, and are separated by slow-wave sleep. This is illustrated in Figure 7.1, in which the peaks of the oscillations represent fast-wave sleep and the troughs represent slow-wave sleep. The tonic phase of REM sleep is characterized by low-amplitude, fast-frequency EEG and muscle atonia, shown in Figure 7.3. At frequent intervals during the period of fast sleep, there are rapid movements of the eyes, increased muscle atonia—with muscle twitches superimposed—and sometimes rapid limb and body movements. This is the phasic component of REM or fast-wave sleep. Associated with the rapid eye movements are regular,

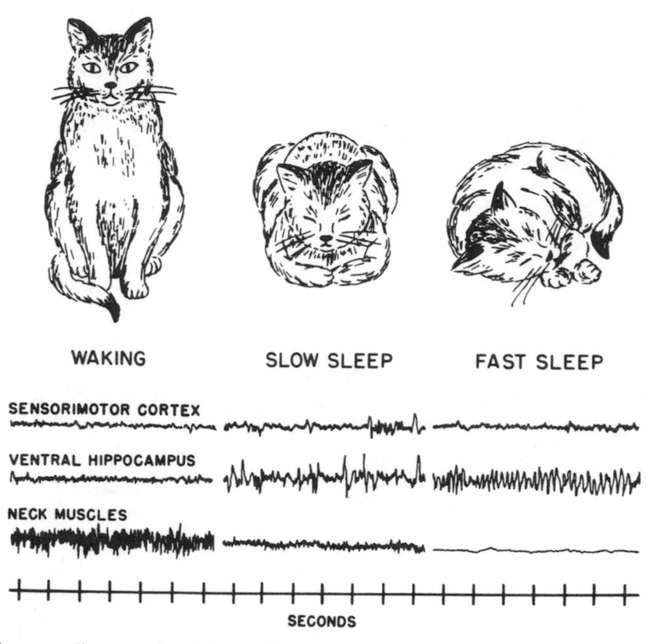

FIG. 7.3 The recordings on the right are from a cat in deep or fast wave sleep and are similar to those in the awake animal on the left, prompting Jouvet to call it paradoxical sleep. Cats usually begin each sleep period with 20–25 minutes of light or slow wave sleep (middle) followed by fast wave sleep. (Adapted from Jouvet, 1967.)

high-amplitude waves recorded from the visual cortex and lateral geniculate that originate in the pontine reticular formation (see Figure 7.3) and have been designated pontogeniculo-occipital or PGO spikes.

Microelectrode recording studies have shown that the pattern of action potentials of neurons in the cerebral cortex is modified during sleep, but the overall activity is not reduced (Evarts, 1967). During REM sleep the total activity may in fact be greater but is irregular, with short bursts of action potentials separated by periods of inactivity (Figure 7.4). Evarts (1967) points out that these observations are inconsistent with the view that the functional significance of sleep is the "restoration of substrates for action potential generation [p. 556]." That sleep has a restorative function is a classic view but one that has not been supported by experimental evidence. Evarts speculates that sleep may be concerned with "relatively long-term plastic structural changes that the nervous system must undergo to make memory and learning possible [p. 556]."

Recently it has been shown that there are electrical correlates of the tonic and phasic components of REM sleep that are similar to those recorded from the neocortex and hippocampus of the waking animal (Robinson, et al., 1977). The tonic component is associated with a noncholinergic neural system that is active during periods of immobility in the waking animal; the phasic component is

TABLE 7.2
Comparison of Slow-Wave Sleep and Fast-Wave Sleep
(After Van Sommers, 1972, p. 94)

Normal or slow-wave sleep	REM or fast-wave sleep
Occurrence:	
75% of total sleep. Predominates in early half of sleep period.	25% of total sleep. Predominates in second half of sleep period.
EEG:	
Sleep spindle bursts and slow waves.	Low-voltage desynchronized pattern.
Motor activity:	
Reduced muscle tone.	Flaccidity in neck and trunk muscles.
No rapid eye movements.	Rapid eye movements. Penile erection.
Physiological states:	
Steady decline in blood pressure, heart rate, respiration, and brain temperature.	Large variability in all physiological indices.
Sleep disorders:	
Somnambulism, enuresis, night terrors, nightmares.	Teeth-grinding.
Sleep-talking common.	Sleep-talking uncommon.
Arousal:	
Confused, amnesic.	Lucid, alert.
Dream states:	
Transient images, thoughts, and reverie predominate.	Narrative "visual" dreams.
Mediator:	
Suggested neurochemical Serotonin.	Norepinephrine.

associated with a cholinergic neural system active during such responses as walking, rearing, and jumping in the waking animal (Whishaw, et al., 1977).[2] These investigators conclude that the motor system of the brain is not shut down during sleep; this has also been suggested by such microelectrode recording studies as shown in Figure 7.4. It appears that the motor system keeps

[2]Vanderwolf and co-workers had previously recorded rhythmical slow waves from the hippocampus and had shown that when this pattern is associated with immobility of the animal, it is not blocked by atropine, but when this pattern is associated with such responses as walking, rearing, and jumping, it is blocked by atropine and presumably mediated by a cholinergic system.

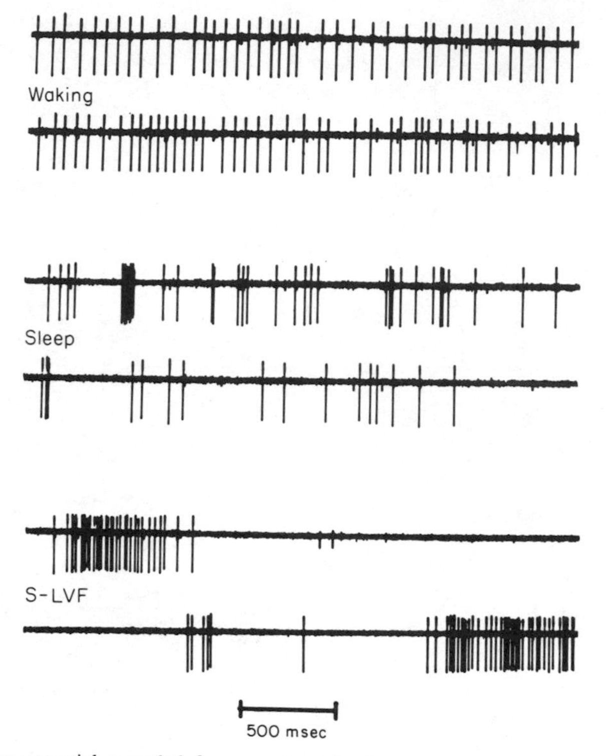

FIG. 7.4 Action potential recorded from a neuron in the motor cortex verified by antidromic activation to be a pyramidal tract neuron. During fast-wave or REM sleep, the overall discharge rate of the neuron is similar or greater than during waking, but the activity occurs in bursts separated by periods of quiescence. (From Evarts, 1967.)

humming—is "free-running"—but is functionally isolated from the muscles because of inhibition of the lower motor neurons of the spinal cord. Robinson and co-workers point out that from the evolutionary point of view, it would be advantageous to keep the motor system running and "ready for action" but to suppress skeletomotor responses to conserve energy supplies and to avoid predators.

ONTOGENY OF SLEEP

The pattern and nature of sleep changes considerably as maturation of the CNS occurs (Figure 7.5). This is the case with humans and also with a number of other mammalian species, and there are striking similarities in the "ontogenesis of the

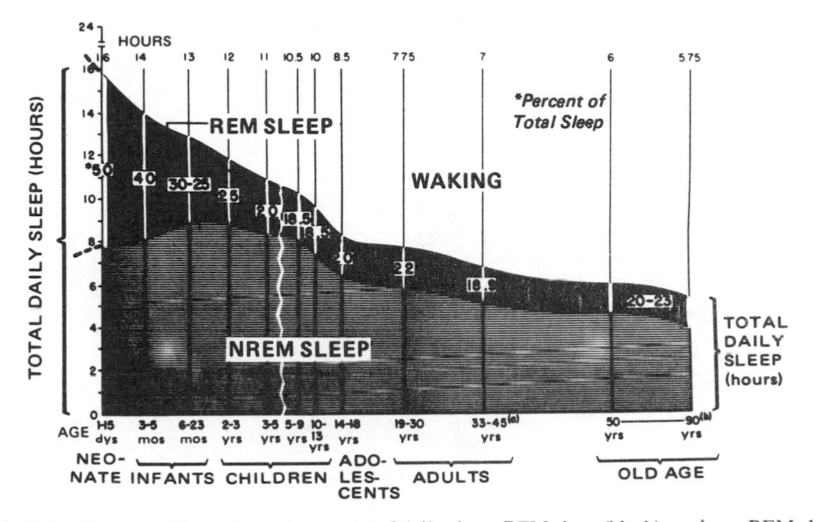

FIG. 7.5 Changes with age in total amounts of daily sleep, REM sleep (black), and non-REM sleep (NREM, gray). The amount of NREM sleep remains fairly constant, decreasing with advanced age. REM sleep falls rapidly from 8 hours at birth to 2–3 hours in the young child. (From Roffwarg, et al., 1966.)

wakefulness–sleep cycle and its EEG correlates in the various species thus far studied'' (Ellingson, 1972, p. 170). Ellingson (1972) summarized some of the major changes as follows:

> (1) Total sleep time is high in the newborn, often increases for a few days after birth, but then decreases gradually with age. (2) The number of wakefulness–sleep cycles per day decreases with increasing age. This occurs first as a result of consolidation of sleep periods, with lengthening of single periods of waking occurring later. (3) Paradoxical sleep as a percentage of total sleep time is high in the newborn and decreases with age . . . and, conversely, quiet sleep as a percentage of total sleep is relatively low in the newborn and increases with age. (4) The number of sleep phases (that is shifts from one stage of sleep to another) is high in the newborn as compared with the adult of a species, and the average durations of sleep phases are low [p. 169].

The changes with age in total sleep time and relative amounts of REM (fast-wave sleep) and slow-wave sleep are shown for the rat, cat, and guinea pig in Figure 7.6.

In some species, such as the rat, cat, and rabbit, in which the CNS is relatively immature at birth, slow-wave sleep may be absent, and REM sleep may occur immediately at the onset of sleep. This is sometimes the case for the human

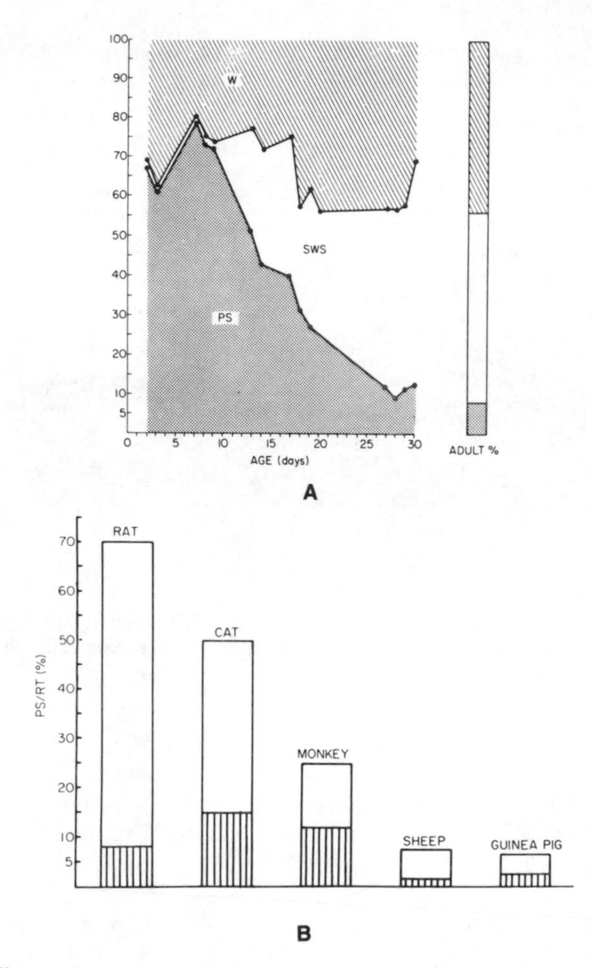

FIG. 7.6 A. Changes in percentages of slow-wave and paradoxical sleep in the rat at various ages. Column on right shows the 24-hour percentages in the adult rat. W = wakefulness; SWS = slow wave sleep; PS = paradoxical sleep. B. Percentages of paradoxical sleep (PS) calculated for total recording time (RT) at 7–8 days of age (white columns) and in the adult (hatched columns) of different mammals. (From Jouvet-Mounier et al, 1970.) The greater relative and absolute amount of paradoxical sleep during infancy has suggested the speculative hypothesis that paradoxical sleep has an important role in the maturational development of the CNS.

infant up to about 3 months of age (Weitzman & Graziani, 1974). Mammals with a more mature CNS, such as the guinea pig, sheep, and goat, display the usual alterations of slow-wave (NREM) and fast-wave (REM) sleep from birth. Two suggestions have been made to explain the absence of slow-wave sleep during early life in some species and the gradual increase both in slow-wave sleep and in the proportion of slow-wave sleep to total sleep time. One proposal is that the critical factor is the presence and the level of serotonin in the brain.[3] The second proposal is that with maturation of the CNS, there is a shift from brainstem to forebrain mechanisms (''encephalization of sleep''). According to McGinty (1971), fast-wave or REM sleep is more primitive and represented in the brainstem, whereas slow-wave sleep, which appears only when the forebrain is sufficiently mature, involves modulatory influences of the forebrain on the brainstem.

Kleitman (1939) was the first investigator to emphasize ''encephalization of sleep and waking'' in order to account for changes in sleep patterns with age (Figure 7.1). He felt that neural systems for sleep initiation and maintenance mature fairly quickly, whereas the neural mechanisms for ''wakefulness capability'' (underlying what he called ''wakefulness of choice'') mature much more slowly. In a recent symposium, Kleitman (1972) commented as follows:

> So, if we are talking about the maturation of sleep, we really should stop at the age of 1 year, when every sleep type of behavior seems completely established. . . . It takes an infant about 4—6 weeks to learn to sleep through the night—8, 9, 10 hours, that is about as much as any adult would sleep. There is very little training required for that purpose. It takes a child five or six years to develop long continuous waking periods, such as when the child gives up the afternoon nap and stays awake from morning to bedtime in the evening. This is what I call wakefulness capability and is something that develops with brain maturation [p. 216].

FACTORS THAT INFLUENCE SLEEP AND WAKING

In this section the various factors that contribute to feeding, drinking, and the other behaviors dealt with in this book are considered in terms of their influence on sleep and waking.

Homeostatic Factors

Sleep is necessary for the survival and well-being of the animal, and from the viewpoint of motivation, it has been classified by some authors as a homeostatic drive (Grossman, 1967; Stern & Morgane, 1974). As with most biological parameters, there is considerable variability in the amount of sleep required by different individuals. Sleep time decreases from infancy to adulthood (Figure 7.5); the typical human adult sleeps about 8 hours per day, but some individuals

[3]Resnick (1972) reported that:
animals such as the rat and rabbit, which are poorly developed at birth, possess low levels of brain 5-HT at birth, whereas the guinea pig and goat, species which are well developed and competent at birth, possess levels of brain 5-HT which approach or, in the case of the goat, exceed the adult levels. Thus, there is presumptive evidence that 5-HT may be involved, directly or indirectly, in the establishment and maintenance of SWS [p. 110].''

sleep considerably more or less. An extreme example is the report of two apparently healthy insomniacs who required less than 3 hours of sleep per day (Jones & Oswald, 1968). The duration of sleep also varies considerably from one species to another. According to Moruzzi (1969): "Hediger... noted that animals like the antelope, which are always in mortal danger when living in their natural state, show an extremely brief duration both of conventional instinctive behaviors, such as drinking and copulation, and of sleep itself. The opposite behavior (longer periods of sleep) is observed in large wild animals, like the bear and lion [p. 183]."

Sleep has been considered from the homeostatic point of view, because following sleep deprivation, there is an increase in drowsiness, a decrease in latency to the onset of sleep, and an increase in total sleep time. Wong (1976), in discussing this point, comments that sleeping is "a consummatory act that is induced by sleep deprivation much in the way that feeding and drinking are induced by food and water deprivation [p. 181]." Sleep, according to many authors, has a "restorative function," and this has been another reason for considering sleep from the viewpoint of physiological regulations and homeostasis (Moruzzi, 1969). According to Schwartz (1973): "The earliest theories of this type postulated that sleep is the result of the accumulation of certain substances, usually metabolic end products, which cause toxic-like depression of brain activity. The removal of these substances during sleep accounts for the return to the waking state [p. 151]."

Over the years there have been a number of attempts to identify a toxin or other homeostatic, deficit signal for sleep. However, as pointed out by Moruzzi (1969) in a scholarly article dealing with the biological significance of sleep, no homeostatic or deficit signals or external feedback loops, which characterize the regulation of body temperature or acid–base balance, have been identified for sleep. According to Moruzzi (1969), sleep is a nonhomeostatic instinct or drive, like sexual or maternal behavior, in which "homeostatic regulation is absent [p. 189]." Moruzzi also suggests that most recent so-called homeostatic theories, such as the hypothesis of "cortical homeostasis" proposed by Ephron and Carrington (1966) and the hypothesis of Stern and Morgane (1974) that REM sleep "repairs" or "maintains the functioning of" catecholamine systems of the brain, are not really homeostatic; he designates them as "cerebrostatic." The proposal of Stern and Morgane can be considered a modern version of the classical view that the CNS undergoes some sort of functional reorganization during sleep.

In summary, whether one regards sleep as homeostatic or not depends on the definition of the term. If the criterion is merely a biological requirement, then sleep has to be considered homeostatic. On the other hand, if homeostatic signals and feedback mechanisms are necessary criteria, then sleep is not homeostatic. Like Moruzzi, I accept the latter view.

Hormones

Certain hormones influence the general excitability of the central nervous system, either directly or indirectly, and because the level of arousal or cerebral activation is altered, there is an influence on alertness, sleep, and waking. The adrenal cortical hormones have such an effect, and so do the thyroid hormones. A person suffering from hypothyroidism may exhibit mental dulling and may sleep more, whereas a hyperthyroid patient may be excessively aroused and not sleep well.

The catecholamines released from the adrenal medulla during stress or emotional excitement may influence reactivity, arousal level, and sleep. Because adrenalin and noradrenaline do not readily cross the blood–brain barrier, these effects may be from peripheral actions of the hormones. However, they may influence the reticular formation, at least in small amounts, and in this way contribute to cortical activation (Rothballer, 1956).

External Stimuli

External stimuli also influence sleep and waking, but, as with the two factors just discussed, there is no evidence of an external stimulus that initiates sleep. In humans and other diurnal species, the absence of auditory, visual, pain, and other stimuli is conducive to sleep. The presence of light, in addition to being a disruptive stimulus that keeps animals awake, may, by its cyclical onset and offset, trigger a "biological clock" that controls circadian rhythms of sleep and activity. For example, laboratory rats and other rodents are less active and sleep a good deal during the day when the lights are on.

The complexity, patterning, and significance of external stimuli, and not merely their occurrence, are important determinants of levels of consciousness.[4] In a monotonous environment, people tend to become bored and drowsy and may

[4]In higher mammals and man, waking is associated with consciousness and sleep, with a temporary, reversible loss of consciousness. Consciousness is not easily defined, but according to Mountcastle (1974), it is an aspect of brain function that has several "observable attributes:"

> Evidence exists that consciousness appears in animals *pari passu* with the development of a complex nervous system as the master regulatory mechanism of the animal. But a nervous system per se is not enough. The attributes of consciousness do not appear when nervous systems containing relatively few neurons control behavior by linking afferent input via "releasers" to neural mechanisms governing innate behavioral patterns. It is difficult to draw any line separating those animals that are conscious from those that are not, and it is perhaps better to regard the different species as distributed along a continuum of an increasing degree and complexity of consciousness. The presence of the conscious control of action may, as an operational definition, be assumed when an organism displays the capacity for choice of action, the ability to set one goal aside in favor of another, the power to withhold action or reaction. Certainly a high order of consciousness is involved in anticipatory planning for action, in modifying action once initiated in terms of then-current events, and in the preparation of

sleep more than usual. For a number of years, a major emphasis was given to the role of external stimuli in the maintenance of waking. Kleitman (1972) put it very simply: "If you have nothing to do, you are likely to sleep much more [p. 216]." The restricting of sensory experience, as in the McGill isolation experiments, has a profound effect on attention and complex problem solving as well as on arousal level and sleep patterns (Hebb, 1966). The view that sleep is a passive process—merely the absence of waking and not subserved by a neural mechanism for its initiation and maintenance—is supported by electrophysiological experiments, discussed in a later section.

Circadian Rhythms and "Biological Clocks"

The effect of light, a stimulus known to have an important influence in the rat on sleep-waking, as well as on rest–activity cycles, was eliminated by blinding the animals (Richter, 1971). The rats, which had been awake and active through most of the night and asleep and inactive in the light, continued to show circadian sleep–waking patterns, although the phase and duration of the rhythm was shifted in some of the animals (Figure 7.7). Richter attributed the continuation of the sleep–waking rhythm after blinding to the existence of a "biological clock." The "biological clock" is normally influenced by the light–dark schedule, and the circadian rhythm is said to be entrained to light. Similar observations have been made in human subjects who lived for several weeks in an underground cave unaware of when daytime and night were occurring in the outside world (Jouvet et al., 1974). Their sleep and waking continued in an approximate circadian manner, but, like Richter's rats, the duration of the rhythm was longer or shorter in some subjects and was no longer determined by the 24-hour day–night sequence.

A number of physiological parameters (e.g., body temperature, see Figure 3.9; heart rate; the release of certain hormones) change in a circadian manner in relation to the circadian rhythm of sleep and waking. When the normal cycle of such rhythms is suddenly changed, as for a "shift worker" who switches from the 8:00 a.m. to the 12:00 p.m. shift or the jet traveller who flys halfway around the world, bodily functions are disrupted, and the person may not "feel right," not only because of lack of sleep initially, but also because it may take several days for these rhythms to become synchronized to the new sleep–waking cycle determined by the day–night sequence.

Little is known about the mechanisms that control the sleep–waking cycle and the circadian rhythms of various physiological parameters. The importance of

---

alternative stratagems to deal with abstract conceptualizations of events that may be encountered. All these latter compose a property of brains called intelligence [p. 254].

Mountcastle suggests that consciousness is associated with attention, the capacity to manipulate abstract ideas, and the capacity for expectancy and self-awareness, and that it presupposes perception and memory.

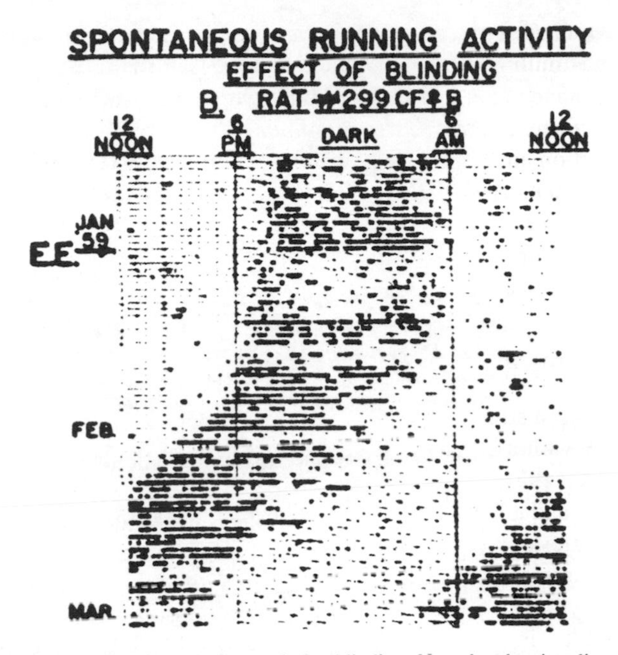

FIG. 7.7 Running activity of a rat before and after blinding. Note that the circadian cycle of activity and rest continues after blinding, but there is a phase shift. (From Richter, 1965.)

"biological clocks" has only recently been recognized, and empirical evidence is limited (Rusak & Zucker, 1975). Speculations about the biological significance of the sleep–waking cycle, which might give some clues about underlying mechanisms, depend largely on whether one considers sleep as having a restorative function or considers it from the ethological point of view. If sleep is involved in "metabolic recovery" or has a restorative function, then the cyclic nature of sleep might be the result of the waxing and waning of some metabolic or homeostatic signals (e.g., a toxin according to classical theories); support for this hypothesis, as indicated previously, is lacking. Alternatively, it has been suggested from the ethological point of view that the function of sleep is to conserve energy during certain periods of the day so that the animal can be active during periods advantageous to its survival (Rusak & Zucker, 1975; and see p. 10 and p. 97).

Emotional and Cognitive Factors

The ease of falling asleep, as well as the maintenance and the duration of sleep, are influenced by the emotional state of the subject. As more and more people live in the cities and participate in complex and competitive industrialized societies, the incidence of sleep disorders has increased. Sleep clinics have been established to deal with these problems, and a new profession of "sleep specialists" has emerged (Kales et al., 1974). Sleep disturbance includes narcolepsy, sleep apnea, somnambulism, nightmares, sleep walking, and enuresis, but the most frequent complaint is insomnia.

Insomnia, being quite prevalent in our society, deserves special comment. According to Kales et al. (1974):

> Insomnia is often a symptom associated with medical disorders, including any condition where pain and physical discomfort are significant symptoms. Where insomnia is associated with medical conditions, both physical and emotional factors must be considered, since it can be assumed that there is always an emotional response to a disease process, varying only in degree. . . . Of course, insomnia may also occur independent of any medical or psychiatric disorder, secondary to a stressful situation such as a death in one's family or financial problems. In these cases, the insomnia is usually transient [pp. 395–396].

A high proportion of "poor" sleepers and insomnia patients have psychological disorders.

Narcolepsy, characterized by excessive sleep of sudden onset, is also associated with psychological factors and usually begins at a point of maturational crisis such as puberty or pregnancy (Guilleminault & Dement, 1974).

Mental activity of a cognitive rather than an emotional nature may also influence sleep. The student who studies right up to the time of going to bed may not sleep very readily and for that matter neither may the professor who works late on his lecture to be delivered to the student's class the next morning. Problem solving, thinking, and cognitive activities, especially when associated with worry, anxiety, or emotional problems, may lead to "tossing and turning" in bed rather than a good night's sleep.

THE NEURAL SUBSTRATES OF SLEEP AND WAKING

In considering the neural mechanisms of sleep and waking, we depart somewhat from the approach of the other chapters. Instead of dealing in turn with each of the factors that contribute to the various motivated behaviors, we take an historical approach and begin with the view that dominated the field for a number of years—a view that emphasized the importance of external stimuli for wakefulness and that regarded sleep as a passive process.

Two major recent developments are then considered. The first is experiments that led to the hypothesis that sleep is an active process, a reversal of the earlier view. The second is the demonstration that there are two kinds of sleep (slow- and fast-wave sleep) and the attempt to associate the two kinds of sleep to the recently demonstrated chemical neuroanatomical pathways projecting from the midbrain and brainstem to forebrain structures.

Sleep as a Passive Process

The modern era of investigating the neural basis of sleep began about 40 years ago shortly after the discovery of the electroencephalogram (Berger, 1929) and the subsequent demonstration of a relationship between brain-wave patterns and level of consciousness (see Figure 7.2). Important pioneering studies were carried out by Bremer (1937), who concluded that sleep was a passive process. Another pioneer investigator of the physiology of sleep, Nathaniel Kleitman (1939), also considered sleep a passive phenomenon; in his words, sleep is the "cessation of an active condition of wakefulness [p. 502]."[5] A few years later, the reticular activating system of the brainstem was demonstrated in experiments by Moruzzi and Magoun (1949), which provided further evidence in support of the passive or deafferentation theory of sleep. The classical studies of Bremer and of Moruzzi and Magoun and their co-workers are described in the following.

Bremer's experiments. Bremer (1937) used an elegantly simple approach. He surgically transected the brainstem of the cat at different levels and observed the effects on "brain waves" and pupil size (Figure 7.8). When the transection was at the level of the midbrain, designated by Bremer as the *cerveau isolé preparation,* the EEG was a high-amplitude, low-frequency (or synchronized) pattern, and the pupil was constricted. These responses were the same as those recorded from an intact sleeping cat, with the exception that they were permanent. When the transection was at the junction of the medulla and spinal cord (encéphale isolé preparation), the EEG was low-amplitude, fast-frequency (desynchronized), and the pupil dilated as in the intact cat when awake (see Figure 7.8). According to Bremer, most of the sensory input to the forebrain was eliminated in the cerveau isolé preparation, and the forebrain remained asleep, whereas in the encéphale isolé preparation, sufficient input through sensory nerves maintained arousal and wakefulness. Bremer concluded that wakefulness depended on sensory input and that sleep was a passive process that occurred when the level of sensory input fell below a certain level.

Moruzzi–Magoun experiments. Moruzzi and Magoun (1949) demonstrated that electrical stimulation of the brainstem reticular formation of anesthetized or sleeping cats produced a desynchronized EEG pattern (high-frequency, low-amplitude) similar to that recorded from an awake animal. When the reticular formation was lesioned, the cats became somnolent, and the re-

[5]The view that sleep was a passive phenomenon and that wakefulness depended on sensory input was not original with Bremer and Kleitman. "Strümpell (1877), who attributed the theory to Pflüger, cited as support the example of a patient who was completely lacking in somatic sensibility and whose only channels of sensory input were one ear and one eye. When those channels were shut off, the man invariably fell asleep in two or three minutes and could be awakened only by removing the blindfold or earplug" (Milner, 1970, p. 261).

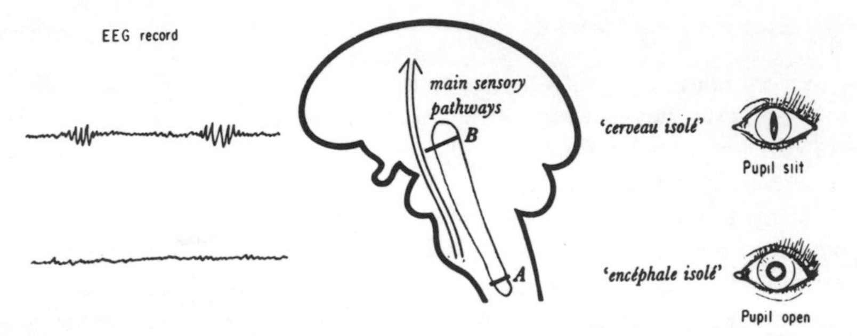

FIG. 7.8 A saggital section of the brain showing where Bremer (1935) made surgical transections. The electrical activity of the brain of a cat transected at *B* was like that of a normal sleeping animal (cerveau isolé). When the transection was more caudal (at *A*), the EEG showed variations typical of sleep and waking (encéphale isolé).

corded EEG was a synchronized pattern typical of what is observed during sleep (Figure 7.9B). In contrast, when the sensory pathways ascending in the brainstem were destroyed, leaving the reticular formation undamaged, the cats were awake and alert, and a desynchronized EEG pattern was recorded (Figure 7.9A). Moruzzi and Magoun suggested that cortical arousal and wakefulness depended on the reticular activating system and not on the sensory pathways as proposed by Bremer. However, the fundamental view was the same: Sleep was considered a passive process attributed to the elimination of the ''waking influence'' of the reticular activating system. Moruzzi (1963) recognized this by commenting: ''From the point of view of the physiology of sleep, therefore, the main result of the new investigations was that Purkinje's (1846) organ of wakefulness was localized, one century later, in the ascending reticular system. However, the validity of all the physiological postulates of the deafferentation theory as developed by Kleitman (1939) and by Bremer (1937) remained unchanged [p. 235].''

The important experiments of Moruzzi and Magoun and their colleagues and the concept of the reticular activating system have had a significant impact on neurophysiology and on our theoretical approaches to motivated and emotional behaviors. For further details, see Mountcastle (1974, pp. 264–268), Thompson (1967, pp. 428–473), and Hebb (1955; 1966).

Sleep as an Active Process

In 1958, Magoun wrote a book called *The Waking Brain;* 13 years later a symposium entitled *The Sleeping Brain* was held in Belgium (Chase, 1972). A number of important developments occurred in the intervening years that revolutionized our views about the neural substrates of sleep and waking.

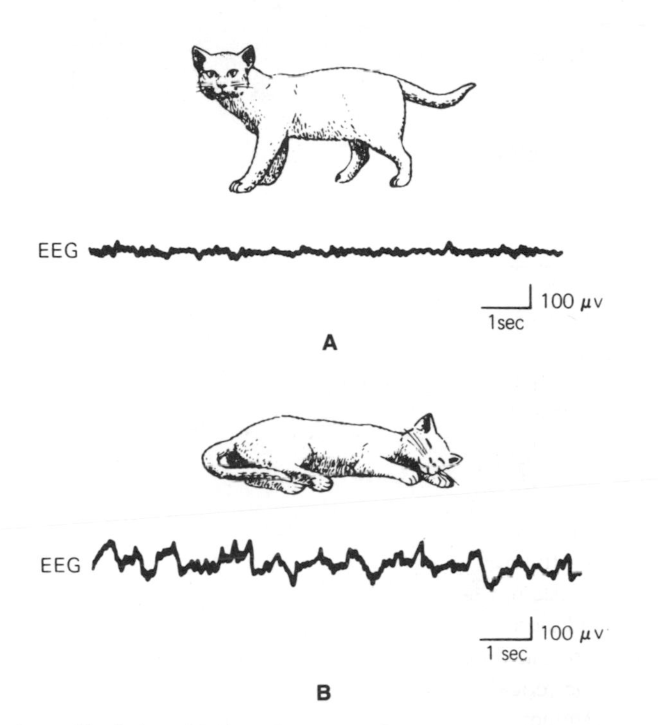

FIG. 7.9 A. A cat with a lesion of the lateral sensory pathways in the midbrain is behaviorally alert, and the EEG is the typical waking pattern. B. A cat with a lesion of the midbrain tegmentum is somnolent and has a desynchronized EEG typical of sleep. The lesion has damaged the reticular activating system, which projects to and regulates the excitability of the cerebrum and the level of consciousness (see Figure 7.2). (From Milner, 1970.)

Using the brain transection technique employed earlier by Bremer, Batini and co-workers (1958; 1959) observed desynchronization of the EEG and sustained behavioral wakefulness following brainstem transection at the midpontine level, just rostral to the trigeminal nerve nucleus (Figure 7.10); these animals showed a marked reduction in the occurrence of sleep following the brainstem transection. The results were interpreted as evidence for a "synchronizing structure" or sleep-inducing mechanism in the lower brainstem (and sleep was considered an active process). Subsequent experiments, in which electrical stimulation of the lower brainstem produced synchronization of the EEG and behavioral sleep, localized the synchronizing or sleep-inducing system to the region of the solitary tract in the medulla (Magnes et al., 1961). Another suggestion is that the brainstem synchronizing system and the reticular activating system exert antagonistic effects on a thalamo-cortical pacemaker system and that sleep–wakefulness and the level of cortical excitability and consciousness depend on

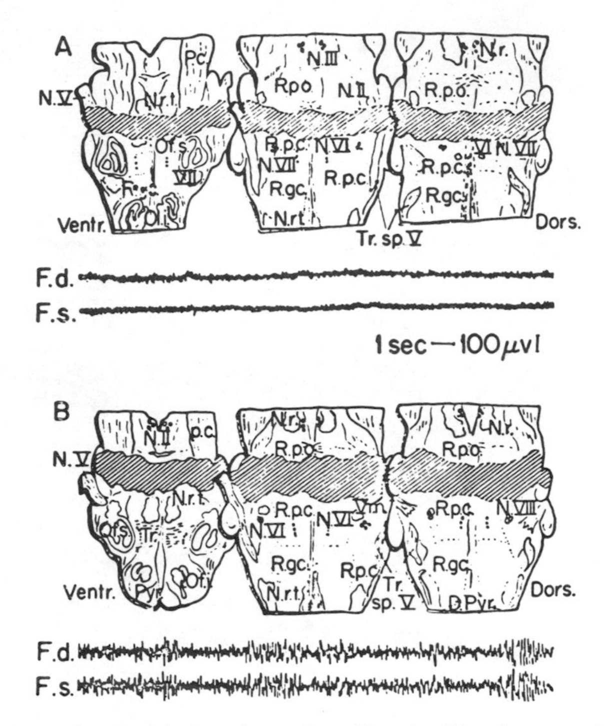

FIG. 7.10 EEG recordings from the frontal area after midpontine (A) and rostropontine (B) transection of the cat's brainstem. Crosshatched areas on brain maps indicate level and extent of damage. F.d. = recording from right frontal area, and F.s. = from left frontal area. Note that recordings in A with midpontine transection are typical of wakefulness. (Batini et al., 1959, as modified by Moruzzi, 1963.)

the relative intensity of these two influences (Mountcastle, 1974). Moruzzi (1963) proposed that the synchronizing system initiates sleep by inhibiting the reticular activating system; ''The presence of the synchronizing structures of the lower brainstem is apparently required only in order to inhibit, or to counteract, the influence of the activating reticular system [p. 244].''

Sleep has been induced by electrical stimulation of other sites in the brain. The original observations were by Hess (1931), who initiated behavioral sleep in the cat by low-frequency stimulation of the thalamus. This early evidence suggested that sleep is an active process, but it was largely ignored because of the impact of the experiments by Bremer and by Moruzzi and Magoun. More recently, electrical stimulation of sites in the preoptic region and basal forebrain (Sterman & Clemente, 1962) and the thalamus (Parmeggiani, 1962) has been reported to induce a synchronized EEG pattern and behavioral sleep (Figure 7.11). The mechanism of these sleep-inducing effects is unclear, but it has been suggested that they may involve inhibitory influences of a basal forebrain system on the

FIG. 7.11 Bilateral electrical stimulation of the basal forebrain in the cat induces sleep. Change from an alert posture on the left to sleep on the right occurred 20 seconds after stimulation began and was accompanied by EEG synchronization. (From Sterman & Clemente, 1962.)

diffuse thalamo-cortical system or on the neurons of origin of the reticular activating system (Mountcastle, 1974).

The Role of Serotonergic, Noradrenergic, and Other Transmitter-Specific Pathways in Sleep and Waking

The important developments discussed in the two previous sections were based on the traditional techniques of lesioning, stimulation, and electrophysiological recording. Although these developments revolutionized the field of sleep research—leading to the views that sleep is the result of an active process and that there are two qualitatively different kinds of sleep (REM or slow-wave sleep and fast-wave sleep)—from the viewpoint of experimental technique or approach, they were a continuation of a tradition dating back to Berger (1929) and Bremer (1937). During the 1960s, in contrast, there was a major new development that resulted in an entirely new approach to investigating the neural mechanisms of sleep and waking. As a consequence, attention shifted to the chemical neuroanatomy and the neurochemistry of sleep.

Using the technique of histofluorescence, a group of Swedish workers visualized and mapped monoaminergic cell bodies (dopaminergic, noradrenergic, and serotoinergic) and their fiber projections (see Figure 2.24). At the same time, the mechanisms for the neurochemical regulation of neurotransmitters were being worked out and drugs developed that could be used as research tools in studying the functional significance of these transmitter-specific neural pathways.

The monoaminergic cell bodies were shown to be clustered in the brainstem, and it was noted that these cells or their fibers were in regions that had previously been implicated, from lesion and stimulation studies, in sleep and waking. Important pioneering studies were made by Jouvet (1972; 1974), who implicated serotonergic neurons in slow-wave sleep; serotonergic, noradrenergic, and cholinergic neurons in REM or fast-wave sleep; and catecholaminergic and cholinergic neurons in waking; and formulated the first neurochemical theory of sleep. The major research strategy used by Jouvet and by a number of other investigators has been to manipulate—by lesions, electrical stimulation, and drugs—the various transmitter-specific neural pathways in order to observe the effects on sleep and waking. The results from such studies have implicated serotonergic neurons of the raphé nuclei located in the midbrain in slow-wave sleep (for review, see Jouvet, 1974). When the raphé nuclei were lesioned, cats showed a marked reduction in slow-wave sleep. Subsequent lesion studies indicated that the critical area was the anterior portion of the raphé. In cats, rats, and monkeys, when the synthesis of serotonin in central serotonergic neurons is blocked by the drug p-chlorophenylalanine, thereby depleting the brain of serotonin, attenuation of slow-wave sleep occurs. However, this drug has a number of other effects, and these results must be interpreted with caution.

The same research strategy has also been used to implicate noradrenergic neurons of the locus coeruleus in arousal and in REM sleep (Jouvet, 1972). However, the attentuation of arousal and of REM sleep from lesions of the locus coeruleus has not been consistently replicated, and because the locus coeruleus also contains cholinergic neurons (Stern & Morgane, 1974), an unequivocal association cannot be made between REM sleep (and arousal) and the noradrenergic neurons of the locus coeruleus that project to the cerebral cortex and hippocampus. The lack of specificity of the techniques has been a shortcoming of this research strategy for studying the chemical anatomical pathways and neurochemical mechanisms that subserve sleep and waking (Holman et al., 1975). This deficiency is especially serious, because it has become apparent that several of the transmitter-specific systems not only partially overlap but also interact functionally (e.g., Jouvet, 1972, has suggested that serotonergic and noradrenergic systems play an important role "in the regulation of the sleep–waking cycle by 'modulating' the activity of other neurons [p. 207].")

A second research strategy is to "observe the changes in CNS neurochemistry during various behavioral states" (Holman et al., 1975, p. 501). Although technically more difficult, it is a more physiological approach. Using this strategy it has been reported that during REM sleep deprivation, there is an increased synthesis and turnover of serotonin (Hery et al., 1970, table 1). This approach should also be very effective for investigating hypotheses formulated to account for results obtained using the first strategy, described in the previous paragraph; for example, it would be effective in testing Jouvet's (1974) hypothesis that the onset of sleep is the result of serotonin being released from the terminals of raphé neurons onto noradrenergic neurons of the midbrain reticular formation that maintain cortical arousal.

Clearly, much work remains to be done on the neurochemistry of sleep. Several transmitter-specific neural pathways have been implicated in slow- and fast-wave sleep and in wakefulness by this exciting new direction of sleep research. The next stage is to test the various hypotheses that have been generated in order to determine the detailed neural mechanisms for the regulation of sleep and waking.

Evidence of a Biological Clock that Controls Sleep and Waking

The first clues about how the brain performs a particular function often come from experiments performed to see whether destroying some part of the CNS will disrupt that function—as, for example, the lesion experiments discussed in Chatper 4, which implicated the hypothalamus in feeding behavior. Such experiments are then followed by others using more sophisticated techniques (e.g., electrophysiological recording, direct administration of neurotransmitters) that attempt to determine the nature of the neural integrative mechanisms. Research on the neural mechanisms that control circadian and other biological rhythms is still in the initial stages.

There have been reports that lesions of the suprachiasmatic nucleus disrupt the circadian rhythm of sleep and waking in the rat (Coindet et al., 1975; Ibuka & Kawamura, 1975). As indicated in Chapters 4 and 5, such lesions also disrupt the nocturnal pattern of feeding and drinking. The suprachiasmatic nucleus receives visual inputs (Moore, 1973), but the disruption of sleep–waking and other circadian rhythms following a lesion of this nucleus does not seem to be the result of eliminating visual signals, because, as indicated in an earlier section, blinding does not eliminate the sleep–waking cycle but only alters the phase and duration of the rhythm. These observations suggest that the suprachiasmatic nucleus is not merely a relay station for visual inputs; rather they suggest that it may be part of the "biological clock." In a recent authoritative review of the literature in this field, Rusak & Zucker (1975) suggest that the suprachiasmatic nucleus has an important and complex role in the "production, coordination, and entrainment" of biological rhythms, but they point out that "whether the suprachiasmatic nucleus serves as a master clock, a central coupler of rhythms, or as a complex oscillator in a multioscillator system is unknown [p. 148]." Further research is needed to investigate this hypothesis and to determine the functional relationships of the suprachiasmatic nucleus and other neural components of the "biological clock" with brainstem structures previously implicated in the control of sleep and waking.

Sleep and waking are altered by lesions and stimulation of other regions of the brain. Waking is severely attenuated or eliminated by lesions of the reticular activating system, and lesions of the "synchronizing system" in the lower brainstem produce continuous wakefulness. It is possible that the suprachiasmatic nucleus and the "biological clock" regulate the sleep–waking cycle by exerting influences on these brainstem systems,but in the absence of any relevant experimental evidence, this is a speculative hypothesis only. Another possibility that should be investigated is whether the "biological clock" regulates the circadian fluctuations of serotonin, noradrenaline, and other neurotransmitters implicated in sleep and waking. Alternatively, the circadian fluctuations of these neurotransmitters may reflect the operation of other neural components of the "biological clock."

It should be emphasized that "biological clock" is merely a hypothetical construct—a term that reflects our ignorance of the neural mechanisms. When the sleep–waking cycle and other biological rhythms can be accounted for in terms of neural integrative mechanisms, this term will no longer be necessary.

SUMMARY

Sleep has been considered from the same point of view as feeding, drinking, and other motivated behaviors. Sleep, consisting of slow- and fast-wave stages, is analogous to the consummatory phase of these behaviors and is preceded by an appetitive phase characterized by drowsiness and, in many cases, species-specific behavioral patterns. The appetitive behavior appears goal-directed, and sleep fulfills some sort of biological requirement—it is purposive. Following sleep deprivation the duration of sleep is increased, and there is a rebound in fast-wave sleep. Sleep also has a higher priority in "motivational time-sharing" following sleep deprivation, further evidence of its biological significance. Sleep is periodic, and in most mammals it is circadian; no evidence has been obtained for a homeostatic signal for sleep or for a sleep hormone. Earlier theories stressed the importance of external stimuli, especially the circadian cycle of light and darkness, in maintaining wakefulness, and sleep was considered a passive process.

The study of the neural mechanisms of sleep is a good example of the interrelationship of experimental techniques, empirical findings, and theoretical concepts. The modern period of sleep research began after the discovery of the electroencephalogram (Berger, 1929). Using electrophysiological recording in conjunction with the technique of brain transections, Bremer (1937) made his classical observations in the cerveau isolé and encéphale isolé preparations and concluded that sleep was a passive process. Moruzzi and Magoun (1949), at the end of the next decade, relied on electrophysiological recording, lesion, and electrical stimulation techniques to demonstrate the role of the reticular activating system in maintaining wakefulness but retained the conceptual orientation that sleep is the absence of waking. In the late 1950s, important new discoveries were made, using the same experimental techniques, which led to a significant conceptual change; the results of this research indicated that sleep is an active process, and two qualitatively different kinds of sleep—REM, or fast-wave sleep, and slow-wave sleep—were identified. Then in the 1960s, an entirely new approach to the neural mechanisms of sleep and waking was undertaken that shifted attention to the chemical neuroanatomy and the neurochemistry of sleep. This was made possible by development of the technique of histofluorescence for visualizing and mapping monoaminergic neurons and their fiber projections, by an understanding of the mechanisms for the neurochemical regulation of neurotransmitters, and by the development of drugs that could be used as research tools for studying the functional significance of these transmitter-specific neural pathways in sleep, waking, and other behaviors.

There has also been an interest in the possibility of a neural oscillatory mechanism, or "biological clock," that controls sleep and waking. The levels of some of the central neurotransmitters implicated in sleep have been shown to vary in a circadian manner. Whether these changes are regulated by a "biological clock" or, indeed, are part of the "machinery" of a "biological clock" is not known. Lesions of the suprachiasmatic nucleus disrupt circadian rhythms of sleep and waking, as well as feeding, drinking, and other behaviors, but whether this nucleus is an important component of a "biological clock" remains to be determined.

8
Brain Self-Stimulation Behavior

E. T. Rolls and
G. J. Mogenson

Brain self-stimulation was first demonstrated when Olds and Milner (1954) observed that rats returned to the place in an open field where they received electrical stimulation of the septal region (Figure 8.1).[1] When tested in a Skinner box, in which the rats could initiate the brain stimulation by pressing down on a lever, they responded for long periods of time at relatively high rates. In subsequent experiments by a number of investigators, various stimulation sites in the hypothalamus and limbic structures were observed to be rewarding, the lateral hypothalamus in the region of the medial forebrain bundle being particularly effective (Olds, 1962). Brain self-stimulation was also reported in a number of other species (e.g., cat, dog, dolphin, gerbil, goat, guinea pig, hamster, monkey, rabbit) including the human (Figure 8.2).

The lever-pressing response (the operant technique) has been the most commonly used behavioral response in studies of brain self-stimulation. Animals will

[1]According to Mogenson and Cioe (1976):

It was nearly 60 years after the brain was first stimulated electrically that it was shown that motivational effects could be elicited by such stimulation. This long delay following the historic experiments of Fritsch and Hitzig, who stimulated the motor cortex in 1870, was due in part to the facts that most of the experiments during this period were in anesthetized animals and that most investigators stimulated the cerebral cortex, which is motivationally neutral (Doty, 1969). In the early 1930s Hess performed important pioneering experiments in unanesthetized, freely moving animals. Although he observed that a variety of motivated behaviors could be elicited by stimulation of the hypothalamus and other subcortical structures, more than two decades passed before further developments occurred. The demonstration of central reinforcement, made possible by the use of stereotaxic surgical procedures for the implantation of chronic stimulating electrodes and by the use of operant techniques, provided the major impetus for the study of brain mechanisms of reinforcement and more generally for the study of brain-behavior relationships [p. 570].

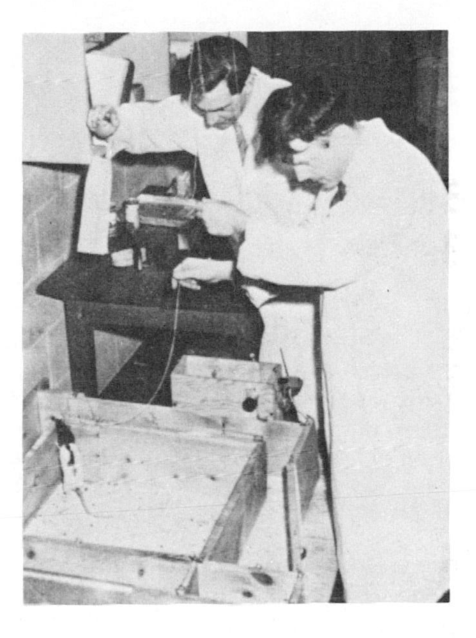

FIG. 8.1 Photograph of Peter Milner and James Olds, co-discoverers at McGill University of brain-stimulation reward. The rat returned to the corner of the open field where brain-stimulation was delivered. (From Hebb, 1966.)

also run along a straight alley, traverse a complex maze, and cross a painful shock grid to obtain brain-stimulation reward (Figure 8.3).

The approach and organizational framework of this chapter is the same as that used in earlier chapters dealing with thermoregulatory behavior, feeding behavior, etc. This approach differs from the usual approach to the topic. The rationale is that if brain-stimulation reward mimics conventional rewards (that is, it involves the direct activation of neural pathways that subserve conventional rewards), as is suggested later in this chapter, then learning about brain-stimulation reward should be relevant to understanding the neural mechanisms for food reward, water reward, etc., and vice versa. Accordingly, we begin with the characteristics of brain self-stimulation and then deal with the factors that influence the behavior before we go on to consider the neural substrates of brain-stimulation reward.

FIG. 8.2 A patient receiving electrical stimulation of the brain. *Left*—Stimulation produces ''pleasurable'' or rewarding effects. *Middle*—Stimulation produces aversive effects. *Right*—No stimulation.

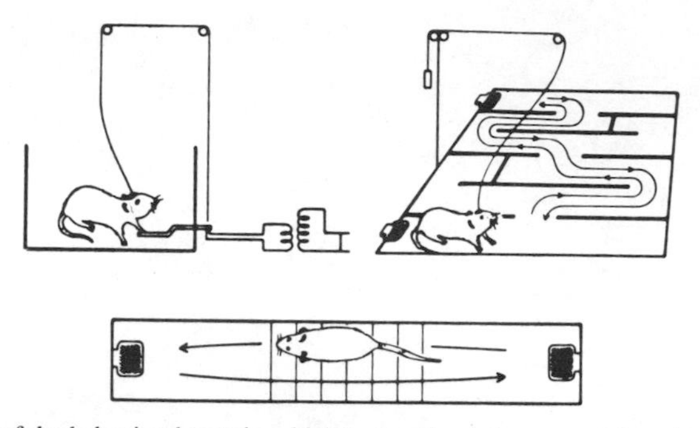

FIG. 8.3 Some of the behavioral tests in which brain-stimulation reward have been investigated: a Skinner box in which pressing a lever initiates the brain-stimulation, a complex maze that must be traversed to reach the lever, and an alley with a grid floor that must be crossed to reach the lever (for details, see Olds & Milner, 1954; Olds, 1958b, 1961).

CHARACTERISTICS OF BRAIN SELF-STIMULATION

Introduction

In Chapter 1, motivated behaviors were characterized as being purposive (or goal-directed), persistent, periodic, and showing priorities in accordance with the internal state of the animal and the external environment. Do the behaviors maintained by brain-stimulation reward exhibit these same characteristics, and are they similar in these respects to feeding, drinking, and other conventional motivated behaviors?

Because animals traverse a complex maze or cross a shock grid to obtain brain-stimulation reward, the behavior would appear to be purposive and goal-directed (Figure 8.3). As demonstrated in numerous lever-pressing experiments, the behavior to obtain brain-stimulation reward is also very persistent; rats will continue to make the operant response for many hours. Indeed, there are cases of continuous responding for more than a day (Figure 8.4), suggesting that the potency of brain-stimulation reward and the persistence of the behavioral responses is greater than that for conventional rewards. The behavioral responding is periodic in the sense that the rate of responding follows a circadian rhythm (Terman & Terman, 1970; 1975), but for the most highly rewarding sites of the brain, the behavior is not periodic in the way that feeding and drinking are (''appetite'' and ''satiety''—the onset and termination of feeding occurring regularly every few hours). Brain self-stimulation may continue for long periods; brain-stimulation reward appears to be relatively insatiable. Also, the animal does not switch readily from responding to obtain brain-stimulation reward in order to respond for food or water (Routtenberg & Lindy, 1965; Morgan &

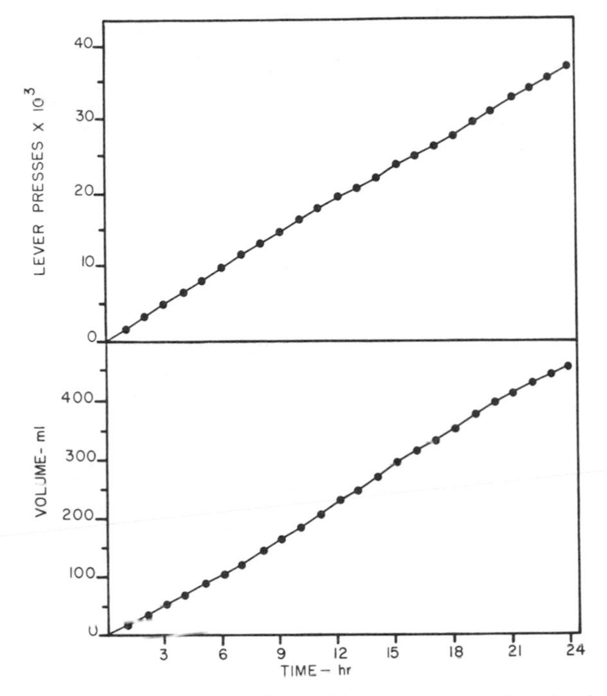

FIG. 8.4 Cumulative records of lever pressing (*top*) to obtain as reward an electrical stimulation of the lateral hypothalamus. The hypothalamic stimulation elicited drinking behavior, and the cumulative water intake for the 24-hour test is shown in the *bottom* panel. (From Mogenson & Stevenson, 1967.)

Mogenson, 1966), suggesting that brain-stimulation reward may not permit the animal to readily switch priorities in accordance with changes in its internal and external environments.

In view of these considerations, it seems appropriate—and even essential—to consider in some detail whether brain-stimulation reward is similar to conventional rewards in terms of its effects on behavior. This is not only relevant to the approach being taken in this chapter but, as indicated in the following, is also relevant from the viewpoint of the historical development of this field.

The Similarity of Brain-Stimulation Reward and Conventional Rewards

A major reason for the initial enthusiasm of behavioral scientists and neuroscientists about the brain self-stimulation experiments was that they seemed to provide an approach that made it possible to deal directly with the neural mechanisms of response reinforcement; it was assumed that brain-stimulation reward and conventional rewards were essentially similar (Olds, 1956). However, there was soon a number of observations that indicated possible differences. Acquisition as well as extinction of an operant response typically occurred more quickly with brain-stimulation reward than with conventional rewards (Olds & Milner, 1954; Seward et al., 1959; Deutsch & Howarth, 1963). Brain-stimulation reward maintained operant responding for long periods of time apparently without satiation (Olds, 1958b; Valenstein & Beer, 1964), and in a choice situation, animals would self-stimulate their brains and ignore food (Routtenberg & Lindy, 1965; Spies, 1965) or water (Falk, 1961; Morgan & Mogenson, 1966) to the point of self-starvation. Animals had difficulty on a partial reinforcement schedule (Brodie et al., 1960; Culbertson et al., 1966), and secondary reinforcement was not established (Seward et al., 1959; Mogenson, 1965) when brain-stimulation reward was used.

Subsequently it was shown that these differences between brain-stimulation reward and conventional rewards were due mainly to differences in experimental procedures, particularly in the delay of reward and the state of food or water deprivation of the animal (Trowill et al., 1969). The tempero–spatial relations between the behavioral response and the reward are not usually the same; typically there is a temporal delay between the operant response and the food, water, or other conventional reward, whereas brain-stimulation reward occurs immediately when the operant response is made. When this factor is equated, either by delaying the brain-stimulation reward after the operant response (Gibson et al., 1965) or by delivering food reward by means of an intraoral fistula (Panksepp & Trowill, 1967b), acquisition is similar for brain-stimulation rewards and conventional rewards. The difference in the delay of brain-stimulation and conventional rewards also contributes to the difference in extinction.

Another factor is the deprivation state of the animal. With conventional rewards the animal is usually food or water deprived, whereas such deprivations are usually not used with brain-stimulation reward. It has been reported that when rats are food deprived, extinction occurs less rapidly using brain-stimulation reward than in the typical procedure with nondeprived animals (Deutsch & DiCara, 1967). On the other hand, extinction occurs more rapidly with food as the reward in nondeprived than in deprived animals (Panksepp & Trowill, 1967a). When the testing conditions were made similar, Gibson and co-workers (1965) obtained similar extinction curves with brain-stimulation reward and a sugar solution reward when either reward followed the animal's response with the same delay.

Failure to obtain secondary reinforcement when using brain-stimulation reward (Seward et al., 1959; Keys, 1964; Mogenson, 1965) may have also been because the animals were not deprived. DiCara and Deutsch (1966) reported that secondary reinforcement could be obtained with electrical stimulation of sites of the brain sensitive to food deprivation when the animals were food deprived.

Difficulties in maintaining behavioral responding when brain-stimulation reward is used on partial reinforcement schedules may be attributed to the inadequate intensity of reward (Sidman et al., 1955). It has been suggested that a standard food pellet used as a conventional reward may be equivalent to several (5–20) brief trains of brain-stimulation reward (Pliskoff & Hawkins, 1967). Brain-stimulation reward has been used successfully for FI-1 minute and VI-1 minute schedules when trains of brain-stimulation were administered, and stable performance on FR-10 minute, FR-200 and DRL-180 second schedules was observed when a two-member chaining procedure was used to produce a temporal delay between response and reward (Brown & Trowill, 1970). If a single presentation of brain-stimulation is preceded by a light, then stable operant behavior is maintained on intermittent reinforcement schedules of FR-200, VR-30, FI-3 minutes and DRL-20 seconds (Cantor, 1971).

Brain-stimulation reward, although sometimes stronger in a competition test than conventional rewards, does not lead to a rigid, inflexible sequence of responding. When the incentive characteristics of the conventional reward are increased or the intensity of brain-stimulation reduced, the preference for brain-stimulation reward disappears. For example, when the alternative in a competition test was a highly palatable saccharine-glucose solution, rats lever-pressed at rates of more than 100 per minute to obtain it, and they showed an equal preference for this solution and electrical stimulation of the lateral hypothalamus following 22 hours of water and food deprivation (Phillips et al., 1970). The relative preference for brain-stimulation reward and conventional rewards depends on the intensity of the stimulation current (Deutsch et al., 1964; Falk, 1961; Morgan & Mogenson, 1966), the length of deprivation (Deutsch et al., 1964; Morgan & Mogenson, 1966), the duration of the test session, and the palatability of the conventional reward (Phillips et al., 1970).

In summary, it appears that any differences between brain-stimulation reward and conventional rewards can be attributed to procedural differences in behavioral testing. This is not to suggest that the two are identical. In fact, there is increasing recognition that the behavioral effects of conventional rewards themselves are not identical (Bolles, 1970; Shettleworth, 1972). How an animal's behavior is altered by a reward or reinforcer, and what and how it learns, is subject to the constraints of its species-specific behavioral organization.

FACTORS THAT INFLUENCE BRAIN SELF-STIMULATION

If brain-stimulation reward and conventional rewards such as food and water are similar, as suggested in the previous section, it seems appropriate to ask whether some of the factors that influence feeding, drinking, and other motivated behaviors also influence brain self-stimulation.

Homeostatic, Deficit Signals

It is well known that the effectiveness of food and water as rewards are enhanced following food and water deprivation. Is brain self-stimulation similarly enhanced by such deprivations?[2] In the first attempt to investigate this question, it was observed that the rate of self-stimulation of the hypothalamus was increased after a period of food and water deprivation (Brady et al., 1957). However, this increase might have resulted from increased motor activity rather than being due to a specific interaction of deprivation and brain-stimulation reward. This interpretation of the results of a more recent study were not confounded by the possibility of such nonspecific changes in activity. Olds (1958a) demonstrated in castrated rats that the rate of self-stimulation of lateral hypothalamic sites was increased by food deprivation, whereas other animals, with electrodes placed somewhat more lateral in the hypothalamus, showed an increase only when injected with androgens. A more recent study also demonstrated that the effect of a particular deprivation or drive state on brain selfstimulation was specific to the site of stimulation. Gallistel and Beagley (1971) observed in rats with two stimulating electrodes that stimulation of one site was preferred when the rats were thirsty (e.g., water-deprived for 24 hours), whereas stimulation of a second site was preferred when the rats were hungry (e.g., food-deprived for 24 hours).

The effects of satiety signals on brain-stimulation reward have been investigated in rats with electrodes in regions of the lateral hypothalamus. Injections of liquid diet or glucose into the stomach (Mount & Hoebel, 1967) and inflating the stomach using an intragastric balloon reduced lateral hypothalamic self-stimulation rates (Hoebel, 1968). The injecting of liquid diet intragastrically did not reduce self-stimulation of the septal region, indicating that it was not a nonspecific effect. Making rats hyperphagic and obese by the administration of

[2]Although we are interested merely in the question of whether internal (hunger, thirst, satiety) signals influence brain self-stimulation, the results of these experiments were considered a few years ago to be of fundamental theoretical interest. When brain-stimulation reward was discovered, the drive-reduction theory (Hull, 1943) was very popular, and a number of investigators attempted to interpret the results of brain self-stimulation experiments by using this theoretical approach (e.g., Miller, 1960). It was assumed that during brain self-stimulation, neurons were activated that were associated with the satisfaction of basic drives such as hunger, sex, and thirst. However, a few years later, there were reports that electrical stimulation of self-stimulation or "reward" sites in the hypothalamus also elicited feeding (Hoebel & Teitelbaum, 1962; Margules & Olds, 1962), drinking (Mogenson & Stevenson, 1966), and other motivated behaviors (Glickman & Schiff, 1967). These observations seemed paradoxical if one assumed that reinforcement resulted from drive reduction, because electrical stimulation of "reward" sites was apparently inducing rather than reducing a drive. Attempts to resolve this paradox were among the important reasons for considering alternatives to the drive-reduction hypothesis of reinforcement and learning (e.g., Glickman & Schiff, 1967) and for emphasizing other determinants of motivated behaviors in addition to homeostatic or deficit signals. During the next few years, there was more interest in the possibility that brain-stimulation reward could be interpreted from the viewpoint of incentive motivation (Pfaffmann, 1960; Bolles, 1972).

insulin (Balagura & Hoebel, 1967) or by force-feeding as a result of elicited feeding by hypothalamic stimulation also attenuated lateral hypothalamic self-stimulation.

External Sensory Stimuli and Cognitive Factors

External stimuli have also been shown to enhance self-stimulation of the brain. Using rats in which electrical stimulation of the lateral hypothalamus initiated water intake, Mogenson and Stevenson (1966) placed a water spout as well as a lever in the test chamber making it possible for the animals to self-stimulate and elicit drinking concurrently. Self-stimulation of the lateral hypothalamus occurred at a faster rate when the animals were induced by the hypothalamic stimulation to drink water, and the self-stimulation rate increased further when saccharin was added to the water (Table 8.1). A similar enhancement of self-stimulation of the hypothalamus was observed in rats that were induced by the stimulation to feed (Coons & Cruce, 1968).

These observations are consistent with the view that brain-stimulation reward is the result of electrical stimulation activating central neural pathways that transmit rewarding or incentive stimuli (Pfaffmann, 1960). By incentive stimuli we refer, for example, to the sensory properties of sweet and salty solutions that are effective rewards even in the absence of homeostatic deficits (Young, 1959). Certain odors are also rewarding in nondeprived animals, and it is of interest that self-stimulation of the olfactory bulb is enhanced by odors (Philips & Mogenson, 1969; Phillips, 1970).

These experiments not only demonstrate that external stimuli can influence self-stimulation behavior but also provide support for an interpretation of reinforcement that emphasizes the incentive properties of nonhomeostatic stimuli.

TABLE 8.1
Self Stimulation Rates* Accompanying Electrically Induced Drinking Behavior in 12 Water-satiated Rats
(After Phillips & Mogenson, 1968)

Condition	Mean±SEM
Control (no fluid available)	517±53.2
Water available	588±49.2**
0.2% saccharin solution available	660±45.6**

*Lever presses/30 minutes
**%Significantly greater than control condition $p < 0.01$

These observations are consistent with the view that feeding and drinking occur because of the incentive or ''appetite-whetting'' characteristics of the food or fluid and that brain-stimulation reward results from the activation of neural pathways that subserve such sensory stimuli (Hoebel, 1969; 1971).[3]

More direct evidence for the role of sensory stimuli in brain-stimulation reward has been obtained recently in experiments that combined electrophysiological recording techniques in unanesthetized animals with chronic electrical stimulation of the hypothalamus. Rolls and co-workers (1976) recorded from single neurons in the lateral hypothalamus in the monkey that were activated by the stimulation of reward sites in the amygdala, nucleus accumbens, and orbitofrontal cortex. The discharge of these neurons was also altered by the sight and taste of food provided that the animal was hungry (for further details, see pp. 112 and Figure 4.10). It appears that such neurons may signal to a hungry monkey that a specific food reward is available. Rolls (1975) suggested that, ''A sufficient condition for electrical stimulation of the brain to provide reward may be that it mimics the effect of a specific natural reward [p. 21].''

It has been suggested that cognitive activities of the brain as well as external stimuli may influence central reward pathways, although direct evidence for this possibility is lacking. After noting the pleasure obtained from science and mathematics and from such pastimes as crossword puzzles or chess, Campbell (1971) suggested that ''Only in the human brain can thinking activate the limbic pleasure areas [p. 22].'' This interesting speculation is worth pursuing, because prefrontal cortex and other regions of the forebrain contribute to cognitive processes.

Hormones

There are a number of reports that the release of adrenaline and noradrenaline from the adrenal medulla and of corticosteroids from the adrenal cortex are increased during self-stimulation of the brain (see Sadowski, 1976). There is also

[3]For further discussion of the interpretation of brain-stimulation reward considered from the viewpoint of incentive motivation, the reader is referred to a recent review article by Mogenson and Cioe (1976):

> Electrical stimulation of the hypothalamus may elicit drinking and feeding not by activating systems for internal deficit signals, . . . but by mimicking incentive stimuli (or ''appetite-whetting'' stimuli) associated with drinking and feeding; it is then not a paradox that animals self-stimulate ''feeding sites'' and ''drinking sites.'' Certain stimuli, such as sweet and salty solutions, are particularly effective reinforcers (Pfaffmann, 1960; Young, 1959), even in the absence of biological deficits and needs; a sensory stimulus ''can function as a reinforcer in its own right'' (Pfaffmann, 1960, p. 255). Pfaffmann has suggested that central reinforcement is due to the activation of neural pathways that normally transmit such incentive motivational stimuli.
>
> The results of experiments by Pfaffmann and Young, as well as those from peripheral self-stimulation experiments (Campbell, 1971), suggest that brain self-stimulation results from activating pathways of the brain that transmit signals from natural reinforcers. These include inputs from smell and taste and other exteroceptive stimuli, but also may include proprioceptive and interoceptive inputs [p. 582].

evidence of increases in thyroxin and growth hormone and decreases in estrogen. However, these endocrine responses seem to be a consequence of arousal and behavioral activation and do not make an essential contribution to brain-stimulation reward. Because the stimulating electrodes are relatively large and activate multiple neural systems, the eliciting of endocrine responses may be merely a concomitant of self-stimulation.

There are also reports that self-stimulation is influenced by the removal of endocrine glands and the administration of hormones. Olds (1958a) demonstrated that the rate of self-stimulation of male rats decreases following castration and that the self-stimulation rate is restored when the animals are administered testosterone. These effects were observed for hypothalamic sites associated with the elicitation of sexual behavior, suggesting that testosterone contributes to the neural excitability of this region. Hoebel (1969) has shown that self-stimulation of hypothalamic sites associated with feeding is increased by the administration of insulin and decreased by glucagon. He attributes these effects to the lowering and raising of blood sugar by these two hormones. Therefore, as with studies in which endocrine responses were measured during self-stimulation, studies involving hormone administration indicate that hormones do not have an essential role in brain-stimulation reward, although they may have indirect effects.

Circadian Rhythms

When rats were able to self-stimulate the hypothalamus or septal area on a 24-hour basis, a circadian rhythm of self-stimulation was observed (Terman & Terman, 1970; 1975). The animals made three or four times as many responses during the 12 hours of darkness as during the 12 hours of light. This circadian rhythm of self-stimulation was entrained to the light, and when the light–dark cycle was reversed, the self-stimulation rhythm was also reversed in a few days.

The mechanisms subserving this circadian rhythm of self-stimulation are not known. It will be important to determine whether the reward potency of the brain stimulation differs between day and night or whether the circadian rhythm of self-stimulation merely reflects the activity cycle of the animal. Because brain-stimulation reward has been related to central catecholamines, it is of interest that the brain noradrenaline levels vary in a circadian fashion (Margules et al., 1972).

THE NEURAL BASIS OF BRAIN-STIMULATION REWARD

To analyze the neural basis of brain-stimulation reward, the sites in the brain where electrical stimulation supports self-stimulation can be determined: Lesions can be made at different brain sites to determine whether any site is essential for brain-stimulation reward; recordings of neuronal activity during brain-stimulation reward can be made; and pharmacological agents can be used to investigate the neurotransmitters and transmitter-specific neurons involved in brain-stimulation reward. These different approaches to the neural basis of brain-stimulation reward, as described in the following, indicate that neural structures involved are present along the general course of the medial forebrain bundle. However, the critical structures and neural mechanisms for brain stimulation have so far not been determined.

Self-stimulation Sites

One group of self-stimulation sites extends along the medial forebrain bundle (MFB), from the ventral tegmental area, through the posterior and lateral hypothalamus and the preoptic area to the septal region. These sites continue further anterior, because self-stimulation can be obtained in the prefrontal cortex in the rat (Routtenberg & Sloan, 1972; Rolls & Cooper, 1974a,b) and in the orbitofrontal cortex in the monkey (Rolls, 1975). They may also continue posteriorly, because self-stimulation can be obtained in certain regions of the brainstem. Some of these sites are near the cells of origin of the catecholamine-containing (e.g., dopaminergic, noradenergic) cell bodies in the brainstem or near their fiber pathways, which ascend in the general region of the MFB (see Figure 2.24; and for review of self-stimulation, see German & Bowden, 1974; Mogenson & Phillips, 1976). Self-stimulation occurs in the region of the noradenergic cell group A6 (Crow, 1972; Ritter & Stein, 1972) and along the course of its ascending fibers in the dorsal ascending noradenergic bundle (Figure 8.5). It has been reported that self-stimulation does not occur in the noradrenergic cell groups A1, A2, and A5 (Clavier & Routtenberg, 1974) but does occur in the region of the ventral ascending noradrenergic bundle, perhaps because of activation of fibers from noradrenergic cell groups A6 and A7 (Ritter & Stein, 1974). Self-stimulation also occurs in the region of the dopaminergic cell group A9 (substantia nigra), in the nigrostriatal pathway, and in the region of the dopaminergic cell group A10 and the mesolimbic dopamine pathway (see German & Bowden, 1974). Some of these ascending catecholamine-containing pathways as well as other ascending and descending pathways are probably activated during self-stimulation of sites along the general course of the medial forebrain bundle. However, the results do not prove that noradrenergic and/or dopaminergic pathways are crucial for brain-stimulation reward because other neural pathways might also be activated. For example, self-stimulation of the locus coeruleus could be due to activation of ''taste neurons'' in this region (Norgren & Leonard, 1973; Norgren, 1974) or of the trigeminal nucleus.

Self-stimulation behavior at this group of sites in or near the MFB is typically rapid, may be accompanied by hyperactivity, and at some sites can be modulated by hunger, thirst, and other drives. For example, self-stimulation of the lateral hypothalamus may be potentiated by food deprivation (see Hoebel, 1969), of the posterior hypothalamus and preoptic area by sex hormones (Caggiula, 1970; Madlafousek et al., 1970), and of the preoptic area by temperature (Wagener, 1973). In addition, eating, drinking, and other stimulus-bound motivational be-

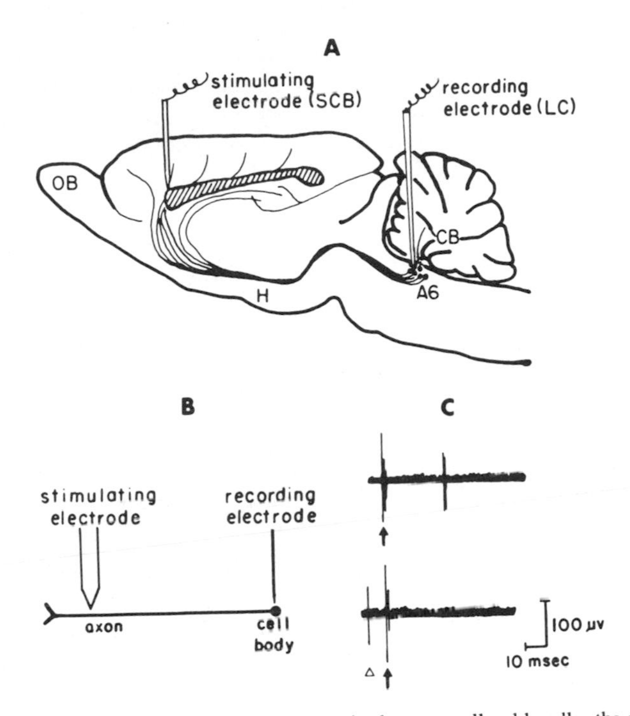

FIG. 8.5 A. Rats self-stimulated with electrodes in the supracallosal bundle, the rostral extension of the noradrenergic fibers from the locus coeruleus (see Figure 2.24). That noradrenergic axons of the LC neurons were being stimulated was shown in subsequent electrophysiological recording experiments with the animal anesthetized with urethane. A microelectrode was lowered to the locus coeruleus, and action potentials from single neurons were recorded. Stimuli were then administered through the electrode in the supracallosal bundle, and action potentials were recorded from the locus coeruleus (see B). The constant latency of the action potentials (top record in C) and the cancellation of the action potential when a spontaneous action potential occurred (shown by triangle in bottom record in C) demonstrated that the electrical stimulus initiated action potentials in the axon that traveled toward the cell body in the locus coeruleus (After Takigawa & Mogenson, 1977). It is concluded from these experiments that fibers from neurons in the locus coeruleus were being activated when the supracallosal bundle was stimulated.

haviors, may be elicited by stimulation of these sites (e.g., Mogenson & Stevenson, 1966). Stimulation at this group of sites may produce a specific reward similar to, for example, food or water, in that a rat may prefer stimulation at one site when hungry and at a different site when thirsty (Gallistel & Beagley, 1971). These findings are consistent with evidence that structures along the general course of the MFB are associated with physiological regulations and with ingestive, thermoregulatory, and other behaviors (see Chapters 3, 4, and 5).

A second main group of self-stimulation sites is in limbic and related structures (see Rolls, 1975), for example in the amygdala, hippocampus, entorhinal, retrosplenial, and cingulate cortex, as well as the septal nuclei and nucleus accumbens septi. Self-stimulation at these sites may be slower than at the sites near the

MFB, is typically not accompanied by hyperactivity, is not usually at sites from which eating and drinking or other stimulus-bound behaviors are elicited (Olds, 1965; Rolls, 1975), and is not modulated by drives such as hunger and thirst. It should be noted, however, that there are major connections between these limbic self-stimulation sites and the self-stimulation sites along the general course of the MFB.

Self-stimulation can occur at other sites, for example, the olfactory bulb (Phillips & Mogenson, 1969; Phillips, 1970), the subfornical organ (Robertson et al., 1976) and the central gray of the midbrain (Cooper & Taylor, 1967), where, although the stimulation may produce analgesia, there is no consistent relation between reward sites and sites at which analgesia is produced (Mayer & Liebeskind, 1974).

Stimulation at the majority of other brain sites, for example in much of the neocortex, thalamus, and cerebellum, is neutral (see Olds, 1961; Routtenberg & Malsbury, 1969), and stimulation in a medial periventricular system, which includes the medial hypothalamus (see Olds & Olds, 1965), and in parts of the reticular formation (Keene & Casey, 1970; Keene, 1973) is aversive.

Thus self-stimulation sites are found along the general course of the MFB as it traverses the hypothalamus and from limbic structures that are activated by this stimulation. The effect of lesions that damage some of these structures is discussed in the following.

Effects of Lesions on Self-stimulation

Lesions that produce damage in the general region of the MFB can disrupt self-stimulation, particularly if the lesions are caudal to the self-stimulation site. The critical damage is not necessarily to the entire MFB but could be to fiber pathways such as dopamine- or noradrenaline-containing pathways that accompany the MFB. In studies of this type, Olds and Olds (1969) and Boyd and Gardner (1967) have shown that lateral or posterior hypothalamic self-stimulation rate is decreased by small lesions in or near the MFB both anterior (in the preoptic area) and posterior (near the interpeduncular nucleus of the midbrain) to the stimulating electrode. A posterior lesion produces the greater decrease in rate, and in contrast to an anterior lesion, there is little recovery of rate over the few days following the lesion. In both studies, ipsilateral lesions were effective, but contralateral lesions were not. Therefore it appears that the organization of the stimulated system is unilateral between the anterior commissure and the ventral tegmental area of Tsai. These experiments receive support from the effects of xylocaine, a local anesthetic, which is injected into sites anterior and posterior to MFB self-stimulation sites (Stein, 1969), which similarly decrease self-stimulation rates.

Lesions that damage limbic sites may produce temporary, but do not usually produce permanent, disruption of self-stimulation at more caudal sites and may even facilitate self-stimulation at more caudal sites (see Rolls, 1975). For example, bilateral lesions to the amygdala or sulcal prefrontal cortex in the rat temporarily disrupt but do not permanently abolish lateral hypothalamic self-stimulation. Thus evidence from lesion experiments shows that these structures may be involved in but are not essential for self-stimulation. The view that these structures may modulate reward because they are normally involved in learning about reward is developed in the following subsection. It is of interest that if a suitably simple response (e.g., tail-raising) is chosen, hypothalamic self-stimulation can still occur when most of the cortex has been removed (Huston & Borbely, 1973). This suggests that the cortex is not essential for basic reward processes.

Thus the available lesion evidence suggests that structures in or caudal to the hypothalamus are critical and that more rostral areas are not essential for brain-stimulation reward. Research in this area would be greatly facilitated if there were a measure of reward that was specific to reward and not affected by such factors as arousal and motor capacity.

Fiber Pathways Related to Self-Stimulation

By making small lesions at self-stimulation sites and tracing the resulting orthograde fiber degeneration, it is possible to draw inferences about the nature of neural systems involved in brain-stimulation reward. Routtenberg (1971) demonstrated that degeneration from rodent prefrontal cortex reward sites occupies the most medial edge of the internal capsule in a course toward the midbrain. It is in this area in the monkey, about 1 mm dorsolateral to the MFB, that Routtenberg and co-workers (1971) obtained good self-stimulation. It may be, therefore, that fibers traveling caudally from the frontal cortex along the general course of the MFB are involved in brain-stimulation reward.

A problem with this type of study is that there is no assurance that the degenerating fibers are actually fired during self-stimulation or, if they are, that their activation is essential for brain-stimulation reward. Also, the lesion at the self-stimulation site may destroy neurons that are not functionally related to the self-stimulation, and no check is possible. Nevertheless, experiments performed with the lesion-degeneration technique are useful in providing an indication about possible reward pathways.

The Activity of Single Neurons During Brain-Stimulation Reward and Natural Reward

The evidence described on self-stimulation sites, the effects of lesions on self-stimulation, and fiber degeneration studies, suggest that structures close to or associated with the MFB are involved in self-stimulation. To find out which structures are activated during the self-stimulation, and to determine whether these structures are involved in natural reward processes such as those that occur when food is given to a hungry animal, recordings can be made from single neurons during brain-stimulation reward.

Recordings have been made in this type of experiment from neurons in the hypothalamus, because this structure is connected to many of the pathways implicated in self-stimulation, and also from neurons in, for example, the prefrontal cortex and the amygdala, which send fibers to or through the self-stimulation sites in the general region of the MFB. In general, electrical stimulation at the lowest current that produces reward is applied to a reward site, and recordings are made from single neurons in different parts of the brain to determine how the neurons are affected by the stimulation. If a single unit is fired with a short, fixed latency by the stimulus pulses, it is probably directly excited by the stimulation. This means that its axon must pass under both the recording and stimulating electrodes. A further test, for collision, can be applied to determine whether the direct excitation is antidromic or orthodromic (see Figure 8.5). If a single unit is fired with a longer variable latency by the stimulus pulses, and if collision cannot be demonstrated, the neurons must be transsynaptically activated by the stimulation. If the firing rate of a neuron is affected by the stimulation yet the action potentials are not in phase with the stimulus pulses, it is likely that polysynaptic activation is occurring. Using these methods, it is possible to show which neural pathways are activated by the rewarding stimulation across synapses through the CNS. Provided that the stimulating current is kept at or below the value that was sufficient for self-stimulation, it can be concluded that the activated neurons were activated by self-stimulation. The specificity of the activation with respect to reward can be assessed by using implanted electrodes that do not support self-stimulation. Also, it should be stressed that in most areas of the brain, single neurons are not activated by brain-stimulation reward.

After neural systems activated in self-stimulation have been traced, further tests can be performed to determine their role in brain-stimulation reward as well as in naturally rewarded behavior. In these further tests, the firing rates of the neurons can be recorded while the animal performs a wide variety of behavioral tasks in order to determine whether the activity of a neuron activated by the brain-stimulation reward is related in some way to natural reward or is instead related to some other type of behavior such as motor movements. In this way it is possible to assess which neural system activated by the electrical stimulation mediates the reward. Of course, it is also true that electrical stimulation that produces reward is a useful tool when searching in the brain for systems concerned with natural reward.

The hypothalamic region. The activity of single neurons in the hypothalamus is altered during self-stimulation of a number of different sites. The influence is usually transsynaptic and may consist of excitation or of inhibition. In the rat, units in the preoptic area and hypothalamus are activated in self-stimulation of the lateral and posterolateral hypothalamus (Rolls, 1971b,

1974; Ito & Olds, 1971; Ito, 1972; Olds, 1974) and the nucleus accumbens (Rolls, 1974), and the latency of the activation is usually in the range of 1–15 msec from the stimulation pulse. In the rhesus monkey and squirrel monkey, units in the lateral hypothalamus and as far forward as the preoptic area are activated in self-stimulation of many different sites (e.g., the orbitofrontal cortex, amygdala, nucleus accumbens, lateral hypothalamus, and the region of the locus coeruleus; see Rolls, 1975). Because single neurons in the lateral hypothalamus are activated by self-stimulation of a number of different sites, this region represents an area of neurophysiological convergence for these sites and thus may have a crucial role in reward.

In the hypothalamus it has been possible to investigate the function of neurons activated by brain-stimulation reward in normal behavior (Rolls, 1975). It has been found in the monkey that some of the hypothalamic neurons are activated when the monkey receives natural rewards, such as water for a thirsty animal and food for a hungry animal. The neuron illustrated in Figure 4.10A decreased its firing rate when glucose was on the tongue but not when other solutions or foods (e.g., water, sodium chloride, or peanuts) were given to the monkey. The firing of this type of hypothalamic unit does not merely respond to taste but appears to be activated when a particular taste is rewarding, as shown by the following evidence. If the monkey was hungry (e.g., after 24-hour food deprivation), then this type of neuron has a large response when food is tasted, but as the animal becomes satiated, the response of the neuron to the stimulus diminishes toward zero. Thus the response of this type of neuron could be equivalent to food or water taste reward in that these neurons do not respond when food or water is not rewarding. It should be noted that hunger does not affect the spontaneous baseline firing rates of these units but does affect their responsiveness to food. Therefore, these neurons could be associated with food reward and not hunger.

Another class of hypothalamic neuron that responds in association with food or water reward responds when food is seen. An example of a hypothalamic unit that responded when the monkey looked at a peanut is shown in Figure 4.10B. The responses of another unit to a wide variety of visual stimuli are shown in Figure 8.6. The unit responded to the monkey's most preferred foods but not to the sight of nonfood or aversive objects or in relation to motor movements, somatosensory stimulation, or swallowing. It was therefore suggested that this was a neuron that responded to the sight of food (Rolls et al., 1976). The response of this type of neuron is associated with the sight of a food reward, because the response is great when the animal is hungry and disappears when the animal is satiated (Burton et al., 1976); there is no response when nonfood objects that the animal wants (i.e., is willing to work for) are shown; and the response is usually greatest to the sight of a food for which the animal has a high preference. These experiments thus show that there are neurons in the hypothalamus that could signal the taste or sight of food reward and that are activated by brain-stimulation reward. This suggests that electrical stimulation of the brain can produce reward, because it mimics the effects of specific natural

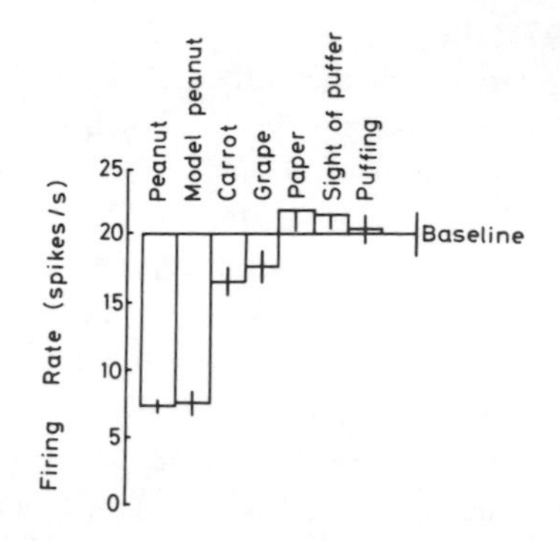

FIG. 8.6 The discharge rate of a neuron in the hypothalamus of a monkey in response to the sight of food and to nonfood stimuli. The neuron responded to the sight of a peanut when the animal was hungry. (From Rolls, Burton, & Mora, 1976.)

rewards. For the hypothalamic units described, the specific natural rewards apparently mimicked by the electrical stimulation at different brain sites were food and water rewards, but it seems likely that at some other self-stimulation sites, other types of natural reward are mimicked. There is a close correlation between the firing of reward neurons and subsequent behavior, and it is thus possible that the reward neurons send signals to a neural system for the motor control of the behavior produced. Feeding behavior and drinking behavior may be controlled by a neural mechanism of this type in the lateral hypothalamus (see Figure 4.11).

To further test the suggestion that the hypothalamic neurons activated by brain-stimulation and by natural rewards do mediate reward, electrical stimulation was applied through the recording microelectrode. Self-stimulation occurred at a low threshold as long as the microelectrode was in this region, but higher currents were required if the microelectrode was moved, and no self-stimulation at all was found if the microelectrode was raised above the region of these neurons (Figure 8.7). This demonstration that self-stimulation occurred in the region that contains neurons activated by brain-stimulation and natural rewards provides further evidence that these neurons mediate reward. It was also possible to show that the electrical stimulation in the region of these neurons mimicked food reward, because self-stimulation through the microelectrode in the region of those neurons was attenuated when the animal was satiated by feeding. Further evidence consistent with the hypothesis that these neurons can mediate reward is that self-stimulation can be attenuated by lesions in this region (see the foregoing).

From the evidence just reviewed, it may be concluded that there are neurons in the hypothalamus that are activated both by natural rewards and by brain-stimulation reward of a number of different sites. Thus a sufficient explanation for self-stimulation at some sites may be that the stimulation excites food- and water-reward neurons. Stimulation at other sites may produce reward because it

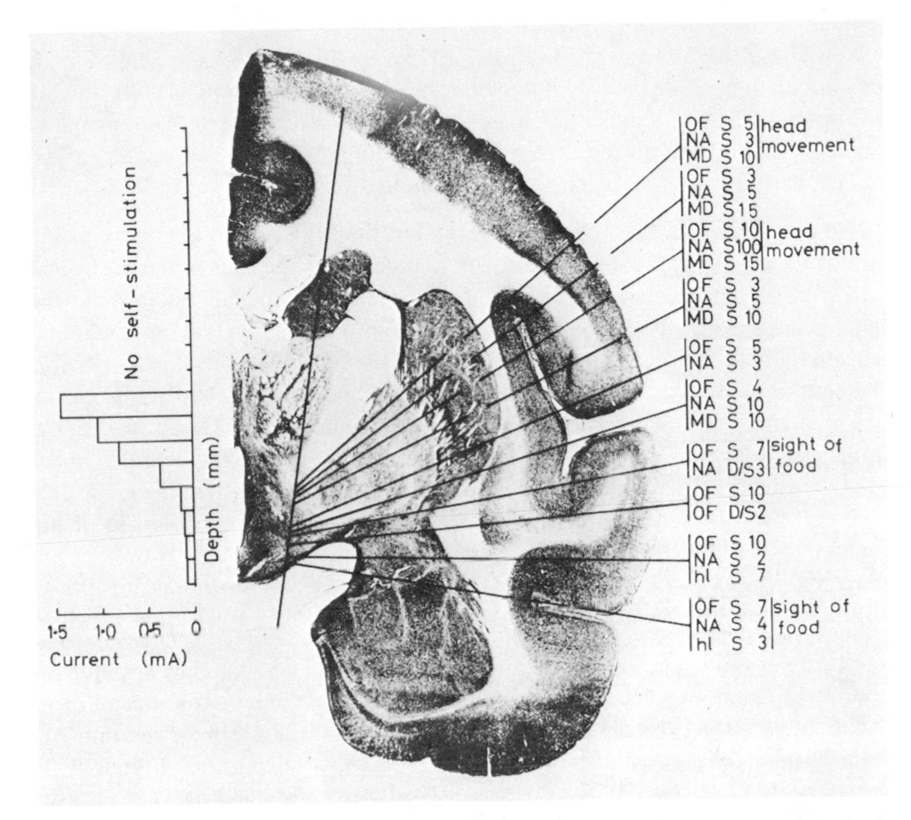

FIG. 8.7 A microelectrode track through the globus pallidus and hypothalamus of a monkey brain. Units activated by both brain-stimulation reward and the sight of food were found at the lower end of this track, in the hypothalamus. Units higher up the track in the globus pallidus were activated by brain-stimulation reward and also by head movements. Self-stimulation through the recording microelectrode occurred in the hypothalamic region, where cells fired to the sight of food and where the self-stimulation was more intense if the animal was hungry. (From Rolls, 1975.)

activates neurons concerned with other types of natural reward. Nevertheless, further investigation of the activity of neurons in the lateral hypothalamus during feeding is required in that the activity of the populations of neurons described here is closely linked to whether reward is given and therefore to the motor behavior of the animal. Because of the possibility that the activity of these units reflects rather than determines the behavior of the animal, it is necessary, as with all analyses of learning and motivation, to perform experiments to distinguish these possibilities. Such experiments might investigate the input and output connections of these neurons, and when in relation to the animal's behavioral response, these neurons are active (Rolls, 1975; 1976).

The amygdala. As indicated previously, self-stimulation is obtained with electrodes in the amygdala, and neurons in the basolateral amygdala are activated by self-stimulation of the MFB (Rolls, 1975). Lesions of the amygdala impair self-stimulation of the MFB, but recovery of self-stimulation usually occurs within a few days. It seems that the amygdala is involved in, but is not essential for, brain-stimulation reward. We return to the amygdala later in this chapter and suggest how it might contribute to brain-stimulation reward.

The prefrontal cortex. Self-stimulation of the MFB and of lower brainstem sites as far caudal as the locus coeruleus also activates neurons in the prefrontal cortex (a region receiving projections from the dorsomedial nucleus of the thalamus; Leonard, 1969) of rats and monkeys. The activation of prefrontal neurons is both antidromic and transsynaptic (Rolls & Cooper, 1973; 1974b). The antidromic activation shows that there are cell bodies in the prefrontal cortex that send fibers through the MFB, where self-stimulation is obtained, and to the brainstem. A recent study using a retrograde technique for tracing neural connections (horseradish peroxidase procedure) has demonstrated that fibers project from cell bodies in the prefrontal cortex to the substantia nigra, the site of the dopaminergic nigrostriatal neurons (Bunney & Aghajanian, 1976). However, the prefrontal cortex, like the amygdala, is apparently not essential for brain-stimulation reward, because ablation of the prefrontal cortex as well as much of the other cerebral cortex does not prevent self-stimulation if an appropriately simple response such as tail movement is used (Huston & Borbely, 1973).

The brainstem. Neurons in the brainstem are activated by self-stimulation of some sites along the general course of the MFB in the rat and monkey (Routtenberg & Huang, 1968), but the activation of the majority of brainstem neurons is frequently associated with the elicitation of locomotor activity (Rolls & Kelly, 1972), with motor side-effects (Rolls, 1971b), and with arousal as measured by EEG desynchronization (Rolls, 1971b). Some neurons in the nucleus reticularis gigantocellularis that respond to stimulation at self-stimulation sites in the lateral hypothalamus also respond to deep pinch (Keene & Casey, 1970).

It should be noted that some neurons in the pons and medulla are directly activated by self-stimulation of hypothalamic sites, but they are not grouped particularly in the locus coeruleus (Rolls, 1975).

Brain-Stimulation Reward Investigated with Neuropharmacological Techniques

As noted in the foregoing, ascending noradrenergic and dopaminergic pathways travel along the MFB, recognized for a number of years as the "hot spot" for brain-stimulation reward (Olds, 1962), and could be activated during self-stimulation of this region and damaged by the lesions that disrupt self-stimulation. By using pharmacological agents that block or mimic the synaptic action of dopamine or noradrenaline (e.g., haloperidol is an antagonist or blocker of dopamine receptors, and apomorphine is an agonist) or that deplete dopamine or noradrenaline, it is possible to obtain evidence about the possible role of these neurotransmitters in brain-stimulation reward. The neuropharmacological approach is most effective when the chemical anatomy of the neural pathways is known (see Figure 2.24). Furthermore, the interpretation of experimental effects of drugs on self-stimulation of the brain is greatly assisted when the mechanisms for the biosynthesis and synaptic action of the neurotransmitters are known. These are major reasons why there has been a keen interest in recent years in the role of catecholamine-containing (dopaminergic, DA; and noradrenergic, NA) neurons in brain-stimulation reward.

There have been a number of studies showing that drugs that inhibit the synthesis of catecholamines, such as α-methyl-p-tyrosine, and drugs that deplete catecholamine stores, such as reserpine, produce a decrement in brain self-stimulation (e.g., Stein, 1962; Poschel & Ninteman, 1966). On the other hand, self-stimulation is facilitated by drugs that increase catecholamine levels, such as monoamine oxidase inhibitors, or that increase the synaptic release and block the reuptake of catecholamines, such as amphetamine (Horowitz et al., 1962; Stein, 1962; Olds, 1970). On the basis of such neuropharmacological evidence, Stein (1971) hypothesized that brain-stimulation reward is subserved by NA neurons.

Self-stimulation electrodes in the region of the MFB are close to the trajectory of ascending NA pathways, and according to Stein's noradrenergic hypothesis, brain-stimulation reward results because these NA neurons are activated. One group of NA fibers arises from the locus coeruleus (A6) in the pons and projects rostrally via the dorsal noradrenergic bundle to innervate the cerebral cortex and hippocampus (see Figures 2.24B and 8.5). Another group arises from other NA cell bodies in the brainstem (A1, A2, A5, A7) and ascend via the ventral NA bundle to innervate ventral telencephalic structures such as the hypothalamus and olfactory bulb. There have been reports of self-stimulation with electrodes in the locus coeruleus (Crow, 1972; Crow et al., 1972; Ritter & Stein, 1973; Cooper & Rolls, 1974), in the supracallosal bundle, a rostral extension of fibers from the locus coeruleus (see Figure 8.5), and in the ventral NA pathway (Ritter & Stein, 1974).

The noradrenergic hypothesis of reward has been further investigated in experiments in which the synthesis of noradrenaline but not of dopamine has been inhibited by drugs (such as disulfiram) that interfere with the enzyme dopamine-β-hydroxylase. Wise and Stein (1969) showed that disulfiram abolished self-stimulation. They also showed that the self-stimulation could be reinstated by administering noradrenaline intraventricularly to restore noradrenaline levels in the brain. This finding suggested that brain noradrenaline was necessary for self-stimulation to occur. However, Roll (1970) suggested that brain self-stimulation was disrupted because the disulfiram made the rats sedated, not necessarily because the disulfiram had interfered with a reward process. This view was supported by the finding that doses of disulfiram that attenuated self-stimulation produced a major attenuation of locomotor activity and were often lethal (Rolls et al., 1974). Thus the noradrenergic hypothesis of reward has not been proven because of the difficulty of showing that reward rather than some other function necessary for self-stimulation behavior is mediated by noradrenergic neurons. However, it should be noted that anatomical evidence of sites from which self-stimulation is obtained and the effects of pharmacological agents on self-stimulation are consistent with the noradrenergic hypothesis of reward (see also German & Bowden, 1974; Stein et al., 1976).

Self-stimulation has also been obtained from sites known to contain DA neurons, such as the substantia nigra (A9) (Routtenberg & Malsbury, 1969; Crow, 1972; Phillips & Fibiger, 1973), the ventral tegmental area (A10) (Dreese, 1966; Crow, 1972), the nucleus accumbens (Rolls, 1971a), and the prefrontal cortex (Rolls, 1975), implicating dopamine systems in brain-stimulation reward. A role for dopamine has also been suggested by pharmacological studies. Treatments such as α-methyl-p-tyrosine and reserpine, which deplete the brain of dopamine, and chlorpromazine and haloperidol, which block dopamine receptors, attenuate self-stimulation. Treatments such as amphetamine, which facilitate the release of dopamine, facilitate self-stimulation. Because all these treatments affect noradrenaline as well as dopamine, it is of interest that pimozide (and also haloperidol and spiroperidol), which blocks dopamine but not noradrenaline receptors, attenuates self-stimulation (Wauquier & Niemegeers, 1972) and that this impairment of self-stimulation is specific, at least with respect to arousal (Rolls et al., 1974).

The attenuation of self-stimulation by dopamine antagonists could be due to motor effects and not because of a specific effect on reward mechanisms. The dopaminergic nigrostriatal pathway (A9) is part of the extrapyramidal motor system, and impairment of this pathway leads to Parkinson's disease in man, which includes a lack of voluntary behavior, and to catalepsy (Hornykiewicz, 1973). It has been reported that a dopamine receptor blockade produced a much greater impairment of the lever-press response to obtain food or water than of simply licking a tube for water or chewing food (Rolls et al., 1974). Because a number of behaviors, including avoidance responding, are impaired by dopamine-receptor blockade, the effects of such drugs are not specific to self-stimulation (Wauquier & Niemegeers, 1972).

The lick response, which is not disrupted by dopamine receptor blockade (Rolls et al., 1974), has been used to obtain brain-stimulation reward. A deficit in self-stimulation using this operant response resulted from the administration of a dopamine-receptor blocking agent, suggesting that dopamine receptors might be involved in reward and not just in the motor responses required to bring about the brain-stimulation reward (Mora et al., 1975).

With unilateral intracranial injections of dopamine-receptor blocking agents it was possible to show that injections into the nucleus accumbens, known to receive an input from the mesolimbic dopaminergic system (A10), produced a severe impairment of self-stimulation with only a mild motor impairment, whereas injections into the caudate nucleus, which receives an input from the nigrostriatal dopaminergic system (A9), produced a smaller attenuation of self-stimulation but a larger motor impairment (Mora et al., 1975). These findings suggest that the mesolimbic DA system may be related to brain-stimulation reward and that there may be a role for dopamine in reward that is distinct from its role in the extrapyramidal motor system. This view is supported by the finding that rats can learn to self-administer apomorphine, which is a dopamine-receptor agonist (Baxter et al., 1974), and that the rate of self-administration of amphetamine can be increased, not decreased, by pimozide, as if the rats were attempting to maintain activation of a dopaminergic system by amphetamine (Yokel & Wise, 1975).

In concluding this section, it should be emphasized that the recent evidence implicating NA and DA systems in brain-stimulation reward does not mean that other neurotransmitters are not involved. Any attempt to explain brain-stimulation reward exclusively in terms of NA and DA pathways—a modern form of phrenology (''chemical'' phrenology)—should be avoided. The goal is to elucidate the entire neural substrates that subserve brain self-stimulation behavior, and other components, including neurons with transmitters other than dopamine and noradrenaline, are no doubt involved. A number of neuropharmacological studies have failed to provide convincing evidence that cholinergic and serotonergic systems have an important role in brain-stimulation reward (for review, see Mogenson & Phillips, 1976). For this reason a recent report that self-stimulation of the subfornical organ is attenuated by atropine sulphate and methysergide, and cholinergic and serotonergic blocking agents, respectively, is of interest (Robertson et al., 1976). The subfornical organ contains cholinergic and serotonergic fibers, but there is no evidence of DA and NA fibers. However, the possibility of transsynaptic activation of DA or NA neurons by cholinergic or serotonergic fibers from the subfornical organ must be considered.

Some Final Speculative Comments About Brain-Stimulation Reward

Brain-stimulation reward mimics natural rewards in that at some sites, it is facilitated by drives such as hunger and thirst and activates neurons that discharge in response to natural rewards such as food and water. This is consistent with and supports the view, presented in Chapter 4, that feeding and other motivated behaviors occur when external sensory inputs (such as the sight, smell, and taste of food) become rewarding when drive (e.g., hunger) is present.[4] The

[4]For further discussion of the view that reinforcement is associated with an interaction of incentive stimuli (e.g., taste) and a ''drive'' or central motive state (''hunger''), see Mogenson and Cioe (1976, pp. 582–583). The hypothesis is that brain-stimulation reward depends on the activation of neural pathways that subserve external, incentive stimuli and internal ''drive'' stimuli.

observation that sites in the lateral hypothalamus at which brain-stimulation reward is modulated by hunger provide information on where this type of control of motivated behavior occurs. At other sites brain-stimulation reward may mimic more general natural rewards (e.g., relaxation or pleasure in man), as indicated by the subjective reports of electrical stimulation of certain sites in humans.

Self-stimulation of the amygdala and prefrontal cortex is of particular interest in relation to evidence implicating these neural structures in learning. Bilateral lesions of the amygdala is followed by deficits in behavioral tests in which animals have to learn the connection between environmental stimuli and rewards and punishments (Jones & Mishkin, 1972). For example, lesioned animals do not form conditioned emotional responses (CER) normally, do not learn to avoid an aversive stimulus normally, and do not learn normally to associate an object with reward. If the amygdala is involved in learning about rewards and punishments, in that it makes a previously neutral stimulus rewarding or punishing (stimulus-reinforcement associations), then self-stimulation of the amygdala may occur, because electrical stimulation excites neural pathways that are related to reward (Rolls, 1975).

Damage to the sulcal prefrontal cortex in the rat or to the orbitofrontal cortex in the monkey leads to increased resistance to extinction, perseveration to previously rewarded stimuli in discrimination reversal tasks, and other deficits that suggest that animals cannot ''unlearn'' the connection between stimuli and rewards (or punishments) (Jones & Mishkin, 1972; Butter, 1969; Rolls, 1975). If this cortex is involved in breaking previously learned associations with reward and punishment or is involved in the execution of behavior under these conditions (see Rolls, 1975, p. 65), then this might explain why this cortical region has connections with reward sites in the brain.

The involvement of the prefrontal cortex in brain-stimulation reward is also of special interest in relation to evidence that this part of the brain is involved in complex cognitive processes related to emotion and receives a dopaminergic input, which may be of crucial importance in the self-stimulation of this site. The association of self-stimulation of the prefrontal cortex with the activation of dopaminergic neurons that project to this region from the midbrain may be relevant to the dopamine hypothesis of schizophrenia. The studies, considered in previous sections, that involve electrophysiological and electrical stimulation techniques have provided new insights and new approaches to the investigation of brain-stimulation reward. These studies suggest that understanding the neural substrates of brain-stimulation reward may help to elucidate the neural mechanisms responsible for disorders of affect and for schizophrenia, both of which involve abnormal reactions to emotional (i.e., rewarding or punishing) stimuli.

There is another interpretation of brain-stimulation reward, recently proposed by Milner (1977), that should be considered briefly. It focuses on the response or motor side of the CNS (in contrast to the emphasis on the sensory side for the hypotheses considered previously) and, from this point of view, suggests a role for dopaminergic and noradrenergic pathways implicated in brain-stimulation reward.

Milner assumes that preprogrammed neural circuits for biting, gnawing, swallowing, head turning, etc., represented in the brainstem, are inhibited by a more rostral mechanism of the CNS. This response-inhibitory mechanism is itself inhibited when motor responses occur. In other words, the occurrence of motor responses depends on disinhibition. Milner reviews experimental evidence that the nigrostriatal pathway and dorsal noradrenergic pathway, which project to the neostriatum, hippocampus, and cortex (see Figure 2.24), inhibit neurons in these forebrain structures. His hypothesis is that the basis of response reinforcement by natural rewards is that these DA and NA pathways are activated to disinhibit the motor system. Direct electrical stimulation of these DA and NA pathways has a similar disinhibitory effect and is, according to Milner, the basis for brain-stimulation reward.

This theoretical proposal is considered in detail by Mogenson and Cioe (1976), who conclude their discussion as follows:

> Milner's ingenious hypothesis suggests a mechanism common to both conventional reinforcement and central reinforcement. Furthermore, it provides a role for drive and incentive stimuli as activators of the catecholamine ''reward'' neurons and at the same time links these neurons to neural systems for motor control. The hypothesis appears to provide a fruitful integration and synthesis of the previous hypotheses of central reinforcement that stressed sensory input (Deutsch and Howarth, 1963; Pfaffmann, 1960) and response elicitation (Glickman and Schiff, 1967) [p. 588].

SUMMARY

Electrical stimulation of certain sites in the hypothalamus, limbic system, and associated neural structures, is rewarding in the sense that animals will work to obtain the brain-stimulation. Rats will run along a straight alley or through a maze, press a lever, or make other operant responses to stimulate their brains. Brain-stimulation reward, as demonstrated in this way, has been observed in a number of species including primates and man.

The characteristics of the behavioral responses that the animal makes to obtain brain-stimulation reward are similar to the appetitive responses made to obtain food, water, or other conventional rewards. The behavior is goal-directed and typically very persistent. The rate of brain self-stimulation varies in a circadian fashion, and because of the strength of brain-stimulation reward, it ''competes'' very effectively with feeding, drinking, and other behaviors. It is suggested in this chapter that brain-stimulation reward involves the activation of neural structures and pathways that subserve conventional rewards. Brain-stimulation rewards may be sometimes more intense than conventional rewards in ''motivational time-sharing,'' because the central pathways are being activated more effectively and/or because the usual satiety influences do not occur or occur only minimally.

There is a great deal of evidence from recent anatomical and neuropharmacological studies that brain-stimulation reward is associated with electrical stimulation of catecholaminergic pathways that originate in the midbrain and lower brainstem and project rostrally to the hypothalamus, the striatum, and other forebrain structures. A controversy has developed concerning the possibilities that brain-stimulation reward depends exclusively on the activation of either noradrenergic neurons or dopaminergic neurons. It seems likely that both noradrenergic and dopaminergic, and possibly other transmitter-specific neurons, can mediate self-stimulation depending on the self-stimulation site. Furthermore, an understanding of the neural mechanisms of brain-stimulation reward depends on the elucidation of the relation of noradrenergic and dopaminergic neurons with stimuli that normally provide reward, such as food and water, and with neural systems for the motor control of behavior. Comprehensive reviews of these topics are now available (see Mogenson & Phillips, 1976; Rolls, 1975; Wauquier & Rolls, 1976).

9
Emotional Behaviors

By emotional behaviors we mean fearful, aggressive, and affectional reactions. For example: In Figure 1.1 is a picture taken from Charles Darwin (1872) of aggressive behavior in a dog and also a picture of a "rage" or "threat" reaction in a baboon; the mountain climber shown in the same figure may on occasion experience fear and ultimately, when he reaches the summit, elation; a bear following hibernation or a child following an afternoon nap may be very irritable. Emotional reactions have the prominent characteristics of the behaviors considered in previous chapters; the distinctive features of emotional behaviors are the intensity and persistence of the behavioral reactions and the strong subjective or affective component experienced in man. It is for these reasons that emotional reactions are sometimes considered unique and a special kind of motivated behavior. However, the intensity and affective component of fear, anger, or love—traditionally regarded as emotions—is only a matter of degree; prolonged hunger or thirst or sleep deprivation may also lead to an intense affective experience as well as a behavioral response.

Emotional reactions are also influenced by the same factors that influence the more traditional motivated behaviors, considered in previous chapters. These factors are discussed before going on to consider what is known about the neural substrates of emotional behaviors.

CHARACTERISTICS OF EMOTIONAL BEHAVIOR

A number of authors have considered emotional reactions as a type of motivated behavior. This seems justified, because emotional reactions or behaviors have, to a greater or lesser degree, the same characteristics as motivated behaviors. Emotional behavior may be distinctive because of the prominence of autonomic responses, particularly those controlled by the sympathetic nervous system, and the greater participation of hormones from the adrenal gland (adrenaline and noradrenaline from the adrenal medulla and corticosteroids from the adrenal cortex) that contribute to the intensity of the affective component. In this section we first consider the characteristics of emotional behavior and then discuss emotional reactions from the viewpoint of the integration of autonomic and endocrine as well as skeletomotor or behavioral responses.

Emotional Reaction Considered as Motivated Behavior

Emotional reactions, like the motivated behaviors considered in previous chapters, are frequently purposive and adaptive. A cat that suddenly encounters a threatening, barking dog displays piloerection of hair, dilation of the pupils, increased heart rate, and increased blood pressure, and it may viciously attack the dog and drive it away. A "frightened" squirrel runs briskly across the yard and climbs a tree to escape from a dog. There are numerous examples of this sort that illustrate, as suggested by Charles Darwin (1872) more than a century ago, that emotional behavior is purposive in the sense that it is adaptive to the circumstances of the animal (Figure 9.1).

On the other hand, a lady finding herself in the path of a rapidly approaching car may "freeze in her tracks" with fright. A bereaved husband may be so distraught with grief following the death of his wife that he cannot function at work or look after the house and children. There was a controversy some years ago as to whether emotions are organizing or disorganizing (Leeper, 1948). They may be either:[1] in some cases strong emotions may be disruptive, and in others they may be organizing or purposive, depending on such factors as arousal level (Hebb, 1955) and the effectiveness of cognitive constraints on the emotional reaction (Hebb, 1966). Hebb proposed that the adequacy and effectiveness of behavioral performance is related to the level of arousal and the nature of the task; there is an optimal level of arousal for effective or adaptive performance, and behavior may deteriorate when arousal is too low or when it is too high, such as during extreme boredom or the terror of a novice soldier in battle for the first time (Figure 9.2).

Emotional reactions, like other motivated behaviors, are characterized by intensity and persistence: Anger and aggressive behavior or fear-like behavior, unlike simple skeletomotor reflex responses, may last for minutes or even hours; hatred, love, and affection in man may persist for years. Strong emotional reactions are typically more intense and persistent than the motivated behaviors considered in previous chapters. This is related to the greater involvement of

[1]Tolman (1923) emphasized the adaptive and purposive nature of emotional reactions. Earlier, Darwin (1872) had recognized that emotional behaviors could be maladaptive as well as adaptive. A definitive article on this issue was written by Leeper (1948), and Hebb (1955) proposed a model that accounted for emotional reactions being both organizing and disorganizing.

FIG. 9.1 Examples of emotional behavior in the dog. *Left*—an aggressive or hostile reaction as it approaches another dog. *Right*—a submissive reaction to another dog. (From Darwin, 1872/1965.)

autonomic and endocrine components and to a greater degree of activation or arousal.

Emotional behaviors may be periodic, as in the "moody" person or the "manic-depressive" individual, but they usually do not show the regular rhythmicity of feeding or sexual behavior. If behavioral arousal is considered an important dimension of emotions and of emotional expression, it should be noted that behavioral arousal does fluctuate in a regular circadian manner (Moruzzi, 1969). However, in general, the characteristic of periodicity is not as prominent for emotional behaviors as for a number of the other well-known motivated behaviors.

Emotional behaviors are also similar to the so-called motivated behaviors in terms of "motivational time-sharing." Because of their intensity and persistence, emotional reactions compete effectively with motivated behaviors that normally have a high priority. The frightened animal may ignore food and water or not seek a shelter from the cold, and the "love-sick" teenager may be disinterested in meals or other normal pursuits.

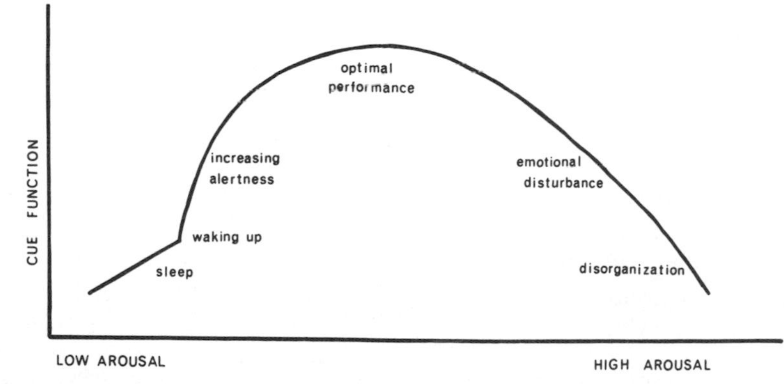

FIG. 9.2 Cue function or the reaction to environmental stimuli is related to arousal level according to an inverted U-shape function. When arousal level is too low or too high, performance suffers; there is an optimal level of arousal for good performance of a variety of responses. (From Hebb, 1966.)

Autonomic, Endocrine, and Skeletomotor Responses in Emotional Reaction

In characterizing emotional reactions, it is necessary to include autonomic and endocrine responses as well as skeletomotor responses (Figure 9.3). For example, aggressive behavior is accompanied by increased heart rate and blood pressure and other responses under the control of the sympathetic nervous system. There may be increased blood flow to skeletal muscles mediated in the cat by cholinergic vasodilator fibers (Abrahams et al., 1964). At the same time, there is activation of the anterior pituitary-adrenal cortical axis and release of catecholamines from the adrenal medulla (Levi, 1975). By raising blood sugar levels, increasing metabolic rate, and increasing blood flow to active muscles, these autonomic and endocrine responses support the "fight or flight" responses of the animal in adapting to an "emergency situation" (Cannon, 1929).

The hypothalamus, as indicated in more detail in a later section (see p. 250), has an important integrative role for the coordination of autonomic, endocrine, and skeletomotor (behavioral) components of emotional reactions (see Figure 9.3). This is clearly demonstrated in the classical experiments of Hess (1954), in which aggressive responses in cats were elicited by electrical stimulation of the hypothalamus. These experiments, which have been extended by other investigators (see Figure 9.8), have been summarized by Folkow and Neil (1971) as follows:

> These hypothalamic response patterns some of which have relay stations of considerable complexity at mesencephalic and lower brainstem levels as well, usually involve somatomotor and hormonal components as well as autonomic pathways so linked together that they adjust the organism as a whole to face particular situations in an appropriate manner. This necessitates an integrated influence, not only of interoreceptors and exteroreceptors, but telereceptors as well, in turn calling for intimate interaction between cortical hypothalamic, lower bulbar and spinal structures via excitatory and inhibitory connections and feedback loops, the entire system of autonomic "centers" works as a well-coordinated unit [p. 342].

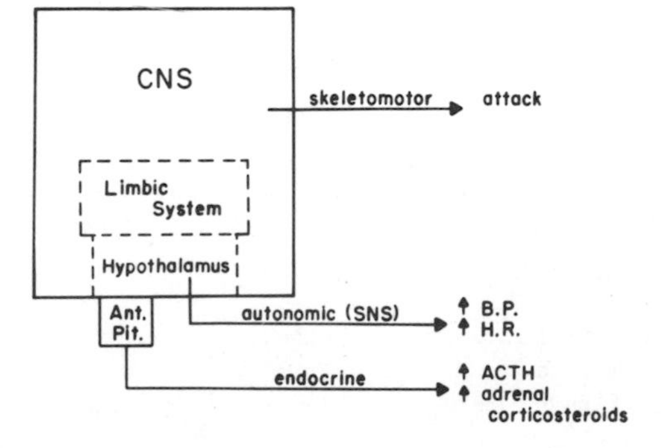

FIG. 9.3 The responses associated with an emotional reaction such as fear or aggression are controlled by the autonomic nervous system (and especially the sympathetic nervous system—SNS), the endocrine system, and skeletomotor system. It is the function of the hypothalamus in association with limbic forebrain structures to coordinate these three types of responses.

FACTORS THAT CONTRIBUTE TO EMOTIONAL BEHAVIOR

Emotional reactions are influenced by some of the same factors that influence the more traditional motivated behaviors. The contribution of these factors to emotional behaviors is considered in this section.

Deficit or Homeostatic Signals

Emotional behaviors are not directly initiated by homeostatic, deficit signals, but such signals may sensitize the animal to certain emotion-provoking stimuli. For example, prolonged hunger or exposure to a cold or hot ambient temperature may lead to increased irritability. This may result, as indicated later, because deficit or homeostatic signals increase the level of behavioral arousal and the reactivity of the person or animal to environmental stimuli. Pain, which may be initiated internally or by an external source, frequently makes the individual more emotionally responsive (Melzack, 1973).

Hormones

Hormones have been shown to act on the CNS and to influence temperament and mood and the reactivity of the animal or person to emotion-provoking stimuli. Hormones may affect the development of neural systems, as in the case of testosterone; they may excite existing neural systems, as in the case of estrogens that induce sexual receptivity in certain species; or they may influence the general excitability of the CNS, as in the case of thyroid and adrenal cortical hormones.

A person with thyroid deficiency exhibits marked changes in temperament and mood and may be depressed, generally unhappy, and dissatisfied with life. The hyperthyroid patient, on the other hand, is irritable and overly sensitive to emotion-provoking stimuli. Ovarian hormones, besides initiating sexual receptivity, also influence emotional reactivity. Women may be irritable and emotionally upset at the beginning of menstruation. The menopause, when the secretion of the estrogens is reduced, may also be a period of emotional turmoil. Testosterone, besides contributing to sexual development and sexual arousal, has effects on emotional behavior. For example, aggressive behavior may fail to occur in the castrated male, especially if castration is performed before sexual maturity. As noted recently by Moyer (1976), castration ''can convert the raging bull into a gentle steer [p. 59]'' and reduce the ''number of asocial acts by individuals convicted of violent sex crimes [p. 187].''

The release of certain hormones, such as those from the adrenal medulla and adrenal cortex, is associated with stress and strong emotional reactions (Frankenhaeuser, 1975; Selye, 1950). The adrenal cortical hormones exert metabolic effects throughout the body as part of the adaptive response to stressful conditions. It has been shown that young animals exposed to mild stressors, such as weak electric shocks, have larger adrenal glands than adults and are more resistant to the breakdown effects of prolonged stress (Levine, 1960). The adaptive response includes an increase in the excitability of the CNS and thereby the reactivity of the animal to stimuli in its environment. A variety of behavioral effects have been reported in relation to blood levels of adrenal cortical hormones. For example, following adrenalectomy, the startle response to a loud sound habituates more rapidly. There is also increased output of the catecholamines from the adrenal medulla during emotional behavior. The effects of these hormones on bodily function complement those of the sympathetic nervous system (e.g., increasing heart rate, blood pressure, and blood sugar levels), preparing the animal, as suggested by Cannon (1929), for "fight or flight." The catecholamines do not readily cross the blood–brain barrier, but there is some evidence that they affect the reticular activating system to increase cortical activation (Rothballer, 1956).

External Stimuli

The role of external stimuli as elicitors of emotional behavior has been recognized for a long time and has been emphasized by a number of theorists. Charles Darwin, who devoted much of his monograph, *The Expression of the Emotions in Man and Animals,* to describing behavioral responses, identified the environmental events or stimuli that initiate fear, anger, anxiety, and other emotional reactions (e.g., Figure 15 on p. 125 of his book shows the defensive–aggressive reaction of a cat that has been frightened by a dog). William James' (1890) classic theory emphasized the importance of the subjective experience of sensory inputs from the viscera, muscles, and other effectors for emotional expression, but it also recognized that the emotional expression or reaction was precipitated by events in the external environment. John Watson (1919), in designating the three innate patterns of emotional reactions in human infants (rage, fear, and love), suggested that the stimuli for these reactions were respectively: restriction of movement, loud sound, and pleasurable tactile stimulation. Hebb (1966), who carried out ingenious studies of fear in chimpanzees, drew attention to the wide variety of stimuli that could elicit this emotional reaction (e.g., a carrot of an unusual shape, a biscuit with a worm in it, a "death mask" of a chimpanzee or man as shown in Figure 9.4). Ethologists have been particularly concerned with identifying the characteristics of environmental stimuli that function as "innate releasers" of emotional and motivated behaviors (e.g., Hinde, 1970; see Figure 1.2).

The task of attempting to relate specific stimulus characteristics to particular species-specific behaviors, which has produced impressive results in certain species (e.g., the stickleback and the herring gull), is more difficult in higher

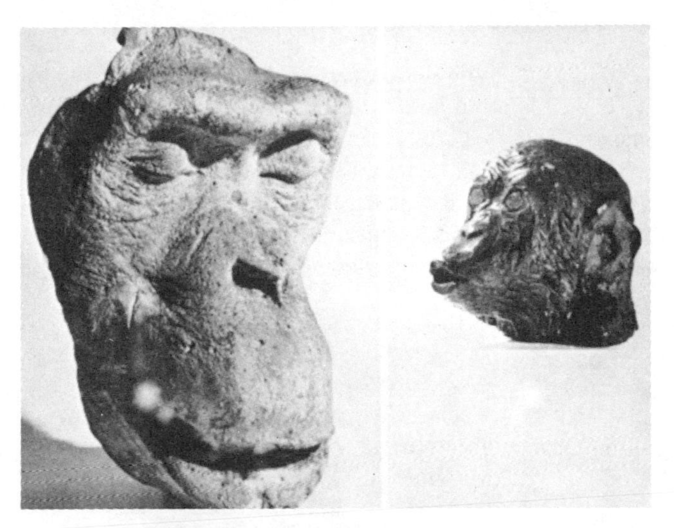

FIG. 9.4 Examples of clay models of a mask of a monkey's face and of a monkey's head that were effective in eliciting fear reactions in the chimpanzee. Hebb emphasized that objects were particularly effective in producing fear reaction if they were presented in a manner that differed from what would be expected from past experience. (From Hebb, 1966.)

mammals (see Figure 1.2). In higher mammals, emotional reactions are characterized by their variability. This has been emphasized by Hebb (1949) in his studies of fear and other emotional behaviors. He pointed out that a greater variety of circumstances elicit an emotional reaction in higher mammals (e.g., fear of the strange, being interrupted from sleep, hunger, and withdrawing a drug from a drug addict, all induce fear reactions), and, on the other hand, the same provocation or stimulus could in the same animal produce different emotional reactions at different times. He also noted that for higher mammals, in comparison to lower mammals, the behavioral expression of a single emotional reaction is more variable. According to Hebb, it is the more complex mediating processes or cell assemblies (or in our terminology, cognitive processes) of the higher mammal that are responsible for the greater variability of the environmental stimuli that elicit or provoke emotional behavior as well as for the greater range of emotional reactions. Young (1973) has also called attention to the complexity of environmental events and stimuli that trigger emotional reactions and has referred to ''a cognitive basis of emotion—in perception, memory and imagination [p. 228].'' This point is discussed more fully in a later section [p. 248].

Biological Rhythms

External stimuli are important determinants of emotional behavior, but emotional reactivity—the probability of occurrence of an emotional response to an evoking stimulus—will vary in relation to circadian rhythms of waking and sleeping and in relation to sexual and seasonal cycles (e.g., hibernation and the breeding season, respectively). A major factor, particulary in circadian rhythms of waking and sleeping, is the "arousal level" of the brain as determined by the reticular activating system. This is discussed in more detail in a later section (p. 256). Rhythms in the blood levels of hormones are also important, as, for example, in the premenstrual irritability and increased emotional responsiveness of women associated with fluctuating plasma hormone levels. Another example is the purported irritability of bears after their long "winter nap"—as with some humans upon waking in the morning.

Experiential Factors

Environmental experience can influence behavioral reactions to emotion-provoking stimuli in a number of ways. An initially neutral stimulus may come to elicit an emotional reaction when it is associated with a potent "emotional

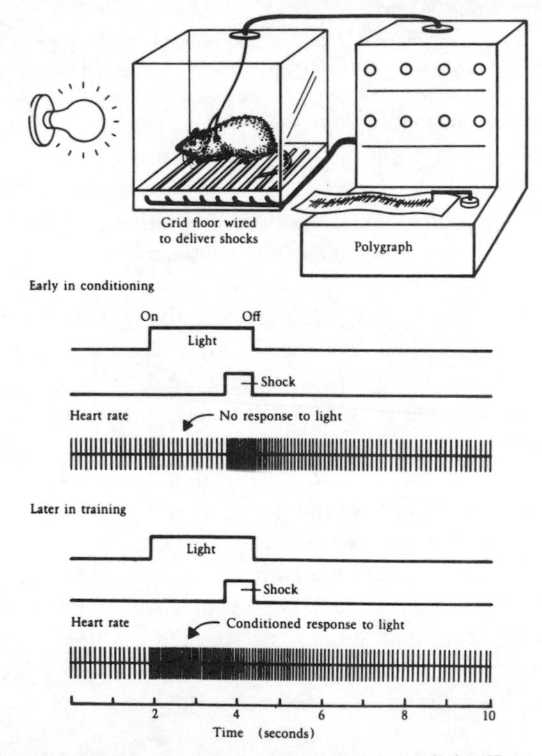

FIG. 9.5 The heart rate of the rat is being conditioned to a light. Heart rate is recorded on a polygraph from electrodes placed on the front shoulders of the animal. At first the light does not increase heart rate, but grid shock does. Later in training, the heart rate increases at the onset of the light prior to the grid shock, and the animal crouches and defecates. (From *A primer of psychobiology: Brain and behavior,* by T. J. Teyler. W. H. Freeman and Company. Copyright ©1975.)

stimulus.'' For example, in the rat a light paired with foot shock will result in conditioning so that increased heart rate (shown in Figure 9.5), crouching, defecation, and other ''fearlike'' responses occur when the light comes on. This has been designated as a *conditioned emotional response*. On the other hand, there may be habituation of an initially strong emotional reaction to the sight of a major surgical operation (the sight of blood perhaps causing nausea and fainting on first exposure) or to a hypodermic needle with repeated exposure to this stimulus. In addition to specific learning and habituation of emotional reactions to stimuli, there may be a more general influence of experiential factors. This has been clearly demonstrated in early sensory restriction studies in Scottish terriers carried out by Donald O. Hebb and co-workers at McGill University and in maternal deprivation studies in monkeys carried out by Harry Harlow and co-workers at the University of Wisconsin.

The most prominent feature of the behavior of the Scottish terriers following the restriction of early sensory experience was hyperactivity manifested by ''clumsy and puppy-like playfulness, frantic nuzzling and hard licking, and sometimes even a curious whirling type of behavior'' (Thompson, 1955, p. 129). Melzack (1965) has attributed the hyperactivity to a high degree of behavioral arousal and emotional excitement, which occurred especially when the dogs encountered a novel environment, but which was sometimes observed to occur spontaneously. According to Melzack, sensory restriction resulted in an impairment in the development of perceptual or cognitive processes (designated by Hebb as *central mediating processes*) so that there was a deficit in filtering environmental stimuli and, as a consequence, higher levels of emotional excitement. Because of the excessive arousal of the restricted dogs, they had ''difficulty in attending and responding selectively to relevant information'' (Melzack, 1965, p. 291).[2]

[2]According to Hebb (1966):

Scottish terriers were reared in partial isolation, from early weaning onward (R. Melzack, W. R. Thompson). They could hear and smell other dogs and the human caretakers in the same room, but otherwise were cut off from all social contacts, in small cages just large enough to allow them to stand and turn around comfortably. They grew well and stayed in excellent health until they were removed for testing at ages between 9 and 12 months. There was no sign that they were unhappy; a dog that is reared normally and then put in isolation is obviously miserable, but these dogs had known no other existence. They were ''happy as larks'' and physically ''as strong as bulls,'' in the words of the Scottie expert who supervised their rearing and who won a number of first class ribbons with them at dog shows. Such prizes are awarded only for physical form and posture; in obedience tests the isolation-reared dogs would have got nowhere, for they turned out to be almost untrainable. The whole picture of their social behavior was aberrant in a way that is hard to describe; they were dominated by normal dogs, would permit another dog to eat simultaneously from the same food dish (the normal Scottie reared outside the laboratory will not permit this) and reacted to familiar or unfamiliar people with a strange combination of approach and avoidance—a sort of diffuse or disorganized emotional behavior—that was never seen in normally reared dogs. The peculiarities diminished with time but did not disappear, and the personalities remained grossly abnormal [pp. 151–152].

Monkeys isolated from their mothers also exhibited striking changes in their behavior (Figure 9.6). Initially there were distress calls, loud screeching, and disoriented running about. The highly agitated infant monkeys struggled to regain contact with the mother and slept poorly. The ''agitation'' phase, which lasted for 24–36 hours after separation, was followed by a period of ''depression'' when there was little play activity or social contacts. The infant monkeys were hunched over with their heads between their legs, making but a few sluggish movements. When reunited with the mother, there was a tendency toward a greater amount of physical contact, designated by Harlow (1962) as ''contact comfort.''

Harlow has observed that monkeys raised in isolation were also abnormal when they grew up. This was very evident in their sexual and maternal behaviors, which was described as follows:

> When the laboratory-bred females were smaller than the sophisticated males, the girls would back away and sit down facing the males, looking appealingly at these would-be consorts. Their hearts were in the right place, but nothing else was. When the females were larger than the males, we can only hope that they misunderstood the males' intentions, for after a brief period of courtship, they would attack and maul the ill-fated male. . . . They (the males) approached the females with a blind enthusiasm, but it was a misdirected enthusiasm. Frequently the males would grasp the females by the side of the body and thrust laterally, leaving them working at cross purposes with reality. [Harlow, 1962, p. 7]

Sensory experience may be especially crucial for the subsequent behavior of some animal species if it occurs during a certain critical period of early development. A good example is ''imprinting'' (Figure 9.7). Normally the gosling or lamb becomes imprinted to its mother and other members of its species, contributing to its group affiliation and development of social and emotional behavior. Young animals imprinted to man or another species because of early exposure behave very differently when they become adults.

FIG. 9.6 Infant monkeys reared in isolation from their mothers displayed abnormal behavior and were ''emotionally upset.'' (From Sackett, 1968).

FIG. 9.7 External stimuli and experiential factors contribute to motivated behaviors as indicated in previous chapters. Early experience is particularly important for behavior and in some species there may be a "critical period" for the effects of visual and other stimuli on later behavior. A good example is imprinting shown here for a kid fed by bottle instead of by the mother goat. (Photo courtesy of Rebecca Woodside).

So the contributions of learning and experiential factors to emotional behavior may be both direct and indirect: direct as in the case of conditioned fear or imprinting; indirect in that species-specific environmental experience may be necessary for the normal appearance of emotional responses. Hebb (1966) cites, as an example of the indirect influence of learning on emotional behavior, the fear of strangers, "which occurs only when the child has first learned to recognize familiar persons, but does not require that he have any previous exposure to strangers, or any unpleasant event associated with strangers [p. 143]." A further example cited by Hebb is the experimental neurosis observed by Pavlov when dogs, conditioned to discriminate between an oval and a circle, showed disturbed "neurotic-like" behavior when the discrimination was made difficult by making the oval and circle more and more similar. Hebb (1966) comments that "the dog's breakdown in behavior was not learned, but it could not have occurred until after the conditioning process had established the discriminative behavior with the test objects (an oval and a circle) [p. 143]."

Cognitive Factors

In a previous section, we referred to the greater variety of stimuli that elicit an emotional reaction in higher mammals and the greater variation in emotional expression. For fear reactions, for example, Hebb (1966) has noted that "pain, sudden loud noise, and sudden loss of support cause fear in any mammal [p. 241]," but going from rat, to dog, to monkey and chimpanzee, a greater variety of environmental stimuli will elicit fear, and the "degree and duration of disturbance is greater [p. 241].'' More recently, Young (1973) has also emphasized the complexity of the conditions that induce an emotional reaction in man and higher animals. Young (1973) says, ''emotional reactions arise within a psychological situation and . . . the kind of emotion aroused depends directly upon the environmental situation. There is a cognitive basis of emotion—in perception, memory and imagination [p. 278].'' This is illustrated by Young (1973) as follows:

> It is common knowledge that emotions are aroused by reading a novel, watching a TV broadcast, following a stage play, hearing an impassioned address, reading a comic strip, hearing jokes and stories, etc. In all such situations it is the *meaning* or *understanding* of the situation that elicits the feeling. We identify ourselves with the characters or the situation and are emotionally aroused. We feel the anxiety, grief, resentment, anger, satisfaction, joy or relief of the characters. There is thus a cognitive identification with others and their situations [p. 353].''

Schachter and co-workers (1967), in a series of well-known experiments, have demonstrated how emotional reactions, as well as the labels attached to the emotional state, are a joint function of cognitive factors and physiological arousal. When human subjects were administered adrenaline in order to produce sympathetic arousal, their behavior as well as their subjective experience (whether fear, or anger, or euphoria, or no emotion at all) depended on what they were told or the nature of the contrived situation. For example, subjects receiving adrenaline behaved in an excited, euphoric manner if they were left with a ''stooge'' who showed manic–euphoric behavior, but subjects showed aggressive behavior if the ''stooge'' responded to a frustrating questionnaire with outbursts of anger. Subjects who received a placebo and subjects who were informed that the drug injection would make their hearts beat faster and make them feel generally nervous and aroused did not show these responses in the presence of a ''stooge.'' Although increased arousal is an important characteristic of emotional behavior, it should be recognized that cognitive factors are also important. Young (1973) cites a number of examples to illustrate this point—for example:

> Many people are emotionally disturbed when they see an animal in danger or in pain. Here is an example: A kitten had climbed up into the bed springs. Jimmy, aged five and a half, tried to catch her but in the attempt pulled her in such a way that the kitten's paw became caught in a spring. She hung by one foot, crying plaintively. Jimmy was much distressed by the outcry and called for help. At last a housemaid came to the rescue and crawled under the bed to extricate the kitten. While the kitten was being freed, Jimmy stood beside the bed crying and jumping excitedly. There was nothing he could do to relieve the distress of his pet. He was clearly disturbed emotionally by the situation. When at last the kitten was freed, Jimmy's excitement quickly subsided [p. 352].

Cognitive factors not only contribute to the initiation of emotional behavior but also are involved in restraining or controlling emotional reactions and in ensuring that emotional behaviors are expressed in a socially acceptable manner. Children are more likely to display temper tamtrums and other emotional outbursts than adults. The higher neural systems involved in cognitive processes are not mature in the child, and the cognitive constraints of emotional expression may be lacking; cognitive processes develop slowly during the first years of life as a result of experience and socialization.

The relative absence of temper tantrums in the adult does not mean that there is less susceptibility to emotional reaction. Hebb (1966) maintains that emotionality increases with age, although the behavioral expression of emotions may be more effectively restrained because of social and experiential factors. Hebb also suggested that for man, our so-called "civilized environment" provides a "protective cocoon" to reduce the emotional provocations to which we are exposed.[3]

The greater susceptibility to emotional reaction in man and monkey can have undesirable consequences and may even contribute to disease. Gastric ulcers, hypertension, and arthritis have an emotional component, and the incidence of psychosomatic disorders is greater in stressful circumstances. An example is a study by Brady (1967) who produced ulcers in monkeys placed in a stressful situation of having to press a lever at frequent intervals in order to avoid painful electric shocks. However, the greater susceptibility may, on occasion, have desirable consequences as well. "Man and ape can feel fear for others and man, at least, can be angered by injury or injustice to others" (Hebb, 1966, p. 246). The previous example of Jimmy's concern for the kitten whose foot was caught in the bed spring illustrates this point as well as the altruism, which is "defined as intrinsically motivated purposive behavior whose function is to help another person or animal" (Hebb, 1966, p. 247) and which often characterizes the behavior of man.

THE NEURAL SUBSTRATES OF EMOTIONAL BEHAVIORS

The first neurological model for emotions was formulated by Cannon (1927) as a reaction—and an alternative—to the popular James–Lange theory, which assumed that sensations from the viscera are experienced as emotional feelings. Cannon associated emotional feelings with the dorsal thalamus and emotional

[3]According to Hebb (1966):

The adults in it have learned elaborate rules of courtesy, good manners, and how to behave in public, so their behavior is predictable and usually will not cause one embarrassment, shame, anger or disgust. All this is achieved by prolonged training in childhood and later enforced by legal penalties (e.g., for slander, indecent exposure, dumping garbage in the street) or by social ostracism. In this environment, in short, one can count most of the time on not being suddenly exposed to the causes of strong emotion without warning and without adequate opportunity to avoid them. All this of course implies also that we as adults have been trained to suppress strong emotion when it does occur, as far as we can. Emotional outbursts are thus rare in the civilized adult on his own ground, but there is no reason to conclude from this that he is less susceptible to attacks of emotion than his five-year-old son, who must live in an environment tailored to adult needs rather than those of a five-year-old [p. 244].

expression with the hypothalamus. His model was supported by the demonstrations of ''rage-like'' reactions in decorticate dogs and cats (Bard, 1934), which had earlier been designated as sham rage because the ''attack'' responses were poorly directed. According to Cannon's model, decortication released the diencephalic integrative centers for emotional expression from cortical inhibition.

Experimental studies during the next few years indicated that the neural mechanisms for aggressive and other emotional behaviors were more complex than suggested by this model. We next consider some of the evidence that has implicated—in addition to the hypothalamus—the limbic forebrain structures, the cerebral cortex, and the reticular activating system.

The Hypothalamus and Emotional Behaviors

Bard (1934) observed that the ragelike reactions in the decorticate animal or following transection of the forebrain just rostral to the hypothalamus disappeared when the transection was made caudal to the hypothalamus. About the same time, more direct evidence implicating the hypothalamus in aggressive behavior was provided by Hess (1954) in pioneering experiments using chronic electrical stimulation of the hypothalamus. The behavior elicited by hypothalamic stimulation has been described by Hess and Akert (1955) as follows:

> The animal assumes a typical defense posture, extends the claws and lashes the tail. The syndrome is further characterized by hissing, spitting and growling; by wide-open pupils and eyes; retraction of the ears; a bushy tail, and piloerection of the back. At this stage of general excitement a slight move on the part of the observer is sufficient to make him the object of a brisk and well-directed assault [p. 127].

The elicited behavior was similar to that observed when a cat is confronted by a barking dog or other threat; and the behavioral response appeared purposive and directed, in contrast to the diffuse, poorly directed, ''sham rage'' response observed by Bard in the decorticate animal. The directed attack behavior observed by Hess may be attributed to the stimulated cats having intact cerebral cortex and limbic forebrain structures enabling the behavior to be guided by environmental stimuli.

Ragelike or aggressive behaviors elicited by electrical stimulation of the hypothalamus of the cat have been studied extensively by Flynn and co-workers, who established that two distinct patterns of behavior may be elicited (Figure 9.8). The first, designated as the *affective–defensive reaction* or threat attack, is characterized by hissing, spitting, growling, pupillary dilitation, and other sympathetic nervous system responses. However, in spite of the highly ''agitated'' appearance of the cat, it rarely makes a direct and injurious attack. The second pattern is characterized by a relative absence of affective display and a directed, stalking attack on the victim (e.g., rat), which usually terminates by biting the

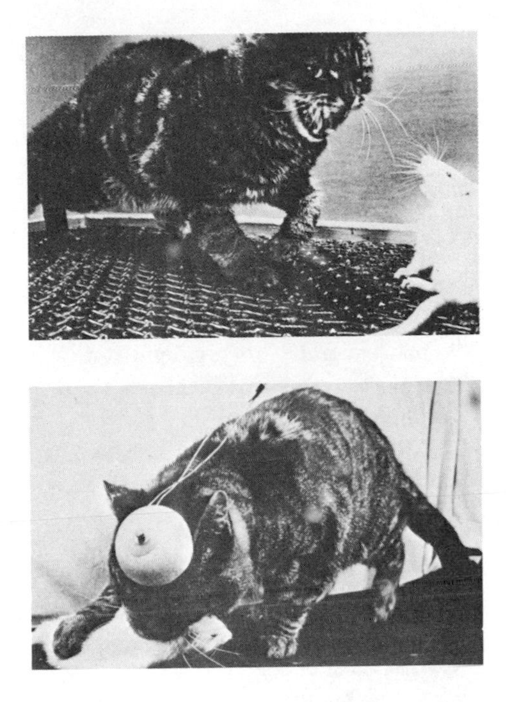

FIG. 9.8 Aggressive behaviors elicited by chronic electrical stimulation of the hypothalamus in the cat. *Top*–threat attack. *Bottom*–biting attack. (From Flynn, 1967.)

neck area of the "prey" and silently killing it. The pattern of responses is similar to normal predatory behavior observed in cats but is usually not followed by eating the victim. The hypothalamic sites for eliciting the affective reaction include much of the medial hypothalamus, whereas the sites for predatory or stalking attack are more lateral (Chi & Flynn, 1971; Wasman & Flynn, 1962). It is of interest that the lateral sites, in the perifornical region of the hypothalamus, from which predatory attack is elicited are in the region from which ingestive behaviors have been induced. In fact, both stalking attack and feeding have been elicited from the same hypothalamic electrode (Hutchinson & Renfrew, 1966).

The Limbic System and Emotional Behaviors

The first experimental evidence implicating limbic forebrain structures in emotional behaviors was obtained by Klüver & Bucy (1937), who observed a reduction in fearfulness and aggression in monkeys following bilateral removal of the temporal lobes. However, it was more than a decade later that MacLean (1949) coined the term "limbic system" and suggested that the amygdala, cingulate gyrus, septum, and other forebrain structures make important contributions to emotional behaviors and visceral regulations. During the next few years, attention shifted to these limbic forebrain structures, and the results of a number of lesion and stimulation studies clearly implicated them in aggressive, fearful, and other emotional behavior as well as in feeding, sexual, and the other motivated behaviors considered in previous chapters.

The amygdala has received the greatest amount of attention. The emotional placidity and reduced fearfulness and aggressiveness observed by Klüver and Bucy following bilateral temporal lobectomy was produced by selectively lesioning the amygdala in the cat (Schreiner & Kling, 1953). One of the most impressive demonstrations of this effect was in the wild cat or lynx, which following bilateral amygdalectomy became as gentle as a house cat (Schreiner & Kling, 1956). Another interesting demonstration is the effect of such lesions on the dominance hierarchy of a colony of rhesus monkeys (Figure 9.9). When bilateral amygdalectomy was performed on the most dominant monkey, Dave, he became submissive to all of the others in the colony. On the other hand, electrical stimulation of the amygdala has been reported to elicit violent and aggressive behavior in the cat (MacLean & Delgado, 1953) and in man (Heath et al., 1955; Stevens et al., 1969) and to influence aggressive behavior elicited by hypothalamic stimulation. Egger & Flynn (1963) reported that electrical stimulation of the basolateral amygdala suppressed hypothalamic attack, whereas stimulation of an area more dorsal and lateral enhanced the hypothalamic attack. These authors suggested that an important function of the amygdala is to modulate hypothalamic activity and that it contains "anatomically distinct regions that suppress or facilitate the hypothalamus" (Egger & Flynn, 1967, p. 165). In a recent monograph entitled *The Limbic System,* Isaacson (1974) suggests that "the extrahypothalamic portions of the limbic system exert regulatory effects on the hypothalamus and midbrain [p. 229]" and concludes that "some portions of the limbic system act to suppress established responses, while others intensify them. In the intact animal, the net effect of all these systems on behavior is probably the resultant, at the hypothalamic level, of the total amount of facilitating and inhibiting influences [p. 237]."

Stimulation of certain limbic structures alone may not elicit aggressive behavior, but when an aggressive reaction is being elicited by hypothalamic stimulation, the elicited responses may be altered by the limbic stimulation. For example, the latency or hypothalamically elicited attack behavior is increased by concurrent stimulation of the anterior cingulate gyrus but not by stimulation of the posterior region of the same gyrus. Because the cingulate stimulation did not influence the latency of the initial movement, these investigators concluded that it was not having a general inhibitory effect but rather that "the anterior cingulate gyrus serves as a modulator of the hypothalamic attack mechanism" (Siegel & Chabora, 1971, p. 175).

The septum also exerts inhibitory effects on aggressive behavior. This was first suggested by a report of hyperirritability, hyperreactivity, and vicious attack behavior in the rat following lesions of the septum (Brady & Nauta, 1953). This septal syndrome of hyperemotionality has been observed in other animal species and, interestingly, in an elderly man with a tumor of this region (Zeman & King,

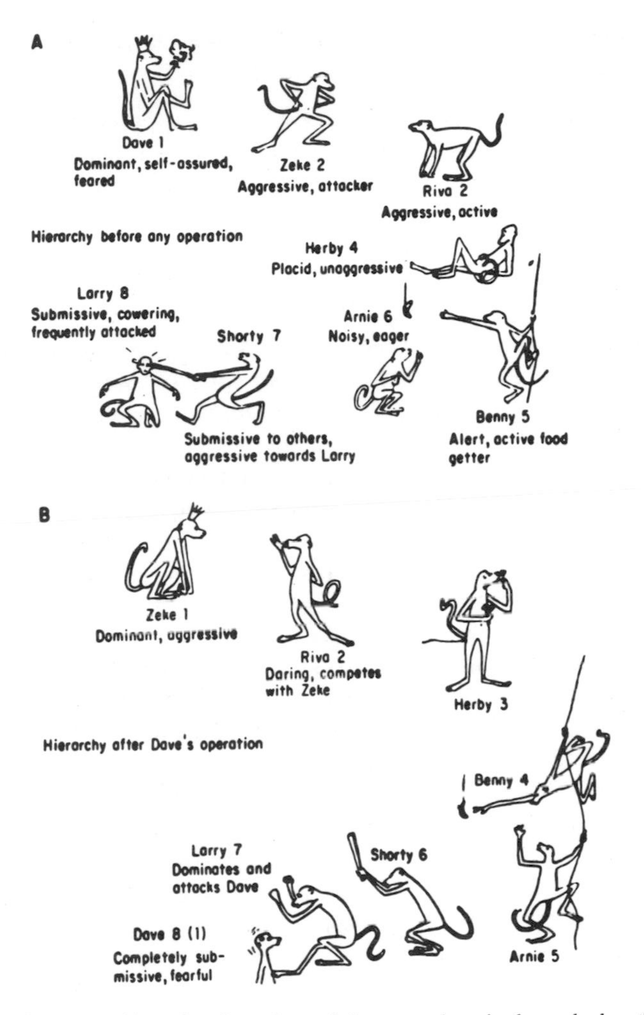

FIG. 9.9 The dominance hierarchy of a colony of rhesus monkeys is changed when the dominant monkey receives bilateral lesions of the amygdala. A. The hierarchy before Dave is lesioned. B. After he is lesioned. (Modified from Rosvold, Mirsky, & Pribram, 1954.)

1958). If the amygdala is lesioned bilaterally after a septal lesion, there is immediate attenuation of the hyperirritable, aggressive behavior (King & Meyer, 1958). In later studies it was shown that electrical stimulation of the septum increased the latency of attack behavior elicited by stimulation of the hypothalamus (Siegel & Skog, 1970). On the other hand, the latency of hypothalamically elicited hissing and flight behavior was decreased by septal stimulation, suggesting that the septum might also exert modulatory effects on hypothalamic systems concerned with aggressive behavior.

Higher-Order Integration of Autonomic, Endocrine, and Behavioral Responses

The hypothalamus, in spite of its small size, is involved in many functions. As indicated in previous chapters, it makes important contributions to temperature regulation, to metabolic regulations, to energy and water balance, and to reproduction. It contributes to feeding, drinking, and other motivated behaviors as well as to aggression and various emotional behaviors. The reason that such a small structure contributes to so many important functions is that it is strategically located in the brain to funnel a variety of neural signals to influence the autonomic nervous system and the endocrine system. The hypothalamus influences cardiovascular and other autonomic responses and controls the release of hormones by the pituitary gland. The unique role of the hypothalamus is not the initiation of discrete, isolated autonomic and endocrine responses but rather to contribute to the overall integration and coordination of autonomic, endocrine, and behavioral responses for adaptation of the animal to the environment and for homeostatic regulation (see Figure 9.3).

Limbic forebrain structures are involved in many of the same functions as the hypothalamus. Electrical stimulation of the amygdala, hippocampus, septal area, and other limbic structures alters blood pressure, heart rate, gastrointestinal mobility, and the release of hormones from the pituitary (Kaada, 1951; McCleary & Moore, 1965). There is a great deal of evidence that limbic structures exert modulatory influences on hypothalamic systems associated with the pituitary gland and the autonomic nervous system, thereby influencing visceral responses associated with the hypothalamus (Figures 1.8B & 9.3). Furthermore, as indicated previously, stimulation and lesions of limbic forebrain structures can have dramatic effects on motivational and emotional behaviors.

Limbic forebrain structures, like the hypothalamus, receive visceral afferents and the feedback influence of hormones. In addition these structures receive rich and varied inputs from vision, hearing, and other exteroreceptors and interact with the frontal cortex (Nauta, 1971). They are in a position, therefore, to contribute to the higher-order integration or ''functional coupling'' of autonomic, endocrine, and skeletomotor responses necessary for adaptation and homeostasis (Mogenson & Huang, 1973). The greater variety of inputs received by limbic forebrain structures make possible physiological and behavioral responses that are appropriate to the exigencies of the external environment.

Psychosomatic disorders, which are known to have an important emotional component, may be considered as a failure of the ''functional coupling'' of autonomic, endocrine, and behavioral responses.

The Cerebral Cortex and Emotional Behaviors

In a previous section the contribution of cognitive factors to emotional behaviors was recognized. According to Schachter (1967), set and perception of the social situation are important in determining both the content of subjective emotional experience and the nature of the emotional reaction when arousal is produced in an artificial manner (e.g., administration of adrenaline). Although the neural mechanisms are poorly understood, it is assumed that the cerebral cortex is also involved in constraints on emotional expression or behavior that appear during maturation of the CNS and as a result of socialization of the individual. Young children are more likely to display temper tantrums and other emotional outbursts than adults, because these higher functions have not developed, and, therefore, the neural systems that subserve the "cognitive constraints" on emotional reactions are lacking.

The frontal cortex in particular, which is closely linked anatomically to limbic forebrain structures and the hypothalamus, is thought to have an important role in emotional as well as motivated behaviors and in physiological regulations involving the autonomic nervous system and endocrine controls. This has been emphasized by Nauta (1971), who has summarized the evidence as follows:

> The frontal lobe is characterized so distinctly by its multiple associations with the hypothalamus, that it would seem justified to view the frontal cortex as the major—although not the only—neocortical representative of the limbic system. The reciprocity in the anatomical relationship suggests that the frontal cortex both monitors and modulates limbic mechanisms [p. 182].

It has been known for some time that frontal lobe pathology is associated with marked changes in emotional as well as motivated behaviors.[4] More recently the effects of frontal lobotomy and of ablations of frontal lobe tumors have confirmed that although complex cognitive processes may be altered, including subtle influences on personality, there are usually prominent changes in drives and emotional reactions, as illustrated by the following summary by Wooldridge (1963):

> A person who has been ambitious, well-adjusted, highly motivated and considerate will, after frontal lobe damage, exhibit lack of drive, insensitivity to the feelings of others, diminished initiative and organizing ability, tactlessness, and frequently a general silliness and lack of responsiveness to ethical and moral considerations. There is often a feeling of euphoria. Physical restlessness and increased talkativeness sometimes occur, but completely unaccom-

[4]According to Ruch (1965):
Since 1848, the date of the famous "crowbar" case of Phineas P. Gage, the relation of the frontal areas to personality has been recognized. The nature of such changes is not the same in all cases of frontal lobe damage, but some form of personality alteration is usually reported. Phineas P. Gage, an "efficient and capable" foreman, was injured on September 13, 1848, when a tamping iron was blown through the frontal region of his brain. He suffered the following change in personality, according to the physician, J. M. Harlow, who attended him. "He is fitful, irreverent, indulging at times in the grossest profanity (which was not previously his custom), manifesting but little deference to his fellows, impatient of restraint or advice when it conflicts with his desires, at times pertinaciously obstinate yet capricious and vascillating, devising many plans for future operation which are no sooner arranged than they are abandoned in turn for others appearing more feasible. . . . His mind was radically changed, so that his friends and acquaintances said he was no longer Gage" [p. 474].

panied by the imaginative thinking that makes such increased activity worth while. Not infrequently are childishness, naiveté, and emotional incontinence displayed in the form of bursts of laughter, or rage [p. 148].

Nauta has suggested that frontal cortex–limbic forebrain relationships may be very important to behavior in general and to emotional behavior in particular, in two ways. First, the frontal lobe–limbic complex contributes to ensuring that the visceral components of the emotional reaction are appropriate to the skeletomotor or behavioral responses. Thus, according to Nauta (1971), "The failure of the affective and motivational responses of the frontal-lobe patient to match environmental situations that he nonetheless can describe accurately could thus be tentatively interpreted as the consequence of a loss of modulatory influence normally exerted by the neocortex or upon the limbic mechanisms via the frontal lobe [p. 182]." Second, the frontal cortex, which is "perhaps the only realm of the neocortex where neural pathways representing the internal milieu converge with conduction systems re-representing the external environment as reported by all exteroreceptive modalities [p. 182–183]," integrates information from the internal environment with information from the external environment. In other words, in the terminology of the previous section, the frontal cortex because of intimate functional relationships with limbic forebrain structures contributes to the "functional coupling" of autonomic, endocrine, and skeletomotor responses for homeostatic regulations in relation to appropriate behavioral adaptations to a variable and changing external environment.

The Reticular Activating System and Emotional Behavior

In Chapter 7, the contributions of the reticular activating system to the varying levels of consciousness and to the regulation of sleep–waking were considered. Lesions of the brainstem reticular formation were shown to produce behavioral somnolence and a synchronized ("sleeplike") pattern in the EEG. On the other hand, electrical stimulation of the brainstem reticular formation of anesthetized animals produced a desynchronized pattern in the EEG and, in the chronic animal, increased alertness and behavioral arousal. Shortly after these classical experiments were reported, Lindsley (1951) formulated the "activation theory" of emotion. He suggested that the reticular activating system served as the neural substrate for the arousal or activation component of emotional behavior.

Recently, after reviewing a large body of electrophysiological and behavioral evidence, Lindsley (1970) described the current status of the theory as follows:

Underlying emotion, which expresses itself through various cortical, visceral, and somatomotor channels, are mechanisms of arousal and activation which involve reticulo-thalamo-cortical systems. Thought, worry, and anxiety reflect emotional arousal at the cortical level; weeping, sweating, intestinal, and other visceral activities regulated by the autonomic nervous system reflect cortical, diencephalic, and brainstem arousal; facial expression, muscle tension, and tremors manifest somatomotor arousal. Mechanisms of arousal and activation are

> especially identified with the reticular formation of the lower brainstem. Upward extensions, including the ascending reticular activating system and its subsystems are closely related to, and interactive with, diencephalic and limbic systems which control emotional expression and emotional–motivational behavior [pp. 183–184).

Hebb (1955) also emphasized the importance of the reticular activating system to emotional behaviors. As indicated at the beginning of this chapter, he suggested that high levels of activation or arousal are associated with strong emotions and that for effective performance of behavioral responses there is an optimal level of arousal (see Figure 9.2).

Brain Neurotransmitters and Emotional Behaviors

Cannon's (1929) emergency theory of emotions first directed attention to the relation of neurotransmitters, or what he designated as *neurohumors,* to emotional behaviors. He stressed the importance of ''sympathin'' [later identified as noradrenaline (Von Euler, 1946)], released by the adrenal medulla, in the emergency functions of the body associated with the ''fight or flight reaction.'' In recent years there has been a great deal of interest in the role of central neurotransmitters, and particularly of the biogenic amines, in emotional behaviors.

Most of the evidence comes from studies of the possible role of noradrenaline and of other biogenic amines in aggressive behavior. Sham rage produced by decerebration has been used as a model of aggression. There have also been studies of changes in central biogenic amines associated with aggressive behavior elicited by electrical stimulation of the brain, with spontaneous and shock-induced fighting, and with predatory aggression (e.g., ''mouse-killing'' by rats). Reis (1972) observed that sham rage in the cat produced by decerebration just rostral to the hypothalamus was associated with a reduction of noradrenaline in the brainstem. The depletion of noradrenaline was correlated directly with the number of occurrences of sham rage. Reis and Gunne (1965) observed a similar fall of brainstem noradrenaline, as well as of noradrenaline in the forebrain and adrenal medulla, when aggressive behavior was elicited by chronic stimulation of the amygdala of the cat; brain dopamine levels were not altered by the elicited aggressive behavior. If the amygdala stimulation did not elicit aggressive behavior, there was no change in brain noradrenaline levels.

Noradrenaline levels are reduced following fighting in mice (Welch & Welch, 1969) and by shock-induced fighting in rats (Eichelman, 1973). For certain types of aggressive behavior, changes in brain noradrenaline have been minimal, and changes in serotonin and/or acetylcholine have been more prominent; mouse-killing in rats, a form of predatory aggression, is associated with the depletion of serotonin and acetylcholine levels in the brain (Eichelman, 1973), and depleting the brain of serotonin induces mouse-killing in rats (Karli et al., 1969). Aggressive behavior observed following lesions of the septal area [''septal rage'' syndrome, (Brady & Nauta, 1953)] is also associated with reduced serotonin (Lints &

Harvey, 1969) and acetylcholine (Sorensen & Harvey, 1971) levels in the brain. The "septal rage" responses are attenuated by p-chlorophenylalanine, a drug that blocks the synthesis of serotonin (Eichelman, 1973).

It appears that the relation between aggressive and other emotional behaviors and brain biogenic amines is complex, and a great deal of research remains to be done. However, the evidence that is available indicates that central biogenic amines are "involved in the modulation of emotional behaviors" (Barchas, et al., 1972, p. 235). In a later section, we consider biogenic amines and CNS monoaminergic pathways in relation to abnormal behaviors and psychiatric disorders.

NEURAL SUBSTRATES OF ABNORMAL BEHAVIOR AND PSYCHIATRIC DISORDERS

Charles Darwin, in his classic monograph, *The Expression of the Emotions in Man and Animals,* suggested that studies of the mentally ill are a valuable source of information about emotional reactions (or what he called *expression*). Darwin (1872) stated: "It occurred to me that the insane ought to be studied, as they are liable to the strongest passions, and give uncontrolled vent to them [p. 13]."[5] Investigations of abnormal behavior and mental illness are also relevant to understanding the neural substrates of emotional behavior. There is a growing body of evidence that aggressive, violent behavior, schizophrenia, and other psychiatric disorders are frequently associated with dysfunction of neural systems involved in normal emotional behavior—in particular limbic forebrain structures and neural pathways having biogenic amines as their neurotransmitters (Girvin, 1975; Goldstein, 1974; Snyder, 1974; Valenstein, 1973). In the first part of this section, the association of limbic forebrain structures with aggressive behavior is considered, and in the second part, the association of the biogenic amines with psychiatric and behavioral disorders is briefly discussed.

[5]According to Darwin (1872/1965):

The insane notoriously give way to all their emotions with little or no restraint; and I am informed by Dr. J. Crichton Browne, that nothing is more characteristic of simple melancholia, even in the male sex, than a tendency to weep on the slightest occasions, or from no cause. They also weep disproportionately on the occurrence of any real cause of grief. The length of time during which some patients weep is astonishing, as well as the amount of tears which they shed. One melancholic girl wept for a whole day, and afterwards confessed to Dr. Browne that it was because she remembered that she had once shaved off her eyebrows to promote their growth. Many patients in the asylum sit for a long time rocking themselves backwards and forwards "and if spoken to, they stop their movements, purse up their eyes, depress the corners of the mouth, and burst out crying" [p. 154].

Aggressive, Violent Behavior in Man

The attenuation of emotional reactivity of monkeys following bilateral temporal lobectomy was previously described (the Klüver–Bucy syndrome), and evidence was considered that indicates that reduced emotionality in monkeys, cats, and other species is associated with lesions of the amygdala. Strikingly similar observations have been made in man; Terzian and Ore (1955) reported that a violent, aggressive young man became much more placid (and also highly hypersexual, as did the Klüver–Bucy monkeys) after bilateral temporal lobe ablations. More recently, Valenstein (1973) described the case of a 23-year-old Danish woman, who for several years had been very emotionally unstable with frequent periods of aggressive and destructive behavior, which included throwing objects, attacking other people, and self-mutilation. Following bilateral surgical removal of the amygdala, she was described as ''quiet and cooperative'' and ''receiving elementary school training'' (Valenstein, 1973, p. 230). Similar effects have been reported by Dr. J. Narabayashi of Tokyo, Japan, who has performed amygdalectomies to treat a number of patients showing symptoms of aggressive and violent behavior (see Valenstein, 1973, pp. 215–223).

Electrical stimulation of the amygdala in patients, like electrical stimulation of this region in experimental animals, has been reported to elicit strong emotional reactions including aggressive and violent behavior (see Figure 9.10 for an interesting example). Valenstein (1973) has reviewed a number of these cases. For example, when Dr. R. Heath of New Orleans stimulated the amygdala of a female patient, she ''became enraged and attempted to strike out, [and] on other occasions she became fearful and felt an impulse to run'' (Valenstein, 1973, p. 107). Another report, referred to previously, in which clinical observations were very similar to those in laboratory animals, involved a tumor of the septal area (Zeman & King, 1958). A 67-year-old man over the period of several months became increasingly irritable and disturbed. He argued with his wife and neighbors and at times was confused and wandered aimlessly about. After the man died, it was found that he had a tumor of the septum pellucidum. Aggressive and often violent behavior has also been reported in some patients with tumors and epileptic foci in the temporal lobes in the region of the amygdala. A well-known case is that of Charles Whitman, responsible for killing several people on the university campus at Austin, Texas, who at autopsy was found to have a malignant tumor in the temporal lobe (Girvin, 1975). Psychiatric disorders have been reported in a series of patients with tumors of the inferomedial region of the temporal lobes (Malamud, 1967). In another study of a series of 2484 patients, temporal lobe seizures were associated with behavioral and psychiatric symptoms in more than 50% of the cases (Figure 9.11).

Studies of the sort described in this section have raised the possibility of using neurosurgical procedures to treat patients who display violently aggressive and

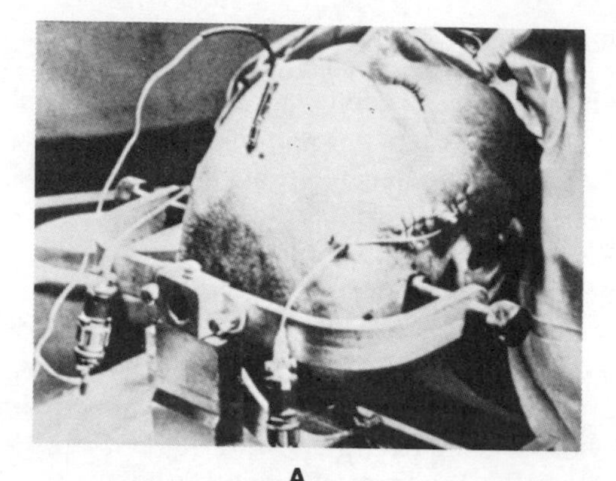

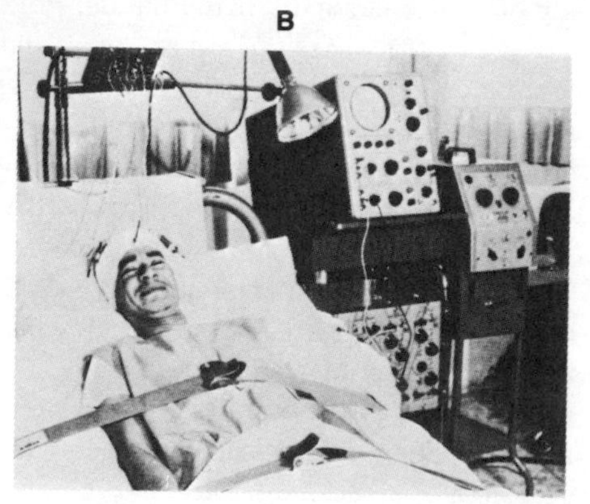

FIG. 9.10 A. The stereotaxic apparatus used for fixation of electrodes under anesthesia. B. The patient after surgery in the recording studio where studies of brain stimulation were carried out. (From Velasco-Suárez, 1970).

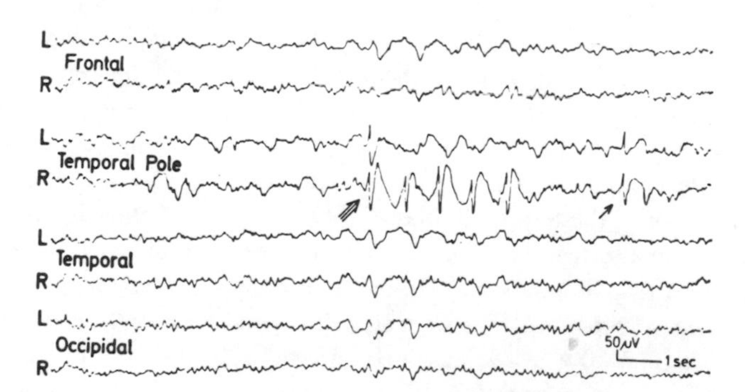

FIG. 9.11 EEG recording during a grand mal seizure in an epileptic. (From Jovanović, U.J., *Psychomotor Epilepsy*, 1974. Courtesy of Charles C Thomas, Publisher, Springfield, Illinois.)

other seriously abnormal behaviors. This procedure, called *psychosurgery,* has become very controversial in recent years. It seems appropriate, therefore, for the student to have an opportunity to consider the issues surrounding the use of psychosurgery in more detail. Accordingly, a concise yet thoughtful account from Thompson (1975) is included as the last section of this chapter (Appendix A).

Biogenic Amines and Psychiatric Disorders

During the last few years, there has been a growing interest in the possible role of the biogenic amines (e.g., serotonin, noradrenaline, dopamine) in abnormalities of mood and affect and in psychotic behavior. In Chapter 8 we dealt with the relation of dopaminergic and noradrenergic systems with brain stimulation-reward and the suggestions of the possible association of these pathways with disorders of affect and behavior.

The biogenic amines were implicated initially in neuropharmacological studies: Lysergic acid diethylamide (LSD), which was observed to have dramatic effects on affect and emotional reactions as well as on perception and thought processes, was shown to interact with serotonin; the tranquilizing drug, reserpine, was subsequently found to markedly deplete serotonin in the brain. Following such findings, a popular view by the mid-1960s according to Kety (1970), was that "serotonin was a hormone which controlled mood. When it was elevated, mood was high; when it was depleted, mood was depressed [p. 63]."

Reserpine also depletes the brain of the catecholamines (dopamine and noradrenaline), and there has been a great deal of evidence in recent years that these neurotransmitters are associated with depression and perhaps even with schizophrenia.

Kety (1970), a major proponent of the view that catecholamines have a critical role in depression, has reviewed the evidence. Hypertensive patients receiving reserpine show a high incidence of depression, and this has been attributed to the drug depleting catecholamines from the synaptic terminals. Drugs that elevate mood (e.g., MAO inhibitors) have the opposite effect and increase the availability of noradrenaline and dopamine in nerve terminals. One of the most potent antidepressant drugs, imipramine, increases the concentration of noradrenaline at the synapse by blocking the reuptake of this neurotransmitter by the presynaptic membrane. The antidepressant effects of electroconvulsive shock therapy, according to Kety (1970), may also be related to effects on central noradrenaline; electroshock therapy, effective in about 80% of depressed patients, increases the synthesis and turnover of noradrenaline in the brain.

The relationship between brain biogenic amines and mood is not simple, and there is likely more to affective disorders than low brain levels of noradrenaline or serotonin during depression and high levels during periods of mania or euphoria. Recently, Mendels (1974), after reviewing much of the relevant literature, concluded that "the depletion of brain norepinephrine and dopamine, or serotonin, is, in itself, not sufficient to account for clinical depression [p. 447]." Mendels, as well as Kety (1970), suggested that other transmitter-specific systems may be involved. Depression could also depend on an interaction between depletion of brain biogenic amines and cognitive and social factors. "The biogenic amines may play the chords of the affective state, but the melody is carried to a large extent by cognitive factors" (Kety, 1970, p. 69). Depletion of catecholamines may be a necessary but not a sufficient condition for depression; cognitive and social factors may also have to contribute.

The status of the catecholamine hypothesis of affective disorders (depressions and elations) has been summarized by Schildkraut (1971) as follows:

> Although many of the clinical findings are compatible with the catecholamine hypothesis of affective disorders, this formulation cannot be definitively established on the basis of existing data, and future investigation with more sophisticated techniques will be required to test this hypothesis. Since it is generally agreed that depressions are not homogeneous clinical or biochemical entities, it is possible that the hypothesized functional deficiency of norepinephrine may ultimately be confirmed only for certain subgroups of this heterogeneous diagnostic classification (e.g., manic-depressives). Even in these subgroups many different environmental and constitutional factors may possibly contribute to the development of such a functional deficiency of norepinephrine and several different biochemical mechanisms may be immediately operative in producing it [p. 218].

There is also a relationship between the biogenic amines and schizophrenia. Following the introduction of reserpine into clinical psychiatry, one of the first theories implicated serotonin and the indoleamines in schizophrenia. Himwich (1971), who made important contributions to this field, has reviewed the relevant evidence. More recently there has been increasing interest in the possible role of the catecholamines in schizophrenia.

The catecholamines were implicated initially in the etiology of schizophrenia on the basis of their chemical similarity to the hallucinogens, known for some time to produce psychotic symptoms. For a number of years, the popular view was that schizophrenia is due to an alteration in the metabolism of adrenaline. The first proponents of this hypothesis were Osmond and Smythies (1952), who proposed that in schizophrenic patients, because of an abnormality in the biosynthesis of adrenaline, there is the formation of adrenochrome, which has some of the properties of mescaline, a psychotomimetic compound. This hypothesis was investigated extensively but has not been substantiated. More recently, it has been proposed that the "aberrant metabolite" in schizophrenia is 6-hydroxydopamine, which destroys noradrenergic and other catecholaminergic neurons (Stein & Wise, 1971). According to the noradrenergic hypothesis of schizophrenia, chlorpromazine is effective in treating schizophrenic patients, because it blocks the uptake of the endogenously formed 6-OH-dopamine into noradrenergic nerve endings and thus protects them from the noradrenaline-depleting effects of this metabolite.

The effects of the amphetamines, the phenothiozines, (e.g., chlorpromazine), and the butyrophenones (e.g., haloperidol) have been particularly important in the implication of the catecholamines in schizophrenia. For example, amphetamine, known to cause the release of catecholamines from nerve fibers, exaggerates the schizophrenic symptoms of patients and produces a psychosis that is clinically very similar to acute paranoid schizophrenia. It has been proposed that "amphetamine psychosis" is a "model" schizophrenia (Kety, 1959; Snyder, 1974). The complementary evidence is that the phenothiozines and butyrophenones (the neuroleptic drugs) are the best antidotes for "amphetamine psychosis," as well as for the treatment of schizophrenia (Snyder, 1974). One possibility is that the phenothiozines and butyrophenones have beneficial effects in the treatment of schizophrenia because of a blockade of postsynaptic dopamine receptors. This view has been summarized by Snyder et al. (1974) as follows:

> Carlsson and Lindquist speculated that the phenothiozines block catecholamine receptor sites, whereupon a message is conveyed by means of a neuronal feedback to the cell bodies: "We receptors are not receiving enough transmitter; send us more catecholamines!" Accordingly, the catecholamine neurons proceed to fire more rapidly and, as a corollary, synthesize more catecholamines and release more metabolites. These speculations have been confirmed in studies showing that phenothiozines and butyrophenones do accelerate catecholamine synthesis in proportion to their clinical efficiency [p. 1246].

Another possibility for the mechanism of action of antipsychotic drugs, suggested more recently, is the presynaptic block of the coupling between nerve impulse in the axon and neurosecretion. Seeman and Lee (1975) have presented evidence of "a direct correlation between the clinical antipsychotic activity of neuroleptic drugs and their ability to block the impulse-coupled release of dopamine [p. 1217]."

The question of the possible relationship of schizophrenia and of other psychiatric disorders to particular monoamine pathways of the brain, mapped in recent years by histofluorescence and other histochemical techniques (Figure 2.24), is receiving active consideration. One proposal by a group of Swedish workers is that the mesolimbic–mesocortical dopamine pathway, which projects from the midbrain ventral tegmental region (A10) to limbic forebrain structures and prefrontal cortex, may be associated with schizophrenia (Hökfelt et al., 1974). This is an attractive hypothesis in that the prefrontal cortex and limbic structures are implicated, because the symptoms of schizophrenia include thought disorders and hallucinations as well as disturbances in affect and goal-directed behaviors.

One of the suggestions of Hökfelt and co-workers (1974) was that schizophrenia might be associated with hyperactivity in mesolimbic and mesocortical dopaminergic (A10) neurons resulting from a lesion somewhere in the brain. In a subsequent report, these investigators suggested that the lesion or pathology might involve GABA neurons, which exert an inhibitory effect on the A10 dopaminergic neurons in the ventral tegmental area (Fuxe et al., 1975). If these

GABA neurons are "diseased," the A10 dopaminergic neurons could be hyperactive because of a reduction in the inhibitory input (Figure 9.12). If this is the mechanism, then the administration of drugs that are GABA agonists should relieve the symptoms of schizophrenia, and if given in combination with a neuroleptic drug, the dose of the latter drug might be reduced. In this way the Parkinsonian symptoms of the neuroleptic drugs might be reduced or avoided. These investigators refer to a clinical study (by Fredriksen, 1975) in which Lioresal (a GABA agonist) was used in combination with neuroleptic drugs. According to Fuxe and co-workers (1975), "out of seven schizophrenic patients studied, six showed substantial improvement after at least one month of treatment with Lioresal, together with a reduced dosage of neuroleptic drugs [p. 182]." These observations must be interpreted with caution, however, and more evidence is needed before these interesting ideas can be assessed.

The speculations and hypotheses concerning the role of noradrenaline, dopamine, and other bionic amines in psychiatric disorders, as well as in normal emotional behaviors, are under active investigation. Important developments are expected during the next few years that will be very relevant to our understanding of motivated behavior in health and disease.

SUMMARY

Emotional behaviors, like the other behaviors considered in previous chapters, are purposive in the sense that they have biological significance in adaptation and survival (Darwin, 1859). Aggressive behavior, which has been studied most extensively, occurs in territorial defense, in predation, and sometimes as a component of sexual behavior. Emotional behaviors are frequently intense and persistent in part because of the prominent contributions of autonomic and endocrine responses and of the reticular activating system. Associated with these changes is a strong subjective or affective component that, in man, makes emotions somewhat unique among the motivated behaviors. If arousal, activation, and the affective components are too intense, emotions may be disorganizing and maladaptive rather than organizing and adaptive. Because of their intensity and

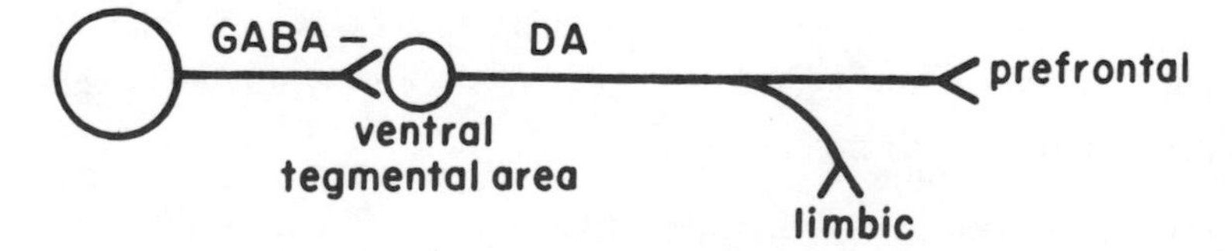

FIG. 9.12 Recent evidence indicates that the dopaminergic neurons of the ventral tegmental areas (VTA), designated A10 by Dahlström and Fuxe (1964), receive a gabaminergic (GABA) input that is inhibitory. According to a recent hypothesis, these GABA neurons become diseased in schizophrenia, resulting in "overactivity" of the mesolimbic and mesocortical DA neurons (A10).

persistence, emotional behaviors have a high priority in "motivational time-sharing."

There is a great deal of evidence from brain transection, lesion, and stimulation experiments that implicates the hypothalamus in emotional behaviors. Limbic forebrain structures, the cerebral cortex, and the reticular activating system are also involved. The hypothalamus is strategically located to receive inputs from these other neural structures as well as signals from the internal environment and feedback effects of hormones. It is in a position, therefore, to make an important contribution to the integration and "functional coupling" of autonomic, endocrine, and skeletomotor components of emotional behaviors. Psychosomatic disorders, in which signals from limbic forebrain structures and the cerebral cortex exert exaggerated effects on autonomic and endocrine responses, may be considered as a failure of this "functional coupling."

The cerebral cortex, especially the frontal lobes, has an important role in emotional behaviors, particularly in man and higher mammals. As a consequence, emotional behaviors in these species are less stereotyped and stimulus-bound than in lower animals and depend much more on experiential and cognitive factors. As cognitive processes become more complex, there is a greater variety of stimuli that elicit emotional reactions and a greater variety and complexity of emotional behaviors. Cognitive factors also contribute to the constraints on emotional expression in man, their role increases in the course of socialization of the individual, and our social environment is structured so as to reduce emotional provocations. Psychiatric disorders may be a disturbance in cognitive functioning as well as a disturbance of affect and emotional expression.

In recent years there has been a great deal of interest in the relationship of emotional behavior and of behavioral and psychiatric disorders to central neurotransmitters and to transmitter-specific pathways that project from the brainstem to the hypothalamus and a number of forebrain structures. The biogenic amines, which may be involved in brain-stimulation reward, have been associated with disorders of affect and mood and with schizophrenia. The neural mechanisms are poorly understood, but there are likely to be significant advances in this important field during the next few years.

APPENDIX A

Human Violence and Psychosurgery

According to Thompson (1975):

> Psychosurgery to control emotional disorders is the subject of much emotional discussion today. The term *psychosurgery* refers to the use of neurosurgical procedures to treat disorders that are primarily behavioral, at least insofar as symptoms are concerned. Such procedures have been used not only to control extreme violence but also for cases of severe terminal cancer

pain, severely debilitating compulsions—the patient whose hands are raw and bleeding because he washes them several hundred times a day—for extreme fear and anxiety that completely incapacitate a person, and for severe psychosis. The problems surrounding the use of psychosurgery do not admit to easy solutions; not only are there scientific and medical problems but also moral and ethical problems. It must be emphasized that our discussion here does not apply to standard neurosurgical procedures, for example, removal of brain tumors, treatment of head injuries, etc., but only to the brain surgery to treat disorders that are first manifested by behavior.

The first modern use of psychosurgery (trephining the skull to let out devils was an ancient use) is instructive. Moniz, a Portugese neurosurgeon, attended a scientific conference in 1935. He heard a paper by an American scientist, Jacobsen, who reported on the effects of removing the prefrontal cortex of monkeys. Jacobsen noted that his previously aggressive and hostile animals became tame and gentle following the operation. On the basis of this evidence, Moniz developed the procedure of frontal lobotomy in humans and was later awarded the Nobel prize for his ''achievement.'' However, what Moniz apparently failed to hear in Jacobsen's talk was the main point of the research: namely that following surgery, the monkeys had severe difficulty in solving tasks involving temporal memory—the so-called *delayed response problem,* in which the animal must remember which of two or more food wells was baited. Much of the subsequent history of psychosurgery has this unfortunate character. Neurosurgeons jump to completely unwarranted scientific conclusions on the basis of inadequate animal experiments, and apply their conclusions to the human brain. This is not meant to be a blanket condemnation of psychosurgery, only a plea for better science.

Among the most extensive experimental studies of human psychosurgery was the Columbia Greystone Project (Mettler, 1952), in which a series of lesions of the prefrontal cortex of systematically varied size were made on a group of ''volunteer'' psychiatric patients. Although one may question their morality, these studies were rather well done from a scientific viewpoint but the results were disappointing. First, the beneficial effects of surgery on psychosis were quite variable. Second, there were no clear effects on psychological test performance. No change occurred in IQ or in any other abilities or personality measures. However, in many cases the relatives of the patients reported marked changes in the patients' personalities. The most common impression was that a patient was only a ''hollow shell'' of his former self. This is a very real and significant effect, even though it was not detected in the psychological tests.

Studies like the Columbia Greystone Project have led to virtual abandonment of prefrontal lobotomy for treatment of psychosis or severe neurosis. However, the procedure still has an important use in terminal pain. For reasons not known, after frontal lobotomy severe chronic pain, as in cancer, no longer seems to bother the patient, at least in many instances. The frontal cortex does have fairly direct connections with the limbic system; the amygdaloid nucleus projects to the dorsomedial nucleus of the thalamus, which in turn projects to the frontal cortex. Whether this limbic input is relevant to altered pain perception or to the changes in personality that often accompany frontal lobotomy remains to be determined. . . .

Psychosurgery to control extreme violence, particularly episodic violence that resembles epilepsy, has focused on the limbic system. Favored targets for lesions are the cingulate gyrus—by placing a lesion here the circuit of Papez can be interrupted without extensive damage to other brain tissue—and the amygdala. . . . The consensus of animal studies does seem to be that destruction of the amygdala tends to reduce aggressive behavior. However, the mechanisms remain unknown and there are many puzzling and contradictory findings. . . .

A most interesting and careful series of clinical–surgical studies on humans have been done by Vernon Mark, Frank Ervin, and their associates at the Massachusetts General Hospital. A readable and interesting account of their work in nontechnical terms is given in the book, *Violence and the Brain* (Harper and Row, 1970), where Mark and Ervin focused on the amygdala. They developed their procedures in several stages on a series of patients. First, recording electrodes were implanted in deep structures like the amygdala to detect abnormal epileptic brain-wave activity. Then, using remote stimulation techniques, they electrically stimulated these regions. Finally, lesions of the amygdala were made. The patients selected had episodic ''attacks'' of extreme and uncontrollable violence which resembled in many ways the clinical epileptic condition of temporal lobe epilepsy. The case history of one of Mark's and Ervin's patients, Julia S., is abstracted here:

> In Julia's case, the relationship between brain disease and violent behavior was very clear. Her history of brain disease went back to the time when, before the age of 2, she had a severe attack of encephalitis following mumps. When she was 10, she began to have epileptic seizures; occasionally these attacks were grand mal seizures. Most of the time, they consisted of brief lapses of consciousness, staring, lip smacking, and chewing. Often after such a seizure she would be overcome by panic and run off as fast as she could without caring about destination. Her behavior between seizures was marked by severe temper tantrums followed by extreme remorse. Four of these depressions ended in serious suicide attempts.
>
> On twelve occasions, Julia had seriously assaulted other people without any apparent provocation. By far the most serious attack had occurred when she was 18. She was at a movie with her parents when she felt a wave of terror pass over her body. She told her father she was going to have another one of her ''racing spells'' and agreed to wait for her parents in the ladies lounge. As she went to it, she automatically took a small knife out of her handbag. She had gotten into the habit of carrying this knife for protection because her ''racing spells'' often took her into dangerous neighborhoods where she would come out of her fuguelike state to find herself helpless, alone and confused. When she got to the lounge, she looked in the mirror and perceived the left side of her face and trunk (including the left arm) as shriveled, disfigured, and ''evil.'' At the same time she noticed a drawing sensation in her face and hands. Just then another girl entered the lounge and inadvertently bumped against Julia's left arm and hand. Julia, in a panic, struck quickly with her knife, penetrating the other girl's heart and then screamed loudly. Fortunately, help arrived in time to save the life of her victim.
>
> The neurological examination made in our hospital showed Julia's ability to assimilate newly learned material was impaired, and, because of her shock treatments, she had a severe deficiency in both recent and remote memory. A brain-wave examination disclosed a typical epileptic seizure pattern with spikes in both temporal regions, in addition to widespread abnormality over the rest of the brain.
>
> Electrodes were placed stereotactically into both temporal lobes, and after she had recovered from the surgical procedure, we recorded epileptic electrical activity from both amygdalas. Electrical stimulation of either amygdala produced symptoms characteristic of the beginning of her seizures. These symptoms were more easily elicited by stimulating her left amygdala, and therefore, a destructive lesion was made in the left temporal lobe in the region of the amygdala and all electrodes were withdrawn. However, the symptoms and seizures persisted and changed to include signs that indicated a small portion of her brain was firing abnormally, and that this area was related to the movement of her left arm (the one that was brushed against by the girl she stabbed). As the motor tract crosses over from one side of the brain to the opposite side of the body, this suggested that her persistent seizures and attack behavior were initiated in her right temporal lobe. Therefore, we again placed electrodes in her right amygdala.
>
> The following records were made from this patient in a hospital room. Both Julia and her parents knew that sometime during the day her brain was going to be recorded from and stimulated, but they had no idea when we were going to do it. Before we had done any stimulating, but while we were recording, the electrical activity recorded from the leads in Julia's amygdaloid nucleus showed a typical epileptic seizure pattern. The behavior that

> accompanied this change in Julia's brain waves involved her getting up and running over to the wall of her bedroom. Once there, she narrowed her eyes, bared her teeth, and clenched her fists—that is, she exhibited all the signs of being on the verge of making a physical attack. (Vernon Mark and Frank Ervin, *Violence and the Brain,* Harper & Row, 1970, abridged from pages 97–99. Copyright 1970 by Harper & Row, Publishers, Inc. By permission of Harper & Row, Publishers, Inc.).

Following these observations, a destructive lesion was made in Julia's right amygdala. She had only two mild rage episodes in the first postoperative year and none in the second. The investigators are properly cautious in noting that two years follow-up is not sufficient fo make a final evaluation of this case. Julia still had generalized brain disease and epileptic seizures following the procedure. However, her episodes of extreme violence were clearly reduced.

Although it is possible to criticize aspects of their work on both scientific and ethical grounds, these clinical studies by Mark and Ervin are well done. Considerable animal experimentation implicates the amygdala in aggressive behavior. Prior to making lesions, electrical activity of the amygdala was first recorded to determine if abnormal activity was present. The behavior of the patients resembled the clinical diagnosis of temporal lobe epilepsy. Finally, they were extremely violent and dangerous to other humans [pp. 399–403].

10

Problems and Strategies for Future Research

The major purpose of this book is to consider what is known about the neural substrates of behavior. This has been done from the viewpoint of the factors that initiate and contribute to thermoregulatory, feeding, drinking, aggressive, and other behaviors, and it has become quite apparent that our knowledge of the neural mechanisms is fragmentary and incomplete. This should not be a cause for pessimism, however, because behavior and the neural mechanisms that control behavior are exceedingly complex and the scientific study of brain-behavior relationships is a relatively recent endeavor. A number of interesting empirical observations have been made that have attracted scientists to this field from a broad spectrum of neuroscience disciplines. The investigation of the neural substrates of behavior promises to be one of the most fascinating and fruitful fields of research in neuroscience in the years to come.

At the outset it should be noted that progress in a research field depends in part on the experimental techniques and research strategies available and in part on the theoretical ideas and conceptual framework used in interpreting the experimental findings.[1] Research concerned with the neural substrates of behavior is no exception. Accordingly we begin by considering how previous efforts and advances have depended on the available experimental techniques—both for studying behavior and for studying brain function—and on the prevailing theoretical views, and we indicate that some of the gaps in our knowledge reflect limita-

[1]According to Grodins (1970):
Someone has defined the goal of science as the compression of the maximum number of observations into the minimum number of conceptual principles. This is a good definition because it emphasizes the duality of science, i.e., it is neither a collection of facts alone (like a telephone directory), nor a set of conceptual schemes alone (like philosophy), but an interdependent combination of both [p. 722].

tions, both in experimental technique and in theory. We also consider the impact of more sophisticated experimental techniques and strategies introduced in recent years and identify current developments and future directions of research in relation to the major emphasis of the book—namely that several factors contribute to motivated behaviors.

PROBLEMS OF MEASUREMENT AND OF EXPERIMENTAL TECHNIQUES

The Problem of Behavioral Analysis and Measurement

One of the shortcomings of much of brain-behavior research and one of the reasons so little is known about the neural substrates of feeding, drinking, aggressive, and other behaviors is that the measures of behavior have frequently been inadequate or inappropriate. Most neuroscientists, including physiological psychologists, who study brain function and behavior are interested primarily in learning about the "secrets of the brain," and unfortunately they have not always been very sophisticated in behavioral analysis. For example, in assessing the behavioral deficits following lesions of the lateral hypothalamus, the earlier conclusions based on "description by consequence" (e.g., reduced food intake therefore reduced hunger motivation) have had to be re-interpreted in light of more recent studies. From more careful and extensive behavioral observations and tests, including responsiveness of lesioned animals to various sensory stimuli and an analysis of response components that are disrupted, it has been shown that aphagia and adipsia are associated with subtle sensory and motor deficits and the disruption of behavioral arousal (Marshall et al., 1974; Turner, 1973; Ungerstedt, 1974). It has become apparent that deficits in feeding and other behaviors, which have usually been attributed to disrupting the appetitive or motivational phase, may be associated with disrupting the consummatory phase. This has been demonstrated in experiments in which lesions of the central trigeminal system (Zeigler & Karten, 1974) or of central dopaminergic pathways (Stricker & Zigmond, 1976; Ungerstedt, 1974) seriously disrupt ingestive behaviors. Such studies emphasize the value of careful behavioral analysis and not merely a reliance on description by consequence when investigating the neural substrates of motivated behaviors.

When behavior is studied in the laboratory, it is possible to perform well-controlled experiments and to obtain quantitative measures that are highly reliable. However, there is the shortcoming that the environment is rather artificial, and certain factors or determinants that normally initiate or influence the behavior operate only to a minimal extent and may even be absent. For example, laboratory studies of thirst mechanisms are typically carried out in water-deprived animals, and secondary drinking—especially important in higher species and associated with experiential factors, circadian rhythms, etc.—has been largely ignored. Similarly, laboratory studies of thermoregulation have stressed physiological responses, and opportunities for behavioral thermoregulation, which in the animal's natural habitat may be the first line of defense against the cold or the heat, have usually been limited or nonexistent. Although studies of behavior in the laboratory enable the investigator to simplify the environment of the animal and to control many relevant variables, there is the limitation that the effects of important "natural" variables may be ignored. This has certainly been the case for feeding, drinking, and thermoregulatory behaviors in which homeostatic or deficit factors have been stressed and the role of experiential and cognitive factors minimized and even disregarded.

The measurement of an operant response is widely used in laboratory studies of behavior, because quantitative and highly reliable measures can be obtained. The animal, typically food- or water-deprived or placed in the cold, makes an arbitrary response such as pressing a lever, and the rate of operant responding is the criterion and quantitative index of motivation (Teitelbaum, 1976). Because the animal's environment is simplified and artificial and the animal severely deprived, factors or determinants of feeding and drinking that operate in the more natural environment may be absent. Ethologists and other critics of this approach point out that in order to understand an animal's behavior, it is necessary to take account of its evolutionary history and its habitat and ecological niche, as well as its physiological state and present environmental circumstances.

THE PROBLEM OF NONSPECIFICITY OF EXPERIMENTAL MANIPULATIONS OF THE CNS

A very serious difficulty in the investigation of the neural substrates of behavior is that the experimental technique may not be sufficiently discrete and selective. This has been a major problem, because the usual research strategy has been to manipulate a neural structure or pathway (e.g., lesioning, electrical stimulation, drugs) to observe effects on the behavior of the animal (e.g., attack response, drinking, sleep). The experimental manipulation or intervention may not be specific to the neural pathway or to a transmitter-specific component of a pathway, and it may not be specific to the behavior or function being investigated. For example, it is well known but all too frequently ignored that electrical stimulation may activate the central nucleus of the amygdala when the objective of the experiment is to stimulate the corticomedial nuclei and that it may activate the fibers of the medial forebrain bundle indiscriminately when an attempt is being made to identify which transmitter-specific fibers are associated with brain-stimulation reward. Similarly, lesions of the lateral hypothalamus may disrupt feeding behavior, but also a number of other behavioral deficits (e.g., drinking behavior, sexual behavior, exploratory behavior, sensory neglect, drowsiness, etc.), and it may be difficult or impossible to ascribe the feeding deficits to damage to integrative systems in the lateral hypothalamus, to fibers of passage in the medial forebrain bundle, or to the nigrostriatal bundle. In order to avoid erroneous conclusions when interpreting experimental observations, it is important to recognize the limitations of experimental techniques being used. Unfortunately, many investigators have not been cautious enough, and much of the earlier evidence involving the effects of lesions and stimulation of the hypothalamus on feeding, drinking, and other behaviors, has been reinterpreted in recent years (Mogenson,1974a; Morgane, 1975; Stricker & Zigmond, 1976).

There may also be a problem of specificity when drugs are used as research tools to study brain mechanisms that subserve behavior. The following examples serve as illustration. Goldthioglucose was used about 20 years ago to identify central glucoreceptors concerned with the control of food intake. It appears that goldthioglucose does not selectively destroy glucoreceptors, rather, as suggested by Epstein and co-workers (1975), the drug destroys neurons indiscriminately. More recently 6-hydroxydopamine has been used widely to destroy catecholaminergic neurons in order to study their role in feeding, brain self-stimulation, and other behaviors; there has been much controversy about the selectivity of the damage produced by this neurotoxin to noradrenergic and/or dopaminergic neurons. Experienced and highly skilled investigators may be able to produce selective and discrete destruction of these neurons, but there are many reports in the literature in which selective damage has not been achieved. Another example is the use of noradrenergic and dopaminergic blocking drugs to investigate the catecholamine hypothesis of brain-stimulation reward (see Chapter 8). The interpretation of these experiments is difficult, because the drugs may have sedative, motor, or other more general behavioral effects in addition to influencing neural mechanisms for reward. It is necessary to monitor behaviors other than the one of primary concern if any claim as to specific effects is to be made.

The limitations of these experimental techniques (e.g., lesions, stimulation, using drugs as research tools) have become more obvious with the emergence of a new chemical neuroanatomy of the brain, based on the use of histofluorescence to visualize monoaminergic neurons in the midbrain and brainstem and to map their fiber projections into the forebrain (see Figure 2.24). It has become clear in recent years that behavioral deficits (e.g., in feeding, brain self-stimulation, copulatory, and aggressive behaviors) following lesions of the hypothalamus and associated structures may not be the result of destroying integrative centers for the initiation of these behaviors but rather to damaging dopaminergic and other transmitter-specific pathways concerned with the reactivity of animals to stimuli and with their ability to initiate movements and to perform consummatory responses.

The Problem of Recording a Relevant Physiological Parameter in a Behaving Animal

Some of the problems of nonspecificity of the techniques used to experimentally manipulate the CNS can be avoided by using a different research strategy—recording or monitoring an appropriate physiological or other biological parameters in a behaving animal. For example, blood sugar levels may be determined in relation to feeding behavior (Steffens, 1970), blood levels of Angiotensin II may be determined in relation to drinking behavior (Abdelaal, Mercer, & Mogenson, 1976), or the activity of single neurons in the hippocampus may be recorded from when an animal explores its environment (Ranck, 1975). However, this approach has problems as well, and the experiments may be technically very difficult to do. It is not easy to monitor the release of noradrenaline in the cerebral cortex or hippocampus or to record the activity of single neurons in the hypothalamus or amygdala in relation to aggressive, copulatory, or feeding behavior. For this reason this strategy has not been used very extensively. There are relatively few studies in which an easily interpretable physiological parameter has been recorded during motivated and emotional behaviors. Furthermore, because of the technical difficulties of this approach, the behavioral test situations have been very simple and often highly artificial.

This research strategy has been used most extensively to relate the activity of single neurons of the hypothalamus and associated structures to motivated behavior. However, even if the enormous technical difficulties of recording in freely moving animals can be overcome, there are also problems of interpretation. Typically, recordings are made from interneurons remote from sensory inputs and effectors, and it is not always easy to decide whether changes in the activity of the recorded neurons is related to "attentional," "associative," or motor processes rather than to motivational processes (for discussion of this difficulty, see Mogenson, 1975, pp. 261–262). With this reservation about possible difficulties of interpretation, it should be recognized, however, that electrophysiological recording techniques seem admirably suited to the identification and investigation of central receptors that contribute to the initiation of thermoregulatory behaviors, drinking, feeding, and other behaviors. There has been some measure of success in using this approach to study central mechanisms for temperature regulation and thermoregulatory behaviors (e.g., Hardy et al., 1970), but studies to identify osmoreceptors and glucoreceptors associated with thirst and hunger have not been definitive (e.g., Mogenson, 1975). Chronic recordings have been made from osmosensitive neurons in the hypothalamus (see Figure 5.15), but it has not been established whether they are osmoreceptors or interneurons concerned with the preparatory or motor aspects of drinking behavior. Recordings have also been made from glucosensitive cells in the hypothalamus, and studies using microiontophoresis indicate that they are probably glucoreceptors (see Figure 2.23), but they may not be concerned with the initiation of feeding behavior as much as with the autonomic and endocrine control of blood sugar levels (see Chapter 4).

In spite of these difficulties, the technique of electrophysiological recording in the chronic animal is a potentially important approach, and some impressive advances have been made. The best example, although somewhat peripheral to the main theme of this book, is the work of Evarts (1975), who has recorded from neurons in the motor system during limb movements in monkeys and revolutionized the field of motor neurophysiology. Olds and co-workers (e.g., Phillips & Olds, 1969) and Rolls (1975) have made promising beginnings in relating the activity of neurons in the hypothalamus and associated structures to feeding behavior and brain-stimulation reward. As described in Chapters 4 and 8, Rolls and co-workers (e.g., Burton et al., 1976) have reported in the unanesthetized monkey that neurons in the hypothalamus respond when the animal is presented with preferred foods but not when nonpreferred foods are presented (see Figure 4.10). Control experiments indicate that these neurons are associated with the appetitive phase of feeding and are not merely motor correlates associated with the consummatory phase. Another example, in which electrophysiological recording from chronic, freely-moving animals has paid off, is the work of Ranck (1975) and of Vanderwolf and co-workers (Vanderwolf et al., 1975), who have shown a relationship of neuronal activity, and in particular the rhythmical slow waves recorded from the hippocampus, to the behavior of rats.

In concluding this section, it should be emphasized that progress in a research field depends very much on the experimental techniques that are available. Research may be handicapped, because available techniques are not appropriate to tackle or to answer certain questions. On the other hand, the development of new techniques may provide opportunities for empirical discoveries and may lead to new ways of thinking about a problem and to significant conceptual advances. Some of the techniques used for studying the neural substrates of motivated behaviors are relatively crude, as indicated previously; this does not mean that they should not be used—they may be the best that are available—but it is important to recognize their limitations and to be very cautious in interpreting the results obtained.

THE VIRTUES AND LIMITATIONS OF THEORETICAL MODELS

The other important ingredient in research in addition to appropriate experimental techniques and research strategies is the theoretical orientation and conceptual framework that is employed. Theoretical models can be very valuable in the choice of experiments and in the provision of a conceptual framework for interpreting experimental observations.[2] The virtue of models, and what makes their use so tempting, is that they help to organize and to simplify a vast amount of what would otherwise be a disparate array of data.

Unfortunately, theoretical models do not always reflect the complexity of the brain and brain function or the complexity of behavior. This was the case for the dual-mechanism model that was popular and influential for nearly two decades. Because the early investigations of the neural substrates of feeding, drinking, thermoregulatory, and other behaviors were undertaken at a time when these behaviors are considered only in terms of their contributions to homeostasis and biological adaptation, the classical studies of Richter (1943) had a major influence on this field and led to an emphasis on homeostatic determinants of behavior. The pioneering lesion and stimulation studies, which implicated the hypothalamus in feeding, drinking, thermoregulation, etc., provided experimental evidence on which the dual-mechanism model was formulated (Fig. 1.8A).

In recent years it has been more widely recognized that understanding the neural substrates of behaviors initiated by homeostatic, deficit signals (although interesting and worthwhile) is only part of the story. A number of behaviors (e.g., aggressive, sexual) are not initiated by homeostatic, deficit signals. Furthermore, even for the so-called self-regulatory behaviors that do contribute to homeostasis (e.g., thermoregulatory behavior, feeding, drinking), there are other determinants that may be the most important. For example, in Chapter 5, it was indicated that rats drink more water than required for body-fluid homeostasis, drinking being associated with circadian rhythms, habit and experience, palatability cues, etc. In Chapter 3, it was suggested that behavioral thermoregulatory responses may anticipate rather than occur in response to deviations in body temperature or even in response to temperature signals from the skin. Experiential and cognitive determinants of motivated behaviors, although clearly demonstrable in the rat and other laboratory animals, become much more important in higher primates and man. For the so-called emotional behaviors (for example, aggressive responding), experiential and cognitive factors are the major determinants, and homeostatic (and hormonal) factors are involved only indirectly in that they may increase the reactivity or responsiveness of the animal to the emotion-evoking situation. Experiential and cognitive factors are also important determinants of sexual behavior, especially in man and primates.

The dual-mechanism model with its emphasis on homeostatic determinants and hypothalamic mechanisms was not only overly simplistic in that it did not recognize sufficiently these other determinants of behavior, but it led us to believe unwittingly that we knew more about brain function and behavior than

[2]According to Hinde (1970), "The crossing of the bridge between behavior and physiology can often be facilitated by the use of models. . . . They serve merely to indicate the job which the physiological machinery must be doing to provide the physiologist with an indication of what he must look for [pp. 8–9]."

was justified from the available experimental evidence. Another example of theoretical ideas and models that are overly simplistic or wrong and that continue for a number of years to influence research and obscure the true complexity of brain function concerns the physiological mechanisms of sleep. For more than 20 years, a widely accepted view was that sleep is a passive process (see Chapter 7). This was during a period of active investigation of the electrophysiology of sleep and waking, and the theoretical model strongly influenced the direction of research and the interpretation of experimental observations, some of which were consistent with the opposite view—that sleep is an active process. Finally after the report of results completely incompatible with the passive sleep model (Batini et al., 1958), a theoretical reorientation took place, and sleep was considered subsequently to result from an active process.

In summary, theoretical models are important and even indispensable as guides in the choice of relevant experiments and as assistance in interpreting experimental results. However, a model has to be viewed in relation to a wider conceptual framework that takes account of all the determinants of the phenomenon under consideration, and it must be continuously revised and updated in light of new experimental evidence. Unfortunately, this has not always been the case in trying to understand the neural substrates of motivated behaviors. For a number of years, emphasis was placed on certain factors or determinants (e.g., homeostatic, deficit signals) and on certain neural components (e.g., hypothalamus), and other factors and neural systems were ignored. This is a major reason for the gaps in our knowledge of the neural mechanisms that subserve motivated behaviors and in particular that subserve experiential and cognitive factors. We return to this point later, but first, in view of the prominent role that the hypothalamus has had in this field for a number of years, it seems appropriate to consider its current status.

THE HYPOTHALAMUS AND MOTIVATED BEHAVIOR: CURRENT STATUS

One of the first neural structures to be implicated in motivated behaviors was the hypothalamus. Using the stereotaxic procedure, which made it possible to produce discrete lesions in deep structures of the brain, it was shown in the 1940s and 1950s that lesions of the hypothalamus dramatically altered a number of motivated behaviors, including feeding, drinking, sexual, and aggressive behaviors. Electrical stimulation of the hypothalamus using electrodes placed stereotaxically were shown to elicit feeding, drinking, thermoregulatory responses, sexual, and aggressive behaviors. Evidence of this sort suggests that the hypothalamus has an important role in motivated behaviors.

In recent years many of these findings have been re-interpreted. The results of stimulation experiments do not necessarily reflect the activation of specific neural systems associated with feeding, drinking, or sexual behavior. Deficits in feeding and drinking behaviors following lesions of the lateral hypothalamus have been attributed to the disruption of motor functions, sensory–motor integration, and behavioral arousal rather than to damage to integrative mechanisms that subserve hunger and thirst motivation. It is appropriate to ask, in view of these developments, what is the role of the hypothalamus in ingestive and other motivated and emotional behaviors, and does the hypothalamus make any contribution at all.

It was suggested previously that in spite of these recent developments and the reassessment of the dual-mechanism model, one should not disregard or neglect the hypothalamus when considering the neural substrates of feeding behavior. This is also the case with some and perhaps all of the other behaviors considered in this book.

By using a thermode to administer a thermal stimulus to the preoptic region behavioral thermoregulatory responses as well as autonomic and endocrine thermoregulatory responses, can be elicited (see Chapter 3). Because these experiments involve a physiologically appropriate stimulus, the results cannot be attributed to nonspecific effects as in the case of experiments in which electrical stimulation is used. There is also evidence that the preoptic region contains osmoreceptors and is responsive to the dipsogenic hormone, angiotensin (see Chapter 5). Furthermore, there are regions of the hypothalamus that are responsive to estrogen and testosterone, and there is evidence that these hormones sensitize or activate neural systems subserving sexual behavior (see Chapter 6). Although the integrative mechanisms by which these thermal, neural, and endocrine signals initiate the various behaviors are poorly understood, an important role for the hypothalamus cannot be ruled out on the basis of existing experimental evidence.

It is well known that there are fiber pathways between the midbrain and brainstem and forebrain structures that pass through or in the region of the hypothalamus. However, to suggest that all the experimental evidence from classical and more recent experiments on the hypothalamus can be attributed merely to damaging, activating, or blocking these pathways is a gross oversimplification. Lesions and electrical stimulation may be too crude to reveal definitively what the contributions of the hypothalamus are to the various motivated behaviors. However, some of the newer techniques may not be anatomically, physiologically, or behaviorally specific either, and partly because of these limitations, the neural substrates of feeding, aggression, brain-stimulation reward, etc., are poorly understood. Although it is proper to be skeptical about experimental evidence that implicates the hypothalamus in motivated behaviors, it would be a serious mistake to conclude that the hypothalamus has no role and to ignore it in future research.

Even in behaviors that depend very much on experiential and cognitive factors, involving more rostral structures of the brain, the hypothalamus may make an indirect contribution. Motivated and emotional behaviors depend upon the coordination of autonomic, endocrine, and skeletomotor systems (see Figure 9.3). The hypothalamus by virtue of its strategic location in the brain—having functional relationships with the pituitary gland, the autonomic nervous system, limbic forebrain structures, and neural systems for motor control—is in a position to contribute to the "functional coupling" of physiological regulations and behavioral responses (Figures 1.8b and 1.9). As indicated in Chapters 1 and 2, the "housekeeping chores" of the body are taken care of at the same time that the animal engages in a variety of behaviors. This vital and distinctive contribution of the hypothalamus and associated limbic structures ensures that physiological responses (e.g., increased heart rate and blood pressure) are coordinated with, and appropriate to, the particular behavioral responses (e.g., attack or escape) as an animal adapts to its complex, continuously changing environment.

The results of an experiment that demonstrates the contribution of the hypothalamus to behavioral and visceral responses are shown in Figure 10.1. Electrical stimulation of one site in the hypothalamus of the cat elicited feeding, and from another site "fearlike" or defensive behavior was elicited. Subsequently, these sites were stimulated when the cat was anesthetized. Stimulation of the "feeding site" resulted in increased intestinal motility, increased blood flow to the intestine, and reduced blood flow to skeletal muscles. In contrast, stimulation of the "defensive site" resulted in reduced intestinal motility and blood flow and increased blood flow to skeletal muscles. In both cases blood pressure was increased. These observations illustrate the role of the hypothalamus in patterns of autonomic responses appropriate to the behavioral responses (see Figures 1.8B, 1.9).

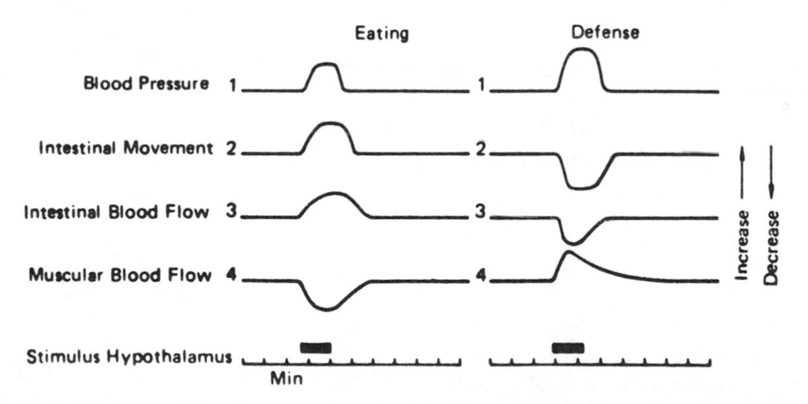

FIG. 10.1 Measures of blood pressure, intestinal motility, and blood flow to intestine and skeletal muscle when sites in the hypothalamus of the anesthetized cat are stimulated. Electrical stimulation in the unanesthetized animal had previously been shown to elicit feeding or a defensive, "fearlike" response. It is concluded that the hypothalamus contributes to autonomic, endocrine, and skeletomotor responses that are coordinated to the behavioral circumstances. (From Folkow & Rubinstein, 1965, as modified by Schmidt, 1975.)

CURRENT AND FUTURE DEVELOPMENTS

In the previous section, we referred to the deemphasis in recent years on the hypothalamus and homeostatic mechanisms and increased emphasis on nonhomeostatic determinants of motivated behaviors, particularly with experiential and cognitive factors. This has been in part because of difficulties with the dual-mechanism model and because of a reinterpretation of the classical hypothalamic lesion and stimulation experiments. Another factor is the influence of ethology and comparative psychology and a growing interest in the study of feeding, thermoregulatory, aggressive, and other behaviors in more natural circumstances. This interest, especially in dealing with the behavior of higher mammals, primates, and man, has led to a greater stress on experiential and cognitive factors. Even in studies of ingestive behaviors, classically regarded as "homeostatic drives," there has been increasing interest in the determinants of spontaneous food and water intakes (in contrast to food and water intakes in the deprived animal) and in neural mechanisms that enable the rat or other animal to drink or feed prior to the occurrence of water or energy deficits. It has been suggested that these nonregulatory or anticipatory mechanisms may be the ones normally utilized by the animal—homeostatic mechanisms having a back-up or emergency role (e.g., Fitzsimons, 1972).

Another major development in recent years has been the investigation of the role of transmitter-specific pathways in motivated behaviors. This has come about because of the introduction of new techniques: Histofluorescence has been used to visualize dopaminergic, noradrenergic, and serotonergic neurons in the midbrain and brainstem and to map their projections to forebrain structures (see Figure 2.24); neuropharmacological and neurochemical techniques used in combination with the classical lesion technique have implicated these pathways in ingestive, sexual, and emotional behaviors and in sleep–waking, thermoregulation, and brain-stimulation reward. The results of these studies have had a major impact on traditional concepts of the hypothalamus and behavior and on the dual-mechanism model. Although the role of monoaminergic and other transmitter-specific pathways in these behaviors is as yet poorly understood, this will continue to be a vigorous field of research. The interest of a number of neuroscientists has been stimulated by recent achievements in the treatment of Parkinson's disease and by the possibility that one or more of these pathways (e.g., mesolimbic-mesocortical dopaminergic pathway) may be involved in schizophrenia.

Behavioral deficits resulting from damage to monoaminergic neurons, particularly to the nigrostriatal dopamine pathway, have directed attention away from hypothalamic integrative mechanisms to neural systems for the motor control of behavior. An important priority for future investigation concerns the interface between the neural integrative mechanisms that subserve the various determinants of behavior and the neural mechanisms for the motor control of behavior

(Mogenson & Phillips, 1976), a previously neglected subject. McGeer (1976), in a lecture to the Society for Neuroscience, aptly reflected current thinking when he spoke of *mood* and *movement* as being "twin galaxies." By this he meant that understanding the neural substrates of affect, drive, and motivation and the neural substrates of action or behavior is a basic problem in neuroscience. He designated them as "twin galaxies" because they are closely related functionally, so that understanding one of these aspects of brain function depends on understanding the other.

At the same meeting, Graybiel (1976) showed that the neural mechanisms subserving mood and movement—or in our terminology, motivation and behavior—are also intimately related anatomically. Her lecture dealt with our current understanding of the anatomical interface between the neural systems implicated in motivation and those implicated in the motor control of behavior (Figure 10.2). Graybiel suggested that the ventral striatum (nucleus accumbens) is a key structure. The ventral striatum receives fiber projections from the dopaminergic neurons (A_{10}) in the ventral tegmental area and in turn sends fiber projections to the dopaminergic neurons (A_9) of the substantia nigra. She proposes that this loop (ventral tegmental area → nucleus accumbens → substantia

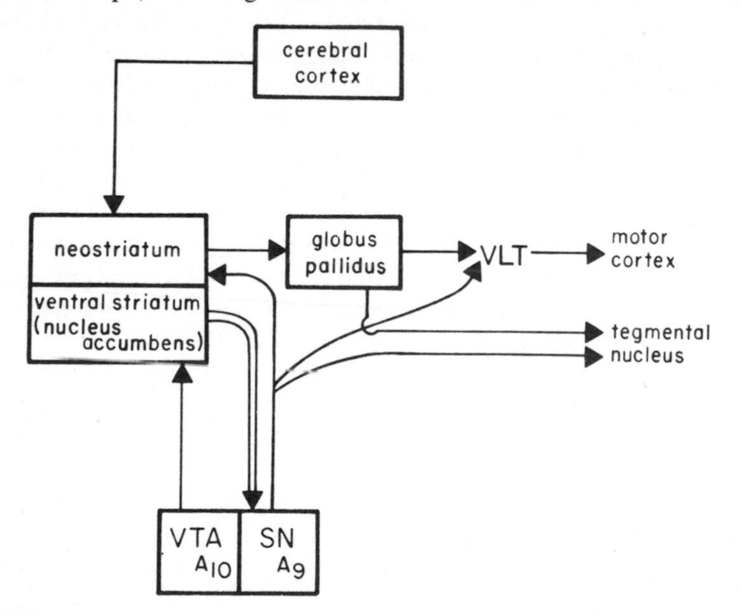

FIG. 10.2 The nucleus accumbens (ventral striatum) receives projections from dopaminergic neurons in the ventral tegmental area (VTA, A_{10}). The nucleus accumbens sends fiber projections to the substantia nigra (A_9), which in turn projects to the neostriatum. According to Graybiel, this loop may have an important role as a link between neural systems for motivation and neural systems for motor control. The neostriatum, which samples the activity of several areas of the brain and sends signals to the globus pallidus, is one of the key structures in the motor control of behavior (see Figure 2.11).

nigra) influences signals to the neostriatum from the substantia nigra. The neostriatum (see Figure 2.11) receives inputs from the neocortex and limbic structures and is in a position to influence "command signals" for the initiation and control of movements. Experimental investigation of this model during the next few years will no doubt help to elucidate the neural mechanisms by which motivation (e.g., or in neurologists' terms, "the intention to respond") gets translated into movement.

Another neglected subject, one that should have top priority in future research, is to investigate the neural mechanisms by which experiential and cognitive factors initiate and contribute to behavior. In going from one chapter to another—from one behavior to another—it has become increasingly apparent, on the one hand, that experiential and cognitive factors are very important determinants and, on the other hand, that the experimental evidence concerning the neural mechanisms that subserve these factors is very limited. This is a difficult challenge and a top priority for future research. There will not be a "neurology of feeding" or a "neurology of aggressive behavior" that is at all adequate until there is a better understanding of the neural basis of learning, memory, and other cognitive processes.[3]

An understanding of the neural substrates of experiential and cognitive factors will be greatly facilitated by developmental studies of brain and behavior. Memory, learning, and other cognitive processes develop gradually as the central nervous system matures and as the animal interacts with its external environment. As proposed by Lindsley (1972), "the developmental approach to behavior and to the study of the brain appears to be one of the most powerful and versatile [p. 446]." The ontogeny of the various behaviors was considered briefly in previous chapters in the hope that this important dimension will not be neglected in future studies of the neural substrates of motivated and emotional behaviors.

[3]According to Lindsley (cited in Clemente, Purpura, & Mayer, 1972):

Brain mechanisms are the neural, structurofunctional patterns by means of which an organism is enabled to interact with its environment. Some patterns, relatively fixed and vital to life processes, represent the maturational unfolding and development of hereditary patterns. These, sometimes referred to as "built in" or "soldered-in" connections, result in so-called "instinctive" or unlearned behaviors. Other, more dynamically oriented, structurofunctional patterns permit a wider and more variable range of interactions with the environment and depend upon a plastic or modifiable record of the interaction and their consequences. The memory storage and utilization, over shorter or longer intervals, of such a dynamic record constitutes learning. The potential for its adaptive use may be called intelligence and the proficiency in learning to adapt, achievement [p. 447].

REFERENCES

Abdelaal, A. E., Mercer, P. F., & Mogenson, G. J. Plasma angiotensin II levels and water intake following β-adrenergic stimulation, hypovolemia, cellular dehydration and water deprivation. *Pharmacology, Biochemistry and Behavior,* 1976, *4,* 317–321.

Abrahams, V. C., Hilton, S. M., & Zbrozyna, A. The role of active muscle vasodilation in the alerting stage of the defense reaction. *Journal of Physiology (London),* 1964, *171,* 189–202.

Adair, E. R., Casby, J. U., & Stolwijk, J. A. J. Behavioral temperature regulation in the squirrel monkey: Changes induced by shifts in hypothalamic temperature. *Journal of Comparative and Physiological Psychology,* 1970, *72,* 17–27.

Adair, E. R., & Hardy, J. D. Posterior hypothalamic thermal stimulation can alter behavioral, but not physiological temperature regulation. *Proceedings of the International Union of Physiological Sciences,* 1971, *9,* 7.

Adair, E. R., Miller, N. E., & Booth, D. A. Effects of continuous intravenous infusion of nutritive substances on consummatory behavior in rats. *Communications in Behavioral Biology,* 1968, *2,* 25–37.

Adolph, E. F. Measurements of water drinking in dogs. *American Journal of Physiology,* 1939, *125,* 75–86.

Adolph, E. F. Urges to eat and drink in rats. *American Journal of Physiology,* 1947, *151,* 110–125.

Adolph, E. F. Thirst and its inhibition in the stomach. *American Journal of Physiology,* 1950, *161,* 374–386.

Adolph, E. F. Regulation of body water content through water ingestion. In M. J. Wayner (Ed.), *Thirst.* New York: Macmillan, 1964.

Adolph, E. F. Regulation of water intake in relation to body water content. In C. F. Code (Ed.) *Handbook of Physiology,* Alimentary Canal (Sect. 6), Food and Water Intake (Vol. 1). Washington, D.C.: American Physiological Society, 1967.

Ahlskog, J. E., & Hoebel, B. G. Overeating and obesity from damage to a noradrenergic system in the brain. *Science,* 1973, *182,* 166–168.

Albert, D. J., & Storlien, L. H. Hyperphagia in rats with cuts between the ventromedial and lateral hypothalamus. *Science,* 1969, *165,* 599–600.

Anand, B. K., & Brobeck, J. R. Hypothalamic control of food intake in rats and cats. *Yale Journal of Biology and Medicine,* 1951, *24,* 123–140.

Anand, B. K., Chhina, G. S., & Singh, B. Effect of glucose on the activity of hypothalamic "feeding centers." *Science,* 1962, *138,* 597–598.

Anderson, K. M., & Liao, S. Selective retention of dihydrotestosterone by prostatic nuclei. *Nature,* 1968, *219,* 277–279.

Andersson, B. The effect of injections of hypertonic NaCl solutions into different parts of the hypothalamus of goats. *Acta Physiologica Scandinavica,* 1953, *28,* 188–201.

Andersson, B. Thirst—And brain control of water balance. *American Scientist,* 1971, *59,* 408–415.

Andersson, B., Brook, A. H., & Ekman, L. Further studies of the thyroidal response to local cooling of the "heat loss center." *Acta Physiologica Scandinavica,* 1965, *63,* 186–192.

Andersson, B., Brook, A. H., Gale, C. C., & Hökfelt, B. The effect of a ganglionic blocking agent on the thermoregulatory response to preoptic cooling. *Acta Physiologica Scandinavica,* 1964, *61,* 393–399.

Andersson, B., Gale, C. C., & Sundsten, J. W. Effects of chronic central cooling on alimentation and thermoregulation. *Acta Physiologica Scandinavica,* 1962, *55,* 177–188.

Aserinsky, E., & Kleitman, N. 2 types of ocular motility occurring in sleep. *Journal of Applied Physiology,* 1955, *8,* 11–18.

Assaf, S. Y., & Mogenson, G. J. Evidence that angiotensin II acts on the preoptic region to elicit water intake. *Pharmacology, Biochemistry and Behavior,* 1977, *5,* 679–699.

Baile, C. A., Bean, S. M., Simpson, C. W., & Jacobs, H. L. Feeding effects of hypothalamic injections of prostaglandins. *Federation Proceedings,* 1971, *30,* 375.

Balagura, S., & Devenport, L. D. Feeding patterns of normal and ventromedial hypothalamic lesioned male and female rats. *Journal of Comparative and Physiological Psychology,* 1970, *71,* 357–364.

Balagura, S., & Hoebel, B. G. Self-stimulation of the lateral hypothalamus modified by insulin and glucagon. *Physiology and Behavior,* 1967, *2,* 337–340.

Baldwin, B. A., & Ingram, D. L. The effect of heating and cooling the hypothalamus on behavioural thermoregulation in the pig. *Journal of Physiology (London),* 1967, *191,* 375–392.

Ball, J. Sex activity of castrated male rats increased by estrin administration. *Journal of Comparative Psychology,* 1937, *28,* 273–283.

Barbeau, A. L-Dopa therapy in Parkinson's disease: A critical review of nine years experience. *Journal of the Canadian Medical Association,* 1969, *101,* 791.

Barchas, J., Ciaranello, R., Stolk, J., Brodie, H. K. H., & Hamburg, D. Biogenic amines and behavior. In S. Levine (Ed.), *Hormones and behavior.* New York: Academic, 1972.

Bard, P. On emotional expression after decortication with some remarks on certain theoretical views. *Psychological Review,* 1934, *41,* 309–329; 424–429.

Barfield, R. J., Wilson, C., & McDonald, P. G. Sexual behavior: Extreme reduction of postejaculatory refractory period by midbrain lesions in male rats. *Science,* 1975, *189,* 147–149.

Barnwell, G. M. A dynamical system theory approach to food intake control. In D. A. Booth (Ed.), *Hunger models: Computable theory of feeding control.* London: Academic, 1977.

Batini, C., Moruzzi, G., Palestini, M., Rossi, G. F., & Zanchetti, A. Persistent patterns of wakefulness in the pretrigeminal midpontine preparation. *Science,* 1958, *128,* 30–32.

Batini, C., Moruzzi, G., Palestini, M., Rossi, G. F., & Zanchetti, A. Effects of complete pontine transection on the sleep–wakefulness rhythm: The midpontine pretrigeminal preparation. *Archives Italiennes de Biologie,* 1959, *97,* 1–12.

Baum, M. J., & Vreeburg, J. T. M. Copulation in castrated male rats following combined treatment with estradiol and dihydrotestosterone. *Science,* 1973, *182,* 283–385.

Baxter, L., Gluckman, M. I., Stein, K., & Scerni, A. Self-injection of apomorphine in the rat: Positive reinforcement by a dopamine receptor stimulant. *Pharmacology, Biochemistry and Behavior,* 1974, *2,* 387–391.

Beach, F. A. Effects of cortical lesions upon the copulatory behavior of male rats. *Journal of Comparative Psychology,* 1940, *29,* 193–244.

Beach, F. A. Analysis of the stimuli adequate to elicit mating behavior in the sexually inexperienced male rat. *Journal of Comparative Psychology,* 1942, *33,* 163–207. (a)

Beach, F. A. Copulatory behavior in prepuberally castrated male rats and its modification by estrogen administration. *Endocrinology,* 1942, *31,* 679–683. (b)

Beach, F. A. Male and female mating behavior in prepuberally castrated female rats treated with androgens. *Endocrinology,* 1942, *31,* 673–678. (c)

Beach, F. A. Effects of injury to the cerebral cortex upon the display of masculine and feminine mating behavior by female rats. *Journal of Comparative Psychology,* 1943, *36,* 169–198.

Beach, F. A. Effects of injury to the cerebral cortex upon sexually-receptive behavior in the female rat. *Psychosomatic Medicine,* 1944, *6,* 40–55.

Beach, F. A. Evolutionary changes in the physiological control of mating behavior. *Psychological Review,* 1947, *54,* 297–315.

Beach, F. A. Cerebral and hormonal control of reflexive mechanisms involved in copulatory behavior. Physiological Reviews, 1967, *47,* 289–316.

Beach, F. A. Coital behavior in dogs. VI. Long-term effects of castration upon mating in the male. *Journal of Comparative and Physiological Psychology Monograph,* 1970, *70* (No. 3, Pt. 2).

Beach, F. A., & Fowler, H. Individual differences in the response of male rats to androgen. *Journal of Comparative and Physiological Psychology.* 1959, *52,* 50–52.

Beach, F. A., Zitrin, A., & Jaynes, J. Neural mediation of mating in male cats: II. Contribution of the frontal cortex. *Journal of Experimental Zoology,* 1955, *130,* 381–401.

Beach, F. A., Zitrin, A., & Jaynes, J. Neural mediation of mating in male cats: I. Effects of unilateral and bilateral removal of the neocortex. *Journal of Comparative and Physiological Psychology,* 1956, *49,* 321–327.

Benkert, O. Pharmacological experiments to stimulate human sexual behavior. In T. A. Ban, J. R. Boissier, G. J. Gessa, H. Heimann, L. Hollister, H. E. Lehmann, I. Munkvad, H. Steinberg, F. Sulser, A. Sundwall, & O. Vinar (Eds), *Psychopharmacology, sexual disorders and drug abuse.* Amsterdam and London: North-Holland, 1973.

Benzinger, T. H. The thermal homeostasis of man. In G. M. Hughes (Ed.), *Homeostasis and Feedback Mechanisms.* Cambridge: Cambridge University Press, 1964.

Benzinger, T. H. Peripheral cold reception and central warm reception, sensory mechanisms of behavioral and autonomic homeostasis. In J. D. Hardy, A. P. Gagge, & J. A. J. Stolwijk (Eds.), *Physiological and behavioral temperature regulation.* Springfield, Ill.: C. C. Thomas, 1970.

Berger, H. Über das Elektroenzephalogramm beim Menschen. *Archiv für Psychiatrie und Nervenkrankheiten (Berlin),* 1929, *87,* 527–571.

Berger, B. D., Wise, C. D., & Stein, L. Norepinephrine: Reversal of anorexia in rats with lateral hypothalamic damage. *Science,* 1971, *172,* 281–284.

Berlyne, D. E. *Conflict, arousal and curiosity.* New York: McGraw-Hill, 1960.

Bermant, G., Glickman, S. E., & Davidson, J. M. Effects of limbic lesions on copulatory behavior of male rats. *Journal of Comparative and Physiological Psychology,* 1968, *65,* 118–125.

Bernard, C. *Leçons de physiologie expérimentale appliquée à la medicine,* (Vol. 2). Cours du semestre d'été. Paris: Baillière, 1856.

Bernard, C. *Leçons sur les Phénomènes de la Vie Communs aux Animaux et aux Végétaux* (Vol. 1). Paris: Bailliere, 1878.

Bernardis, L. L. Disruption of diurnal feeding and weight gain cycles in weanling rats with ventromedial and dorsomedial hypothalamic lesions. *Physiology and Behavior,* 1973, *10,* 855–861.

Berthoud, H., & Mogenson, G. J. Consumatory behavior after intracerebral and intracerebroventricular infusions of glucose and 2-deoxy-D-glucose. *American Journal of Physiology, Integrative & Comparative Physiology,* 1977 (in press).

Beyer, C., Morali, G., & Vargas, R. Effect of diverse estrogens on estrous behavior and genital tract development in ovariectomized rats. *Hormones and Behavior,* 1971, 2, 273–277.

Bindra, D. Neuropsychological interpretation of the effects of drive and incentive-motivation on general activity and instrumental behavior. *Psychological Review,* 1968, *75,* 1–22.

Blandau, R. J., Boling, J. L., & Young, W. C. The length of heat in the albino rat as determined by the copulatory response. *Anatomical Record,* 1941, *79,* 453–463.

Blass, E. M. Evidence for basal forebrain thirst osmoreceptors in the rat. *Brain Research,* 1974, *82,* 69–76. (a)

Blass, E. M. The physiological, neurological and behavioral bases of thirst. In J. K. Cole & T. B. Sonderegger (Eds.) *Nebraska Symposium on Motivation* (Vol. 22). Lincoln: University of Nebraska Press, 1974. (b)

Blass, E. M., & Epstein, A. N. A lateral preoptic osmosensitive zone for thirst in the rat. *Journal of Comparative and Physiological Psychology,* 1971, *76,* 378–394.

Bligh, J. *Temperature regulation in mammals and other vertebrates.* Amsterdam: North-Holland, 1973.

Blundell, J. E. Anorexic drugs, food intake and the study of obesity. *Nutrition (London),* 1975, *29,* 5–18.

Blundell, J. E., Latham, C. J., & Leshem, M. B. Differences between the anorectic actions of amphetamine and fenfluramine—Possible effects on hunger and satiety. *Journal of Pharmacology and Pharmacy,* 1976, *28,* 471–477.

Boling, J. L., & Blandau, R. J. The estrogen–progesterone induction of mating responses in the spayed female rat. *Endocrinology,* 1939, *25,* 359–364.

Bolles, R. C. *Theory of motivation.* New York: Harper & Row, 1967.

Bolles, R. C. Species-specific defense reactions and avoidance learning. *Psychological Review,* 1970, *77,* 32–48.

Bolles, R. C. Reinforcement, expectancy and learning. *Psychological Review,* 1972, *79,* 394–409.

Booth, D. A. *Hunger Models: Computable Theory of Feeding Control.* London: Academic, 1977.

Booth, D. A., Toates, F. M., & Platt, S. V. Control system for hunger and its implications in animals and man. In D. Novin, W. Wyrwicka, & G. A. Bray (Eds), *Hunger: Basic Mechanisms and Clinical Implications.* New York: Raven, 1976.

Boyd, E. S., & Gardner, L. C. Effect of some brain lesions on intracranial self-stimulation in the rat. *American Journal of Physiology,* 1967, *213,* 1044–1052.

Brady, J. V. Emotion and the sensitivity of psychoendocrine systems. In D. G. Glass (Ed.), *Neurophysiology and emotion.* New York: Rockefeller University Press, 1967.

Brady, J. V., Boren, J. J., Conrad, D., & Sidman, M. The effect of food and water deprivation upon intracranial self-stimulation. *Journal of Comparative and Physiological Psychology,* 1957, *50,* 134–137.

Brady, J. V., & Nauta, W. J. H. Subcortical mechanisms in emotional behaviour: Affective changes following septal forebrain lesions in the albino rat. *Journal of Comparative and Physiological Psychology,* 1953, *46,* 339–346.

Bray, G. A. Peripheral metabolic factors in the regulation of feeding. In T. Silverstone (Ed.), *Appetitie and food intake.* Dahlem Konferenzen, Berlin: Abakon Verlagsgesellschaft, 1976.

Bray, G. A., & Campfield, L. A. Metabolic factors in the regulation of caloric stores. *Metabolism,* 1975, *24,* 99–117.

Breese, G. R., Smith, R. D., Cooper, B. R., & Grant, L. D. Alterations in consummatory behavior following intracisternal injection of 6-hydroxy-dopamine. *Pharmacology Biochemistry and Behavior,* 1973, *1,* 319–328.

Bremer, F. Cerveau isolé et physiologie du sommeil. *Comptes Rendus des Séances de la Societé de Biologie et de Ses Filiales (Paris),* 1935, *118,* 1235–1242.

Bremer, F. L'activité cérébrale au cours du sommeil et de la narcase. Contribution à l'étude du mécanisme du sommeil. *Bulletin de l'Academie Royale de Medécine de Belgique (Bruxelles),* 1937, *4,* 68–86.

Bremer, J. *Asexualization: A follow-up study of 244 cases.* New York: Macmillan, 1959.

Brimley, C. C., & Mogenson, G. J. Deficits in ingestive behavior after 6-hydroxydopamine administration to the A_9 and A_{10} dopamine pathways. *Canada Physiology,* 1977, *8,* 31.

Brobeck, J. R., Tepperman, J., & Long, C. N. H. Experimental hypothalamic hyperphagia in the albino rat. *Yale Journal of Biology and Medicine,* 1943, *15,* 831–853.

Brodie, D. A., Moreno, O. M., Malis, J. L., & Brodie, J. J. Rewarding properties of intracranial stimulation. *Science,* 1960, *131,* 929–930.

Brookhart, J. M., Dey, F. L., & Ranson, S. W. Failure of ovarian hormones to cause mating reactions in spayed guinea pigs with hypothalamic lesions. *Proceedings of the Society for Experimental Biology and Medicine,* 1940, *44,* 61–64.

Brookhart, J. M., Dey, F. L., & Ranson, S. W. The abolition of mating behavior by hypothalamic lesions in guinea pigs. *Endocrinology,* 1941, *28,* 561–565.

Brooks, C. McC. The role of the cerebral cortex and of various sense organs in the excitation and execution of mating activity in the rabbit. *American Journal of Physiology,* 1937, *120,* 544–553.

Brooks, V. B. Roles of cerebellum and basal ganglia in initiation and control of movements. *Canadian Journal of Neurological Sciences,* 1975, *2,* 265–277.

Brown, J. L., Hunsperger, R. W., & Rosvold, H. E. Defence, attack, and flight elicited by electrical stimulation of the hypothalamus of the cat. *Experimental Brain Research,* 1969, *8,* 113–129.

Brown, S., & Trowill, J. A. Lever-pressing performance for brain stimulation on F-I and V-I schedules in a single-lever situation. *Psychological Reports,* 1970, *26,* 699–706.

Bunney, B. S., & Aghajanian, G .The precise localization of nigral afferents in the rat as determined by a retrograde tracing technique. *Brain Research,* 1976, *117,* 423–435.

Burton, M. J., Rolls, E. T., & Mora, F. Effects of hunger on the responses of neurones in the lateral hypothalamus to the sight and taste of food. *Experimental Neurology,* 1976, *51,* 668–677.

Butter, C. M. Perseveration in extinction and in discrimination reversal tasks following selective frontal ablations in *Macaca mulatta. Physiology and Behavior,* 1969, *4,* 163–171.

Cabanac, M. Physiological role of pleasure. *Science,* 1971, *173,* 1103–1107.

Cabanac, M. Thermoregulatory behavior. In J. Bligh & R. Moore (Eds.), *Essays on temperature regulation.* New York: Elsevier, 1972.

Caggiula, A. R. Analysis of the copulation-reward properties of posterior hypothalamic stimulation in male rats. *Journal of Comparative and Physiological Psychology,* 1970, *70,* 399–412.

Campbell, H. J. Pleasure-seeking brains: Artificial tickles, natural joys of thought. *Smithsonian,* 1971, *2,* 14–23.

Cannon, W. B. The physiological basis of thirst. *Proceedings of the Royal Society of London,* Series B, 1918, *90,* 283–301.

Cannon, W. B. The James Lange theory of emotions: A critical examination and an alternative theory. *American Journal of Psychology,* 1927, *39,* 106–124.

Cannon, W. B. *Bodily changes in pain, hunger, fear and rage: An account of recent researches into the function of emotional excitement.* New York: Appleton-Century-Crofts, 1929.

Cannon, W. *The wisdom of the body.* New York: W. W. Morton & Co., 1932.

Cantor, M. B. Signaled reinforcing brain stimulation facilitates operant behavior under schedules of intermittent reinforcement. *Science,* 1971, *174,* 610–613.

Carr, W. J., Loeb, L. S., & Dissinger, M. L. Responses of rats to sex odors. *Journal of Comparative and Physiological Psychology,* 1965, *59,* 370–377.

Chase, M. H. *The sleeping brain, perspectives in the brain sciences* (Vol. 1). Los Angeles: Brain Information Service/Brain Research Institute, 1972.

Chi, C. C., & Flynn, J. P. Neuroanatomic projections related to biting attack elicited from hypothalamus in cats. *Brain Research,* 1971, *35,* 49–66.

Clark, T. K., Caggiula, A. R., McConnell, R. A., & Antelman, S. M. Sexual inhibition is reduced by rostral midbrain lesions in the male rat. *Science,* 1975, *190,* 169–171.

Clavier, R. M., & Routtenberg, A. Ascending monoamine-containing fiber pathways related to intracranial self-stimulation: Histochemical fluorescence study. *Brain Research,* 1974, *72,* 25–40.

Clemens, L. G., Wallen, K., & Gorski, R. A. Mating behavior: Facilitation in the female rat following cortical application of potassium chloride. *Science,* 1967. *137,* 1208–1209.

Clemente, C. D., Purpura, D. P., & Mayer, F. E. *Sleep and the maturing nervous system*. New York: Academic, 1972.

Cofer, C. N., & Appley, M. H. *Motivation: Theory and Research*. New York: Wiley, 1963.

Coindet, J., Chouvet, G., & Mouret, J. Effects of lesions of the suprachiasmatic nuclei on paradoxical sleep and slow wave sleep circadian rhythms in the rat. *Neuroscience Letters*, 1975, *1*, 243–247.

Collier, G., Hirsch, E., & Kanarek, R. The operant revised. In W. K. Honig & J. E. R. Staddon (Eds.), *Handbook of operant behavior*. Englewood Cliffs, N.J.: Prentice-Hall, 1976.

Comhaire, F., & Vermeulen, A. Plasma testosterone in patients with varicocele and sexual inadequacy. *Journal of Clinical Endocrinology and Metabolism*, 1975, *40*, 824–829.

Coons, E. F., & Cruce, J. A. F. Lateral hypothalamus: Food current intensity in maintaining self-stimulation of hunger. *Science*, 1968, *159*, 1117–1119.

Cooper, K. E., Cranston, W. I., & Honour, A. J. Observation on the site and mode of action of pyrogens in the rabbit brain. *Journal of Physiology (London)*, 1967, *191*, 325–337.

Cooper, R. M., & Taylor, L. H. Thalamic reticular system and central grey: Self-stimulation. *Science*, 1967, *156*, 102–103.

Cooper, S. J., & Rolls, E. T. Relation of activation of neurones in the pons and medulla to brain-stimulation reward. *Experimental Brain Research*, 1974, *20*, 207–222.

Corbit, J. D. Behavioral regulation of body temperature. In J. D. Hardy, A. P. Gagge, & J. A. J. Stolwijk (Eds.), *Physiological and behavioral temperature regulation*. Springfield, Ill.: C. C. Thomas, 1970.

Cowgill, U. The season of birth in man. *Man*, 1966, *1*, 232–239.

Craig, W. Appetites and aversions as constituents of instincts. *Biological Bulletin*, 1918, *34*, 91–107.

Crow, T. J. A map of the rat mesencephalon for electrical self-stimulation. *Brain Research*, 1972, *36*, 265–273.

Crow, T. J., Spear, P. J., & Arbuthnott, G. W. Intracranial self-stimulation with electrodes in the region of the locus coeruleus. *Brain Research*, 1972, *36*, 275–287.

Culberton, J. L., Kling, J. W., & Berkley, M. A. Extinction responding following ICS and food reinforcement. *Psychonomic Science*, 1966, *5*, 127–128.

Dahlström, A., & Fuxe, K. Evidence for the existence of monoamine-containing neurons in the central nervous system. Demonstration of monoamines in the cell bodies of brainstem neurons. *Acta Physiologica Scandinavica*, 1964, *62*, Suppl. 232, 1–55.

Darwin, C. *The origin of species*. Chicago: Donohue, Henneberry & Co., 1859.

Darwin, C. *The expression of the emotions in man and animals*. Chicago: University of Chicago Press, 1872/1965.

Darwin, C. *The descent of man*. New York: H. M. Caldwell, 1874.

Davidson, J. M. Activation of the male rat's sexual behavior by intracerebral implantation of androgen. *Endocrinology*, 1966, *79*, 783–794. (a)

Davidson, J. M. Characteristics of sex behaviour in male rats following castration. *Animal Behaviour*, 1966, *14*, 266–272. (b)

Debons, A. F., Krimsky, I., Likuski, H. J., From, A., & Cloutier, R. J. Gold thioglucose damage to the satiety center: Inhibition in diabetes. *American Journal of Physiology*, 1968, *214*, 652–658.

Debons, A. F., Krimsky, I., From, A., & Cloutier, R. J. Rapid effects of insulin on the hypothalamic satiety center. *American Journal of Physiology*, 1969, *217*, 1114–1118.

Dement, W. The occurrence of low voltage, fast electroencephalogram patterns during behavioral sleep in the cat. *Electroencephalography and Clinical Neurophysiology*, 1958, *10*, 291–296.

Dement, W., & Kleitman, N. Incidence of eye motility during sleep in relation to varying EEG pattern. *Federation Proceedings*, 1955, *14*, 216.

Dement, W., & Kleitman, N. Cyclic variations in EEG during sleep and their relation to eye movements, body motility, and dreaming. *Electroencephalography and Clinical Neurophysiology*, 1957, *9*, 673–690.

DeRuiter, L. Feeding behavior of vertebrates in the natural environment. In C. F. Code (Ed), *Handbook of Physiology,* Alimentary Canal (Sect. 6), Food and Water Intake (Vol. 1). Washington, D.C.: American Physiological Society, 1967.

DeRuiter, L., Wiepkema, P. R., & Veening, J. G. Models of behavior and the hypothalamus. *Progress in Brain Research,* 1974, *41,* 481–507.

Deutsch, J. A., Adams, D. W., & Metzner, R. J. Choice of intracranial stimulation as a function of delay between stimulations and strength of competing drive. *Journal of Comparative Physiological Psychology,* 1964, *57,* 241–243.

Deutsch, J. A., & DiCara, L. Hunger and extinction in intracranial self-stimulation. *Journal of Comparative Physiological Psychology,* 1967, *63,* 344–347.

Deutsch, J. A., & Howarth, C. I. Some tests of a theory of intracranial self-stimulation. *Psychological Review,* 1963, *70,* 444–460.

DiCara, L. V., & Deutsch, J. A. Secondary reinforcement as a function of drive in intracranial self-stimulation. *Proceedings of the 74th Annual Convention of the American Psychological Association, 1966, 105–106.*

Doty, R. W. Conditioned reflexes formed and evoked by brain stimulation. In D. E. Sheer (Ed.), *Electrical stimulation of the brain.* Austin: University of Texas Press, 1961.

Doty, R. W. Electrical stimulation of the brain in behavioral context. *Annual Review of Psychology,* 1969, 20, 289–320.

Dreese, A. Importance du système mésencéphalo-telencephalique noradrenergique comme substratum anatomique du comportement d'autostimulation. *Life Sciences,* 1966, *5,* 1003–1014.

Dreifuss, J. J., Murphy, J. T., & Gloor, P. Contrasting effects of two identified amygdaloid efferent pathways on single hypothalamic neurons. *Journal of Neurophysiology,* 1968, *31,* 237–248.

Eccles, J. C. *The understanding of the brain.* New York: McGraw-Hill, 1973.

Edholm, O. G., Fletcher, J. G., Widdowson, E. M., & McCance, R. A. The energy expenditure and food intake of individual men. *British Journal of Nutrition (London),* 1955, *9,* 286–300.

Egger, M. D., & Flynn, J. P. Effects of electrical stimulation of the amygdala on hypothalamically elicited attack behavior in cats. *Journal of Neurophysiology,* 1963, *26,* 705–720.

Egger, M. D., & Flynn, J. P. Further studies on the effects of amygdaloid stimulation and ablation on hypothalamically elicited attack behavior in cats. In W. R. Adey & T. Tokizane (Eds.), *Structure and function of the limbic system, progress in brain research* (Vol. 27). Amsterdam: Elsevier, 1967.

Eibl–Eibesfeldt, I. *Ethology: The biology of behavior.* New York: Holt, Rinehart & Winston, 1970.

Eichelman, B. The catecholamines and aggressive behavior. *Neurosciences Research,* 1973, *5,* 109–128.

Ellingson, R. J. Development of wakefulness–sleep cycles and associated EEG patterns in mammals. In C. D. Clemente, D. P. Purpura, & F. Mayer (Eds.), *Sleep and the maturing nervous system.* New York: Academic, 1972.

Ephron, H. S., & Carrington, P. Rapid eye movement sleep and cortical homeostasis. *Psychological Review,* 1966, *73,* 500–526.

Epstein, A. N. Reciprocal changes in feeding behavior produced by intrahypothalamic chemical injections. *American Journal of Physiology,* 1960, *199,* 969–974.

Epstein, A. N. Oropharyngeal factors in feeding and drinking. In C. F. Code (Ed.), *Handbook of physiology Alimentary Canal* (Vol. 1), (Sect. 6). Washington, D.C.: American Physiological Society, 1967.

Epstein, A. N. The lateral hypothalamic syndrome: Its implications for the physiological psychology of hunger and thirst. In E. Stellar & J. M. Sprague (Eds.), *Progress in physiological psychology* (Vol. 4). New York: Academic, 1971.

Epstein, A. N. Feeding and drinking in suckling rats. In D. Novin, W. Wyrwicka, & G. Bray (Eds.), *Hunger: Basic mechanisms and clinical implications.* New York: Raven, 1976. (a)

Epstein, A. N. The physiology of thirst: Fourth J. A. F. Stevenson lecture. *Canadian Journal of Physiology and Pharmacology,* 1976, *54,* 639–649. (b)

Epstein, A. N., Fitzsimons, J. T., & Rolls, B. J. Drinking induced by injection of angiotensin into the brain of the rat. *Journal of Physiology (London),* 1970, *210,* 457–474.

Epstein, A. N., Nicolaidis, S., & Miselis, R. The glucoprivic control of food intake and the glucostatic theory of feeding behaviour. In G. J. Mogenson & F. R. Calaresu (Eds.), *Neural integration of physiological mechanisms and behaviour.* Toronto: University of Toronto Press, 1975.

Epstein, A. N., Spector, N. D., Samman, A., & Goldblum, C. Exaggerated prandial drinking in the rat without salivary glands. *Nature,* 1964, *201,* 1342.

Epstein, A. N., & Teitelbaum, P. Regulation of food intake in the absence of taste, smell and other oropharyngeal sensations. *Journal of Comparative and Physiological Psychology,* 1962, *55,* 753–759.

Epstein, A. W. The relationship of altered brain states to sexual psychopathology. In J. Zubin & J. Money (Eds.), *Contemporary sexual behavior: Critical issues in the 1970's.* Baltimore: Johns Hopkins University Press, 1973.

Evarts, E. V. Unit activity in sleep and wakefulness. In F. C. Schmitt (Ed.), *The neurosciences, first study program.* New York: Rockefeller University Press, 1967.

Evarts, E. V. Changing concepts of central control of movement: Third J. A. F. Stevenson lecture. *Canadian Journal of Physiology and Pharmacology,* 1975, *53,* 191–201.

Evered, M. D., & Mogenson, G. J. Regulatory and secondary water intake in rats with lesions of the zona incerta. *American Journal of Physiology,* 1976, *230,* 1049–1057.

Everitt, B. J., & Herbert, J. The effects of implanting testosterone propionate into the central nervous system on the sexual behavior of adrenalectomized female rhesus monkeys. *Brain Research,* 1975, *86,* 109–120.

Everitt, B. J., Herbert, J., & Hamer, J. D. Sexual receptivity of bilaterally adrenalectomized female rhesus monkeys. *Physiology and Behavior,* 1972, *8,* 409–415.

Falk, J. L. Septal stimulation as a reinforcer of an alternative to consummatory behavior. *Journal of Experimental Analysis of Behavior,* 1961, *4,* 213–217.

Feinier, L., & Rothman, T. Study of a male castrate. *Journal of the American Medical Association,* 1939, *113,* 2144–2146.

Feldberg, W., & Myers, R. D. A new concept of temperature regulation by amines in the hypothalamus. *Nature (London),* 1963, *200,* 1325.

Fenton, P. P., & Dowling, M. T. Studies on obesity. 1. Nutritional obesity in mice. *Journal of Nutrition,* 1953, *49,* 319.

Fibiger, H. C., Zis, A. P., & McGeer, E. G. Feeding and drinking deficits after 6-hydroxydopamine administration in the rat: Similarities to the lateral hypothalamic syndrome. *Brain Research,* 1973, *55,* 135–148.

Filler, W., & Drezner, N. The results of surgical castration in women under forty. *American Journal of Obstetrics and Gynecology,* 1944, *47,* 122–124.

Fisher, A. E. Maternal and sexual behavior induced by intracranial chemical stimulation. *Science,* 1956, 124, 228–229.

Fisher, A. E. Chemical stimulation of the brain. *Scientific American,* 1964, *210,* 60–68.

Fisher, A. E., & Coury, J. N. Cholinergic tracing of a central neural circuit underlying the thirst drive. *Science,* 1962, *138,* 691–693.

Fisher, C., Magoun, H. W., & Ranson, S. W. Dystocia in diabetes insipidus: The relation of pituitary oxytocin to parturition. *American Journal of Obstetrics and Gynecology,* 1938, *36,* 1–9.

Fishman, J. Appetite and sex hormones. In T. Silverstone (Ed.), *Appetite and food intake.* Dahlem Konferenzen, Berlin: Abakon Verlagsgesellschaft, 1976.

Fitzsimons, J. T. Drinking by nephrectomized rats injected with various substances. *Journal of Physiology (London),* 1961, *155,* 563–579. (a)

Fitzsimons, J. T. Drinking by rats depleted of body fluid without increase in osmotic pressure. *Journal of Physiology (London),* 1961, *159,* 297–309. (b)

Fitzsimons, J. T. The effects of slow infusions of hypertonic solutions on drinking and drinking thresholds in rats. *Journal of Physiology (London,* 1963, *167,* 344–354.

Fitzsimons, J. T. Drinking caused by constriction of the inferior vena cava in the rat. *Nature (London),* 1964, *209,* 479–480.

Fitzsimons, J. T. The renin-angiotensin system in the control of drinking. In L. Martini, M. Motta, & F. Fraschini (Eds.), *The hypothalamus.* New York: Academic, 1970.

Fitzsimons, J. T. Thirst. *Physiological Reviews,* 1972, *52,* 468–561.

Fitzsimons, J. T., & Le Magnen, J. Eating as a regulatory control of drinking in the rat. *Journal of Comparative and Physiological Psychology,* 1969, *67,* 273–283.

Fitzsimons, J. T., & Nicolaidis, S. Mécanisme de l'action dipsogénique intracrãnienne de l'angiotensine II. *Journal de Physiologie (Paris),* 1976, *72,* 41A.

Fitzsimons, J. T., & Simons, B. J. The effect of drinking in the rat of intravenous infusion of angiotensin, given alone or in combination with other stimuli of thirst. *Journal of Physiology,* 1969, *203,* 45–57.

Flourens, P. *Récherches experimentales sur les proprietes et les functions du system nerveux dans les animaux vertebres* (2nd ed.). Paris: J. B. Bailliere, 1842.

Flynn, J. P. The neural basis of aggression in cats. In D. C. Glass (Ed.), *Neurophysiology and emotion.* New York: Rockefeller University Press, 1967.

Folk, G. E., Jr. *Textbook of environmental physiology* (2nd ed.). Philadelphia: Lea & Febiger, 1974.

Folkow, F. & Neil, E. *Circulation.* London: Oxford University Press, 1971.

Folkow, B., & Rubinstein, E. H. Behavioural and autonomic patterns evoked by stimulation of the lateral hypothalamic area in the cat. *Acta Physiologica Scandinavica,* 1965, *65,* 292–299.

Fonberg, E. The inhibitory role of amygdala stimulation. *Acta Biologica Experimentalis (Warsaw),* 1963, *23,* 171–180.

Fonberg, E. The role of the hypothalamus and amygdala in food intake, alimentary motivation and emotional reactions. *Acta Biologica Experimentalis (Warsaw),* 1969, *29,* 335–358.

Fonberg, E., & Delgado, J. M. R. Avoidance and alimentary reactions during amygdala stimulation. *Journal of Neurophysiology,* 1961, *24,* 651.

Fowler, S. J., & Kellogg, C. Ontogeny of thermoregulatory mechanisms in the rat. *Journal of Comparative and Physiological Psychology,* 1975, *89,* 738–746.

Frankenhaeuser, M. Experimental approaches to the study of catecholamines and emotion. In L. Levi (Ed.), *Emotions—Their parameters and measurement.* New York: Raven, 1975.

Fredriksen, P. K. Preliminary note on Lioresal (Baclofen) on treatment of schizophrenia. *Lakartidn,* 1975, *72,* 456–458.

Fritsch, G., & Hitzig, E. Über die elektrische Erregbarkeit des Grasshirns. *Archiv fur Anatomie Physiologie und wissenschaftliche Medicin,* 1870, *37,* 300–332.

Fuller, J. L., Rosvold, H. E., & Pribram, K. H. The effect on affective and cognitive behavior in the dog of lesions of the pyriform-amygdala-hippocampal complex. *Journal of Comparative and Physiological Psychology,* 1957, *50,* 89–96.

Fuxe, K., Ganten, D., Hökfelt, T., & Bolme, P. Immunohistochemical evidence for the existence of angiotensin II-containing nerve terminals in the brain and spinal cord in the rat. *Neuroscience Letters,* 1976, *2,* 229–234.

Fuxe, K., Hökfelt, T., Ljungdahl, A., Agnati, L., Johansson, O., & Perez de la Mora, M. Evidence for an inhibitory gabergic control of the mesolimbic dopamine neurons: Possibility of improving treatment of schizophrenia by combined treatment with neuroleptics and gabergic drugs. *Medical Biology,* 1975, *53,* 177–183.

Gale, C. C., Matthews, M., & Young, J. Behavioral thermoregulatory responses to hypothalamic cooling and warming in baboons. *Physiology and Behavior,* 1970, *5,* 1–6.

Gallistel, C. R., & Beagley, G. Specificity of brain stimulation reward in the rat. *Journal of Comparative and Physiological Psychology,* 1971, *76,* 199–205.

Ganten, D., Minnich, J. L., Granger, P., Hayduk, K., Brecht, H. M., Barbeau, A., Boucher, R., & Genest, J. Angiotensin-forming enzyme in brain tissue. *Science,* 1971, *173,* 64–65.

Garattini, S., & Samanin, R. Anorectic drugs and brain neurotransmitters. In T. Silverstone (Ed.), *Appetite and food intake*. Dahlem Konferenzen, Berlin: Abakon Verlagsgesellschaft, 1976.

Gardner, E. *Fundamentals of Neurology*. Philadelphia: W. B. Saunders Co., 1968 (Fifth edition).

Gauer, O. H., Henry, J. P., & Behn, C. The regulation of extracellular fluid volume. *Annual Review of Physiology,* 1970, *32,* 547–595.

German, D. C., & Bowden, D. M. Catecholamine systems as the neural substrate for intracranial self-stimulation: A hypothesis. *Brain Research,* 1974, *73,* 381–419.

Gibson, W. E., Reid, L. D., Sakai, M., & Porter, P. B. Intracranial reinforcement compared with sugar-water reinforcement. *Science,* 1965, *148,* 1357–1359.

Girvin, J. P. Clinical correlates of hypothalamic and limbic system function. In G. J. Mogenson & F. R. Calaresu (Eds.), *Neural integration of physiological mechanisms and behaviour*. Toronto: University of Toronto Press, 1975.

Glickman, S. E., & Schiff, B. B. A biological theory of reinforcement. *Psychological Review,* 1967, *74,* 81–109.

Gloor, P. The amygdala. In J. Field, H. W. Magoun & V. E. Hall (Eds.), *Handbook of Physiology* (Vol. 2), *Neurophysiology (Sect. 1)*. Baltimore: Williams & Wilkins, 1960.

Gloor, P., Murphy, J. T., & Dreifuss, J. J. Anatomical and physiological characteristics of the two amygdaloid projection systems to the ventromedial hypothalamus. In C. H. Hockman (Ed.), *Limbic system mechanisms and autonomic function*. Springfield, Ill.: C. C. Thomas, 1972.

Gold, R. M. Hypothalamic hyperphagia: Males get just as fat as females. *Journal of Comparative and Physiological Psychology,* 1970, *71,* 347–356.

Goldstein, M. Brain research and violent behavior. *Archives of Neurology,* 1974, *30,* 1–35.

Gorzalka, B. B. *Neural mechanisms of progesterone action*. Unpublished doctoral dissertation, University of California, Irvine, 1974.

Gorzalka, B. B. Unpublished observations, 1976.

Gorzalka, B. B., Rezek, D. L., & Whalen, R. E. Adrenal mediation of estrogen-induced ejaculatory behavior in the male rat. *Physiology and Behavior,* 1975, *14,* 373–376.

Gorzalka, B. B., & Whalen, R. E. Genetic regulation of hormone action: Selective effects of progesterone and dihydroprogesterone on sexual receptivity in mice. *Steroids,* 1974, *23,* 499–505.

Gorzalka, B. B., & Whalen, R. E. Effects of genotype on differential behavioral responsiveness to progesterone and dihydroprogesterone in mice. *Behavior Genetics,* 1976, *6,* 7–15.

Goy, R. W., & Phoenix, C. H. Hypothalamic regulation of female sexual behavior: Establishment of behavioral oestrus in spayed guinea pigs following hypothalamic lesions. *Journal of Reproduction and Fertility,* 1963, *5,* 23–40.

Graybiel, A. M. Input–output anatomy of the basal ganglia. In A. Barbeau (Chair), *Basal Ganglia in Health and Disease*. Symposium Lecture, Society for Neuroscience, Toronto, Canada, 1976.

Green, J. D., Clemente, C. D., & De Groot, J. Rhinencephalic lesions and behavior in cats: An analysis of the Kluver–Bucy syndrome with particular reference to normal and abnormal sexual behavior. *Journal of Comparative Neurology,* 1957, *108,* 505–545.

Greer, M. A. Suggestive evidence of a primary "drinking center" in hypothalamus of the rat. *Proceedings of the Society for Experimental Biology and Medicine,* 1955, *89,* 59.

Grodins, F. S. Theory and models in regulatory biology. In J. D. Hardy, A. P. Gagge, & J. A. J. Stolwijk (Eds.), *Physiological and behavioral regulation*. Springfield, Ill.: C. C. Thomas, 1970.

Grossman, S. P. Direct adrenergic and cholinergic stimulation of hypothalamic mechanisms. *American Journal of Physiology,* 1962, *202,* 872–882.

Grossman, S. P. The VMH: A center for affective reactions, satiety, or both? *Physiology and Behavior,* 1966, *1,* 1–10.

Grossman, S. P. *A textbook of physiological psychology*. New York: Wiley, 1967.

Grossman, S. P., & Grossman, R. Food and water intake following lesions or electrical stimulation of the amygdala. *American Journal of Physiology,* 1963, *205,* 761–765.

Grunt, J. A., & Young, W. C. Consistency of sexual behavior patterns in individual male guinea pigs following castration and androgen therapy. *Journal of Comparative and Physiological Psychology,* 1953, *46,* 138–144.

Guilleminault, C., & Dement, W. Pathologies of excessive sleep. In E. D. Weitzman (Ed.), *Advances in sleep research* (Vol. 1). New York: Spectrum Publications, 1974.

Guyton, A. C. *Textbook of medical physiology* (5th ed.). Philadelphia: Saunders, 1976.

Hagamen, W. D., Zitzmann, E. K., & Reeves, A. G. Sexual mounting of diverse objects in a group of randomly selected unoperated male cats. *Journal of Comparative and Physiological Psychology,* 1963, *56,* 298–301.

Hainsworth, F. R., & Epstein, A. N. Severe impairment of heat-induced saliva-spreading in rats recovered from lateral hypothalamic lesions. *Science,* 1966, *153,* 1155–1157.

Hamilton, C. L. Rat's preference for high fat diets. *Journal of Comparative and Physiological Psychology,* 1964, *58,* 459–460.

Hardy, J. D., Gagge, A. P., & Stolwijk, J. A. J. *Physiological and behavioral temperature regulation.* Springfield, Ill.: C. C. Thomas, 1970.

Harlow, H. F. Mice, monkeys, and motives. *Psychological Review,* 1953, *60,* 23–32.

Harlow, H. F. Heterosexual affectional system in monkey. *American Psychologist,* 1962, *17,* 1–9.

Harlow, H. F. A behavioral approach to psychoanalytic theory. *Science and Psychoanalysis,* 1964, *7,* 93–111.

Harlow, H. F. Sexual behavior in the rhesus monkey. In F. A. Beach (Ed.), *Sex and behavior.* New York: Wiley, 1965.

Harper, A. E., Benevenga, N. J., & Wohlhueter, R. M. Effects of ingestion of disproportionate amounts of amino acids. *Physiological Reviews,* 1970, *50,* 428–558.

Harper, A. E., & Boyle, P. C. Nutrients and food intake. In T. Silverstone, (Ed.), *Appetite and food intake.* Dahlem Konferenzen, Berlin: Abakon Verlagsgesellschaft, 1976.

Harris, G. W., & Michael, R. P. The activation of sexual behavior by hypothalamic implants of oestrogen. *Journal of Physiology (London),* 1964, *171,* 275–301.

Harris, V. S., & Sachs, B. D. Copulatory behavior in male rats following amygdaloid lesions. *Brain Research,* 1975, *86,* 514–518.

Hart, B. L. Sexual reflexes and mating behavior in the male dog. *Journal of Comparative and Physiological Psychology,* 1967, *64,* 388–399.

Hart, B. L. Alteration of quantitative aspects of sexual reflexes in spinal male dogs by testosterone. *Journal of Comparative and Physiological Psychology,* 1968, *66,* 726–730.

Hart, B. L. Gonadal hormones and sexual reflexes in the female rat. *Hormones and Behavior,* 1969, *1,* 65–71.

Hart, B. L. *Experimental neurophysiology: A laboratory manual.* San Francisco: W. H. Freeman & Company, 1969.

Hart, B. L. Mating behavior in the female dog and the effects of estrogen on sexual reflexes. *Hormones and Behavior,* 1970, *2,* 93–104.

Hart, B. L. Facilitation by estrogen of sexual reflexes in female cats. *Physiology and Behavior,* 1971, *7,* 675–678.

Hart, B. L., & Haugen, C. M. Activation of sexual reflexes in male rats by spinal implantation of testosterone. *Physiology and Behavior,* 1968, *3,* 735–738.

Heath, R. G. Pleasure response of human subjects to direct stimulation of the brain: Physiologic and psychodynamic considerations. In R. G. Heath (Ed.), *The role of pleasure in behavior.* New York: Harper & Row, 1964.

Heath, R. G., Monroe, R. R., & Mickle, W. A. Stimulation of the amygdaloid nucleus in a schizophrenic patient. *American Journal of Psychiatry,* 1955, *111,* 862–863.

Hebb, D. O. *The organization of behavior: A neuropsychological theory.* New York: Wiley, 1949.

Hebb, D. O. Drives and the C.N.S. (Conceptual Nervous System). *Psychological Review,* 1955, *62,* 243–254.

Hebb, D. O. *A textbook of psychology* (2nd ed.). Philadelphia: Saunders, 1966.

Heimer, L., & Larsson, K. Drastic changes in mating behavior of male rats following lesions in the junction of diencephalon and mesencephalon. *Experientia,* 1964, *20,* 1–4.

Heimer, L., & Larsson, K. Impairment of mating behavior in male rats following lesions in the preoptic-anterior hypothalamic continuum. *Brain Research,* 1966/1967, *3,* 248–263.

Heller, A., & Moore, R. Y. Effect of central nervous system lesions on brain monoamines in the rat. *Journal of Pharmacology and Experimental Therapeutics,* 1965, *150,* 1–9.

Henneman, E. Organization of the motor systems—A preview. In V. B. Mountcastle (Ed.), *Medical physiology* (13th Ed.). St. Louis: Mosby, 1974.

Hervey, G. R. Regulation of energy balance. *Nature (London),* 1969, *223,* 629–631.

Hervey, G. R. Physiological mechanisms for the regulation of energy balance. *Proceedings of the Nutrition Society (London),* 1971, *30,* 109–116.

Hery, F., Pujol, J. F., Lopez, M., Macon, J., & Glowinski, J. Increased synthesis and utilization of serotonin in the central nervous system of the rat during paradoxical sleep deprivation. *Brain Research,* 1970, *21,* 391–403.

Hess, W. R. Le sommeil. *Comptes Rendus de la Société de Biologie,* 1931, *107,* 1333–1360.

Hess, W. R. *Das Zwischenhirn* (2nd ed.). Basel: Schwabe, 1954.

Hess, W. R., & Akert, E. Experimental data on role of hypothalamus in mechanism of emotional behavior. *Archives of Neurology and Psychiatry,* 1955, *73,* 127–129.

Himwich, H. E. *Biochemistry, schizophrenias, and affective illnesses.* Baltimore: Williams & Wilkins, 1971.

Hinde, R. A. *Animal behaviour—A synthesis of ethology and comparative psychology* (2nd ed.). New York: McGraw-Hill, 1970.

Hoebel, B. G. Inhibition and disinhibition of self-stimulation and feeding: Hypothalamic control and post-ingestional factors. *Journal of Comparative and Physiological Psychology,* 1968, *66,* 89–100.

Hoebel, B. G. Feeding and self-stimulation. *Annals of the New York Academy of Sciences,* 1969, *157,* 758–778.

Hoebel, B. G. Feeding: Neural control of intake. *Annual Review of Physiology,* 1971, *33,* 533–568.

Hoebel, B. G., & Teitelbaum, P. Hypothalamic control of feeding and self-stimulation. *Science,* 1962, *135,* 357–377.

Hoebel, B. G., & Teitelbaum, P. Weight regulation in normal and hypothalamic hyperphagic rats. *Journal of Comparative and Physiological Psychology,* 1966, *61,* 189–193.

Hökfelt, T., Ljungdahl, A., Fuxe, K., & Johansson, O. Dopamine nerve terminals in the rat limbic cortex: Aspects of the dopamine hypothesis of schizophrenia. *Science,* 1974, *184,* 177–179.

Holman, B. R., Elliott, G. R., & Barchas, J. D. Neuroregulator and sleep mechanisms. *Annual Review of Medicine,* 1975, *26,* 499–520.

Holmes, J. H., & Gregersen, M. I. Observations on drinking induced by hypertonic solutions. *American Journal of Physiology,* 1950, *162,* 326–337.

Hornykiewicz, O. Dopamine in the basal ganglia. Its role and therapeutic implications (including the clinical use of L-DOPA). *British Medical Bulletin,* 1973, *29,* 1972–178.

Horowitz, Z. P., Chow, M. I., & Carlton, P. L. Self-stimulation of the brain by cats: Effects of imipramine, amphetamine, and chlorpromazine. *Psychopharmacology,* 1962, *3,* 455–462.

Houpt, K. A., & Epstein, A. N. Ontogeny of controls of food intake in the rat: GI fill and glucoprivation. *American Journal of Physiology,* 1973, *225,* 58–66.

House, E. L., & Pansky, B. *A functional approach to neuroanatomy* (2nd ed.). New York: McGraw-Hill, 1967.

Hudson, J. W. Water metabolism in desert mammals. In M. J. Wayner (Ed.), *Thirst.* New York: Pergamon, 1964.

Hull, C. L. *Principles of behavior*. New York: Appleton-Century-Crofts, 1943.

Hull, D. Thermoregulation in young mammals. In G. C. Whittow (Ed.), *Comparative physiology of thermoregulation* (Vol. III), *Special Aspects of Thermoregulation*. New York: Academic, 1973.

Huston, J. P., & Borbely, A. A. Operant conditioning in forebrain ablated rats by use of rewarding hypothalamic stimulation. *Brain Research,* 1973, *50,* 467–472.

Hutchinson, R. R., & Renfrew, J. W. Stalking attack and eating behaviours elicited from the same sites in the hypothalamus. *Journal of Comparative and Physiological Psychology,* 1966, *61,* 360–367.

Ibuka, N., & Kawamura, H. Loss of circadian rhythm in sleep–wakefulness cycle in the rat by suprachiasmatic nucleus lesions. *Brain Research,* 1975, *96,* 76–81.

Isaacson, R. L. *The limbic system*. New York: Plenum Press, 1974.

Ito, M. Excitability of medial forebrain bundle neurons during self-stimulating behavior. *Journal of Neurophysiology,* 1972, *35,* 652–664.

Ito, M., & Olds, J. Unit activity during self-stimulation behavior. *Journal of Neurophysiology,* 1971, *34,* 263–273.

Iversen, S. D. & Iversen, L. L. *Behavioral pharmacology*. New York: Oxford University Press, 1975.

Jacobsen, C. F. Studies of cerebral function in primates. 1. The functions of the frontal association areas in monkeys. *Comparative Psychology Monographs,* 1936, *13,* 3–60.

James, W. *The principles of psychology*. New York: Holt, 1890.

Janowitz, H. D., & Hollander, F. The time factor in the adjustment of food intake to varied caloric requirement in the dog: A study of the precision of appetite regulation. *Annals of New York Academy of Sciences,* 1955, *63,* 56–67.

Janowsky, D. S., Fann, W. E., & Davis, J. M. Monoamines and ovarian hormone linked sexual and emotional changes: A review. *Archives of Sexual Behavior,* 1971, *1,* 205–218.

Johnson, A. K., & Buggy, J. Angiotensin (AII) and intracellular dehydration induced drinking: Mediation by tissue surrounding antero-ventral third ventricle (AV3V). *Federation Proceedings,* 1976, *35,* 814. (Abstract #3359)

Johnson, A. K., & Epstein, A. N. The cerebral ventricles as the avenue for the dipsogenic action of intracranial angiotensin. *Brain Research,* 1975, *86,* 399–418.

Johnston, P., & Davidson, J. M. Intracerebral androgens and sexual behavior in the male rat. *Hormones and Behavior,* 1972, *3,* 345–357.

Jones, B., & Mishkin, M. Limbic lesions and the problem of stimulus-reinforcement associations. *Experimental Neurology,* 1972, *36,* 362–377.

Jones, H., & Oswald, I. Two cases of healthy insomnia. *Electroencephalography and Clinical Neurophysiology,* 1968, *24,* 378–380.

Jouvet, M. The states of sleep. *Scientific American,* 1967, *216,* 62–72.

Jouvet, M. The role of monoamines and acetylcholine-containing neurons in the regulation of the sleep–waking cycle. *Ergebnisse der Physiologie Biologischen Chemie und Experimentellen Pharmakologie,* 1972, *64,* 166–307.

Jouvet, M. The role of monoaminergic neurons in the regulation and function of sleep. In O. Petre–Quadens & J. D. Schlag (Eds.), *Basic sleep mechanisms*. New York: Academic, 1974.

Jouvet, M., Mouret, J., Chouvet, G., & Siffre, M. Toward a 48-hour day: Experimental bicircadian rhythm in man. In F. O. Schmitt & F. G. Worden (Eds.), *The neurosciences, third study program*. Cambridge, Mass.: MIT Press, 1974.

Jouvet–Mounier, D., Astic, L., & Lacote, D. Ontogenesis of the states of sleep in rat, cat, and guinea pig during the first postnatal month. *Developmental Psychobiology,* 1970, *2,* 216–239.

Jovanović, U. J. *Psychomotor epilepsy*. Springfield, Ill.: C. C. Thomas, 1974.

Kaada, B. R. Somato-motor, autonomic and electrocorticographic responses to electrical stimulation of "rhinencephalic" and other structures in primates, cat and dog. *Acta Physiologica Scandinavica Supplement (Stockholm),* 1951, *24,* Supp. 83, 1–285.

Kakolewski, J. W., Deaux, E., Christensen, J., & Case, B. Diurnal patterns in water and food intake and body weight changes in rats with hypothalamic lesions. *American Journal of Physiology,* 1971, *221,* 711–718.

Kales, A., Bixler, E. O., & Kales, J. D. Role of sleep research and treatment facility: Diagnosis treatment and education. In E. D. Weitzman (Ed.), *Advances in sleep research* (Vol. 1). New York: Spectrum Publications, 1974.

Karli, P., Vergnes, M., & Didiergeorges, F. Rat–mouse interspecific aggressive behaviour and its manipulation by brain ablation and by brain stimulation. In S. Garattini & E. B. Sigg (Eds.), *Biology of aggressive behaviour.* Amsterdam: Exerpta Medica Foundation, 1969.

Kayser, C. *The physiology of natural hibernation.* Oxford: Pergamon, 1961.

Keene, J. J. Reward-associated inhibition and pain-associated excitation lasting seconds in single intralaminar thalamic units. *Brain Research,* 1973, *64,* 211–224.

Keene, J. J., & Casey, K. L. Excitatory connection from lateral hypothalamic self-stimulation sites to escape sites in medullary reticular formation. *Experimental Neurology,* 1970, *28,* 155–166.

Keller, A. D., & Hare, W. K. The hypothalamus and heat regulation. *Proceedings of the Society for Experimental Biology and Medicine,* 1932, *29,* 1069–1070.

Kennedy, G. C. The role of depot fat in the hypothalamic control of food intake in the rat. *Proceedings of the Royal Society of London,* Series B: Biological Sciences, 1953, *140,* 578–592.

Kety, S. S. Biochemical theories of schizophrenia. *Science,* 1959, *129,* 1528–1532.

Kety, S. S. Neurochemical aspects of emotional behavior. In P. Black (Ed.), *Physiological correlates of emotion.* New York: Academic, 1970.

Keys, N. W. Secondary reinforcement and reinforcing intracranial stimulation. *Dissertation Abstracts,* 1964, *28,* 3436B.

King, F. A., & Meyer, P. M. Effects of amygdaloid lesions upon septal hyperemotionality in the rat. *Science,* 1958, *128,* 655–656.

Kinsey, A. C., Pomeroy, W. B., & Martin, C. E. *Sexual behavior in the human male.* Philadelphia: Saunders, 1948.

Kinsey, A. C., Pomeroy, W. B., Martin, C. E., & Gebhard, P. H. *Sexual behavior in the human female.* Philadelphia: Saunders, 1953.

Kissileff, H. R. Food-associated drinking in the rat. *Journal of Comparative and Physiological Psychology,* 1969, *67,* 284–300.

Kissileff, H. R. Nonhomeostatic controls of drinking. In A. N. Epstein, H. R. Kissileff, & E. Stellar (Eds.), *The neuropsychology of thirst.* Washington, D.C.: V. H. Winston, 1973.

Kleitman, N. *Sleep and wakefulness.* Chicago: University of Chicago Press, 1939.

Kleitman, N. Invited discussion. In C. D. Clemente, D. P. Purpura, & F. E. Mayer (Eds.), *Sleep and the maturing nervous system.* New York: Academic, 1972.

Kling, A. Effects of amygdalectomy and testosterone on sexual behavior of male juvenile macaques. *Journal of Comparative and Physiological Psychology,* 1968, *65,* 466–471.

Klüver, H., & Bucy, P. C. "Psychic blindness" and other symptoms following bilateral temporal lobectomy in Rhesus monkeys. *American Journal of Physiology,* 1937, *119,* 352–353.

Kolb, B., & Ninneman, A. J. Prefrontal cortex and the regulation of food intake in the rat. *Journal of Comparative and Physiological Psychology,* 1975, *88,* 806–815.

Konorski, J. *Integrative activity of the brain.* Chicago: University of Chicago Press, 1967.

Kucharczyk. J., & Mogenson, G. J. Separate lateral hypothalamic pathways for extracellular and intracellular thirst. *American Journal of Physiology,* 1975, *228,* 295–301.

Kucharczyk, J., & Mogenson, G. J. Specific deficits in regulatory drinking following electrolytic lesions of the lateral hypothalamus. *Experimental Neurology,* 1976, *53,* 371–385.

Kucharczyk, J., Assaf, S., & Mogenson, G. J. Differential effects of brain lesions on thirst induced by the administration of angiotensin II to the preoptic region, subfornical organ and anterior third ventricle. *Brain Research,* 1976, *108,* 327–337.

Larsson, K. Mating behavior in male rats after cerebral cortex ablation: II. Effects of lesions in the frontal lobes compared to lesions in the posterior half of the hemispheres. *Journal of Experimental Zoology,* 1964, *155,* 203–213.

Lashley, K. S. Experimental analysis of instinctive behavior. *Psychological Review,* 1938, *45,* 445–471.

Laties, V. G., & Weiss, B. Thyroid state and working for heat in the cold. *American Journal of Physiology,* 1959, *197,* 1028–1034.

Law, T., & Meagher, W. Hypothalamic lesions and sexual behavior in the female rat. *Science,* 1958, *128,* 1626–1627.

Leeper, R. W. A motivational theory of emotion to replace ''Emotion as disorganized response.'' *Psychological Review,* 1948, *55,* 5–21.

LeMagnen, J. Advances in studies on the physiological control and regulation of food intake. In E. Stellar & J. M. Sprague (Eds.), *Progress in physiological psychology* (Vol. 4). New York: Academic, 1971.

Le Magnen, J. Current concepts in energy balance. In G. J. Mogenson & F. R. Calaresu (Eds.). *Neural integration of physiological mechanisms and behavior.* Toronto: University of Toronto Press, 1975.

Leonard, C. M. The prefrontal cortes of the rat: I. Cortical projection of the mediodorsal nucleus: II. Efferent connections. *Brain Research,* 1969, *12,* 321–343.

Levi, L. *Emotions—Their parameters and measurement.* New York: Raven, 1975.

Levin, R., & Stern, J. M. Maternal influences on ontogeny of suckling and feeding rhythms in the rat. *Journal of Comparative and Physiological Psychology,* 1975, *89,* 711–721.

Levine, S. Stimulation in infancy. *Scientific American,* 1960, *202,* 80–86.

Lewin, R. *Hormones: Chemical communicators.* London: Geoffrey Chapman, 1972.

Lindsley, D. B. Emotion. In S. S. Stevens (Ed.), *Handbook of experimental psychology.* New York: Wiley, 1951.

Lindsley, D. B. The role of nonspecific reticulo-thalamo-cortical systems in emotion. In P. Black (Ed.), *Physiological correlates of emotion.* New York: Academic, 1970.

Lindsley, D. B. Summary and concluding remarks. In C. D. Clemente, D. P. Purpura, & F. E. Mayer (Eds.), *Sleep and the maturing nervous system.* New York: Academic, 1972.

Lints, C. E., & Harvey, J. A. Altered sensitivity to foot shock and decreased brain content of serotonin following brain lesions in the rat. *Journal of Comparative and Physiological Psychology,* 1969, *67,* 23–31.

Lipton, J. M. Behavioral temperature regulation in the rat: Effects of thermal stimulation of the medulla. *Journal de Physiologie (Paris),* 1971, *63,* 325–328.

Lisk, R. D. Diencephalic placement of estradiol and sexual receptivity in the female rat. *American Journal of Physiology,* 1962, *203,* 493–496.

Lisk, R. D. Reproductive potential of the male rat: Enhancement of copulatory levels following lesions of the mammillary body in sexually non-active and active animals. *Journal of Reproduction and Fertility,* 1969, *19,* 353–356.

Louis-Sylvestre, J. Preabsorptive insulin release and hypoglycemia in rats. *American Journal of Physiology,* 1976, *230,* 56–60.

Luttge, W. G., Hall, N. R., Wallis, C. J., & Campbell, J. C. Stimulation of male and female sexual behavior in gonadectomized rats with estrogen and androgen therapy and its inhibition with concurrent anti-hormone therapy. *Physiology and Behavior,* 1975, *14,* 65–73.

Lyon, M., Halpern, N. M., & Mintz, E. Y. The significance of the mesencephalon for coordinated feeding behavior. *Acta Neurologica Scandinavica,* 1968, *44,* 323–346.

Lytle, L. D., Moorcroft, W. H., & Campbell, B. A. Ontogeny of amphetamine anorexia and insulin hyperphagia in the rat. *Journal of Comparative and Physiological Psychology,* 1971, *77,* 388–393.

MacKay, E. M., Calloway, J. W., & Barnes, R. H. Hyperalimentation in normal animals produced by protamine insulin. *Journal of Nutrition,* 1940, *20,* 59–66.

MacKinnon, P. C. B., ter Haar, M. B., & Burton, M. J. Preliminary observations of serum LH levels and of protein metabolism in the brain and anterior pituitary in the rodent around the time of puberty. *Journal of Psychosomatic Research,* 1972, *16,* 271–278.

MacLean, P. D. Psychosomatic disease and the "visceral brain," recent developments bearing on the Papez theory of emotion. *Psychosomatic Medicine,* 1949, *11,* 338–353.

MacLean, P. D., & Delgado, J. M. R. Electrical and chemical stimulation of frontotemporal portion of limbic system in the waking animal. *Electroencephalography and Clinical Neurophysiology,* 1953, *5,* 91–100.

MacLean, P. D., & Ploog, D. W. Cerebral representation of penile erection. *Journal of Neurophysiology,* 1962, *25,* 29–55.

Madlafousek, J., Freund, K., & Grofova, I. Variables determining the effect of electro-stimulation in the lateral preoptic area on the sexual behavior of male rats. *Journal of Comparative and Physiological Psychology,* 1970, *72,* 28–44.

Magalhoes, H. Hamsters. In E. S. E. Hafez (Ed.), *Reproduction and breeding techniques for laboratory animals.* Philadelphia: Lea & Fibiger, 1970.

Magnes, J., Moruzzi, G., & Pompeiano, O. Synchronization of the EEG produced by low frequency electrical stimulation of the region of the solitary tract. *Archives Italiennese de Biologie (Pisa),* 1961, *99,* 33–67.

Magoun, H. W. *The waking brain.* Springfield, Ill.: C. C. Thomas, 1958.

Malamud, N. Psychiatric disorder with intracranial tumors of limbic system. *Archives of Neurology,* 1967, *17,* 113–123.

Malmo, R. B. Osmosensitive neurons in the rat's dorsal midbrain. *Brain Research,* 1976, *105,* 105–120.

Malmo, R. B., & Mundl, W. J. Osmosensitive neurons in the rat's preoptic area: Medial lateral comparison. *Journal of Comparative and Physiological Psychology,* 1975, *88,* 161–175.

Malsbury, C. W. Facilitation of male rat copulatory behavior by electrical stimulation of the medial preoptic area. *Physiology and Behavior,* 1971, *7,* 797–805.

Manshardt, J., & Wurtman, R. J. Daily rhythm in the noradrenaline content of rat hypothalamus. *Nature,* 1968, *217,* 574–575.

Margules, D. L., Lewis, M. J., Dragovich, J. A., & Margules, A. S. Hypothalamic norepinephrine: Circadian rhythms and the control of feeding. *Science,* 1972, *178,* 640–643.

Margules, D. L., & Olds, J. Identical "feeding" and "rewarding" systems in the lateral hypothalamus of rats. *Science,* 1962, *135,* 374–375.

Marshall, J. F., Richardson, J. S., & Teitelbaum, P. Nigrostriatal bundle damage and the lateral hypothalamic syndrome. *Journal of Comparative and Physiological Psychology,* 1974, *87,* 808–830.

Marshall, J. F., & Teitelbaum, P. A comparison of the eating in response to hypothermic and glucoprivic challenges after nigral 6-hydroxydopamine and lateral hypothalamic electrolytic lesions in rats. *Brain Research,* 1973, *55,* 229–233.

Martin, G. E., & Myers, R. D. Evoked release of (^{14}C)norepinephrine from the rat hypothalamus during feeding. *American Journal of Physiology,* 1975, *229,* 1547–1555.

Maslow, A. H. *Motivation and personality.* New York: Harper, 1954.

Masters, W. H., & Johnson, V. E. *Human sexual response.* Boston: Little, Brown, 1966.

May, K. K., and Beaton, J. R. Metabolic effects of hyperphagia in the hypothalamic-hyperphagic rat. *Can. J. Physiol. Pharmacol.* 1966, *44,* 641–650.

Mayer, D. J., & Liebeskind, J. C. Pain reduction by focal electrical stimulation of the brain: An anatomical and behavioral analysis. *Brain Research,* 1974, *68,* 73–93.

Mayer, J. The glucostatic theory of regulation of food intake and the problem of obesity. *New England Medical Center Bulletin,* 1952, *14,* 43–49.

Mayer, J., & Marshall, N. B. Specificity of goldthioglucose for ventromedial hypothalamic lesions and hyperphagia. *Nature (London),* 1956, *178,* 1399–1400.

McCleary, R. A., & Moore, R. Y. *Subcortical mechanisms of behavior*. New York: Basic Books, 1965.

McDonald, P., Beyer, C., Newton, F., Brien, B., Baker, R., Tan, H. S., Sampson, C., Kitching, P., Greenhill, R., & Pritchard, D. Failure of 5α-dihydrotestosterone to initiate sexual behavior in the castrated male rat. *Nature,* 1970, *227,* 964–965.

McFarland, D. J. Recent developments in the study of feeding and drinking in animals. *Journal of Psychosomatic Research,* 1970, *14,* 229–237.

McFarland, D. J. Time-sharing as a behavioral phenomenon. In D. S. Lehrman, R. A. Hinde & E. Shaw (Eds.), *Advances in the study of behavior* (Vol. 5). New York: Academic, 1974.

McFarland, D. J. Personal communication. December 11, 1975.

McGeer, P. L. Mood and movement: Twin galaxies in the inner universe. *Invited lecture, Society for Neuroscience,* Toronto, Canada, November 1976.

McGinty, D. J. Encephalization and the neural control of sleep. In M. B. Sterman, D. J. McGinty, & A. M. Adinolfi (Eds.), *Brain development and behavior*. New York: Academic, 1971.

Mellinkoff, S. M., Frankland, M., Boyle, D., & Greipel, M. Relationship between serum amino-acid concentration and fluctuations in appetite. *Journal of Applied Physiology,* 1956, *8,* 535–538.

Melzack, R. Effects of early experience on behavior: Experimental and conceptual considerations. In P. H. Hoch & J. Zubin (Eds.), *Psychopathology of perception*. New York: Grune & Stratton, 1965.

Melzack, R. *The puzzle of pain*. Harmondsworth, England: Penguin, 1973.

Mendels, J. Biological aspects of affective illness. In S. Arieti (Ed.), *American handbook of psychiatry* (2nd Ed.), (Vol. III) *Adult Clinical Psychiatry*. New York: Basic Books, 1974.

Mettler, F. A. *Psychosurgical Problems* (The Columbia Greystone Associates, Second Group). New York: Blakiston, 1952.

Michael, R. P., Herbert, J., & Welegalla, J. Ovarian hormones and grooming behaviour in the Rhesus monkey (*Macaca mulatta*) under laboratory conditions. *Journal of Endocrinology,* 1966, *36,* 263–279.

Michael, R. P., & Saayman, G. Differential effects on behavior of the subcutaneous and intravaginal administration of oestrogen in the Rhesus monkey *(Macaca mulatta)*. *Journal of Endocrinology,* 1968, *41,* 231–246.

Michael, R. P., Saayman, G. M., & Zumpe, D. Inhibition of sexual receptivity by progesterone in Rhesus monkey *(Macaca mulatta)*. *Journal of Endocrinology,* 1967, *39,* 309–310.

Michal, E. K. Effects of limbic lesions on behavior sequences and courtship behavior of male rats *(Rattus norvegicus)*. *Behaviour,* 1973, *44,* 264–285.

Mickelsen, O., Takahashi, S., & Craig, C. Experimental obesity. 1. Production of obesity in rats by feeding high-fat diets. *Journal of Nutrition,* 1955, *57,* 541–554.

Miller, N. E. Motivational effects of brain stimulation and drugs. *Federation Proceedings,* 1960, *19,* 846–854.

Miller, N. E. Chemical coding of behavior in the brain. *Science,* 1965, *148,* 328–338.

Milner, P. M. *Physiological psychology*. New York: Holt, Rinehart, & Winston, 1970.

Milner, P. M. Theories of reinforcement, drive and motivation. In L. L. Iversen, S. D. Iversen, & S. D. Snyder (Eds.), *Handbook of psychopharmacology* (Vol. 7). New York: Plenum Press, 1977.

Miselis, R. R., & Epstein, A. N. Feeding induced by intracerebroventricular 2-deoxy-D-glucose in the rat. *American Journal of Physiology,* 1975, *229,* 1438–1447.

Mizutani, S., Sonoda, T., Matsumoto, K., & Iwasa, K. Plasma testosterone concentration in paraplegic men. *Journal of Endocrinology,* 1972, *54,* 363–364.

Modianos, D. T., Hitt, J. C., & Popolow, H. B. Habenular lesions and feminine sexual behavior of ovariectomized rats: Diminished responsiveness to the synergistic effects of estrogen and progesterone. *Journal of Comparative and Physiological Psychology,* 1975, *89,* 231–237.

Mogenson, G. J. An attempt to establish secondary reinforcement with rewarding brain stimulation. *Psychological Reports,* 1965, *16,* 163–167.

Mogenson, G. J. Changing views of the role of the hypothalamus in the control of ingestive behaviors. In K. Lederis & K. Cooper (Eds.), *Recent studies of hypothalamic function*. Basel: S. Karger, 1974(a).

Mogenson, G. J. Cerebral electrophysiology of thirst. In G. Peters & J. T. Fitzsimons (Chair), *Thirst (La Soif)*. Journées Internationales de Néphrologie, Vittel, France, 1974 (b).

Mogenson, G. J. Electrophysiological studies of the mechanisms that initiate ingestive behaviours with special emphasis on water intake. In G. J. Mogenson & F. R. Calaresu (Eds.), *Neural integration of physiological mechanisms and behaviour*. Toronto: University of Toronto Press, 1975.

Mogenson, G. J. Neural mechanisms of hunger: Current status and future prospects. In D. Novin, W. Wyrwicka, & G. Bray (Eds.), *Hunger: Basic mechanisms and clinical implications*. New York: Raven, 1976.

Mogenson, G. J., & Calaresu, F. R. Food intake considered from the viewpoint of systems analysis. In D. A. Booth (Ed.), *Hunger models: Computable theory of feeding control*. London: Academic, 1977.

Mogenson, G. J., & Cioe, J. Central reinforcement: A bridge between brain function and behavior. In W. K. Honig & J. E. R. Staddon (Eds.), *Handbook of operant behavior*. Englewood Cliffs, N. J.: Prentice-Hall, 1976.

Mogenson, G. J., & Huang, Y. H. The neurobiology of motivated behavior. In G. A. Kerkut & J. W. Phillis (Eds.), *Progress in Neurobiology*, (Vol. 1, Part 1). Oxford: Pergamon Press, 1973.

Mogenson, G. J., & Kucharczyk, J. Evidence that the lateral hypothalamus and midbrain participate in the drinking response elicited by intracranial angiotensin. In J. T. Fitzsimons & G. Peters (Eds.), *Control mechanisms of drinking*. Heidelberg: Springer-Verlag, 1975.

Mogenson, G. J., Kucharczyk, J., & Assaf, S. Evidence for multiple receptors and neural pathways which subserve water intake initiated by angiotensin II. In J. P. Buckley & C. Ferrario (Eds.), *International symposium on the central actions of angiotensin and related hormones*. New York: Pergamon Press, 1977.

Mogenson, G. J., Mok, A., Grace, J. E., & Stevenson, J. A. F. Behavioral and alimentary response of the rat to acute and chronic changes in environmental temperature. *Journal of Physiologie (Paris)*, 1971, *63*, 346–349.

Mogenson, G. J., & Phillips, A. G. Motivation: A psychological construct in search of a physiological substrate. In A. N. Epstein & J. M. Sprague (Eds.), *Progress in psychobiology and physiological psychology* (Vol. 6). New York: Academic, 1976.

Mogenson, G. J., & Stevenson, J. A. F. Drinking and self-stimulation with electrical stimulation of the lateral hypothalamus. *Physiology and Behavior*, 1966, *1*, 251–254.

Mogenson, G. J., & Stevenson, J. A. F. Drinking induced by electrical stimulation of the lateral hypothalmus. *Experimental Neurology*, 1967, *17*, 119–127.

Moguilevsky, J. A., Kalbermann, L. E., Libertun, C., & Gomez, C. J. Effect of orchiectomy on the amino acid incorporation into proteins of anterior pituitary and hypothalamus in rats. *Proceedings of the Society for Experimental Biology and Medicine*, 1971, *136*, 1115–1118.

Mok, A. C. S., & Mogenson, G. J. Effects of electrical stimulation of the lateral habenular nucleus and lateral hypothalamus. *Brain Research*, 1974, *78*, 425–435.

Money, J. Sex hormones and other variables in human eroticism. In W. C. Young (Ed.), *Sex and internal secretions*. Baltimore: Williams & Wilkins, 1961.

Money, J., & Ehrhardt, A. A. *Man and woman, boy and girl*. Baltimore: Johns Hopkins University Press, 1972.

Montemurro, D. G., & Stevenson, J. A. F. Adipsia produced by hypothalamic lesions in the rat. *Canadian Journal of Biochemistry and Physiology*, 1957, *35*, 31–37.

Mook, D. G., Kenney, N. J., Roberts, S., Nussbaum, A. I., & Rodier, W. I. Ovarian-adrenal interactions in regulation of body weight by female rats. *Journal of Comparative and Physiological Psychology*, 1972, *81*, 198–211.

Moore, R. Y. Retinohypothalamic project projections in mammals: A comparative study. *Brain Research*, 1973, *49*, 403–409.

Mora, F., Sanguinetti, A. M., Rolls, G. T., & Shaw, S. G. Differential effects on self-stimulation and motor behavior produced by microintracranial injections of a dopamine-receptor blocking agent. *Neuroscience Letters,* 1975, *1,* 179–184.

Morgan, C. T. *Physiological Psychology*. New York: McGraw-Hill, 1943.

Morgan, C. W., & Mogenson, G. J. Preference of water-deprived rats for stimulation of the lateral hypothalamus rather than water. *Psychonomic Science,* 1966, *6,* 337–338.

Morgane, P. J. Anatomical and neurobiochemical bases of the central nervous control of physiological regulations and behaviour. In G. J. Mogenson & F. R. Calaresu (Eds.), *Neural integration of physiological mechanisms and behavior*. Toronto: University of Toronto Press, 1975.

Morishima, M. S., & Gale, C. C. Relationship of blood pressure and heart rate to body temperature in baboons. *American Journal of Physiology,* 1972, *223,* 387–395.

Morrison, S. D. The constancy of the energy expended by rats on spontaneous activity, and the distribution of activity between feeding and non-feeding. *Journal of Physiology (London),* 1968, *197,* 305–323.

Moruzzi, G. Active processes in the brainstem during sleep. *The Harvey Lecture Series,* 1963, *58,* 233–297.

Moruzzi, G. Sleep and instinctive behavior. *Archives Italiennes de Biologie (Pisa),* 1969, *108,* 175–216.

Moruzzi, G., & Magoun, H. W. Brainstem reticular formation and activation of the EEG. *Electroencephalography and Clinical Neurophysiology,* 1949, *1,* 455–473.

Mount, G., & Hoebel, B. G. Lateral hypothalamic reward decreased by intragastric feeding: Self-determined "threshold" technique. *Psychonomic Science,* 1967, *9,* 265–266.

Mount, L. E. *The Climatic Physiology of the Pig*. London: Edward Arnold, 1968.

Mountcastle, V. B. *Medical physiology* (Vol. 1, 13th ed.). Saint Louis: C. V. Mosby & Co., 1974, chap. 6.

Mountcastle, V. B. The world around us: Neural command functions for selective attention. *Neurosciences Research Program Bulletin,* 1976, *14* (Suppl.), 1–47.

Moyer, K. E. (Ed.), *Physiology of aggression and implications for control*. New York: Raven, 1976.

Mrosovsky, N. *Hibernation and the hypothalamus*. Englewood Cliffs, N.J.: Prentice-Hall, 1971.

Murgatroyd, D., & Hardy, J. D. Central and peripheral temperatures in behavioral thermoregulation of the rat. In J. D. Hardy, A. P. Gagge, & J. A. J. Stolwijk (Eds.), *Physiological and behavioral temperature*. Springfield, Ill.: C. C Thomas, 1970.

Myers, R. D. *Methods in psychobiology*. London: Academic Press, 1971.

Myers, R. D. The role of hypothalamic transmitter factors in the control of body temperature. In J. D. Hardy, A. P. Gagge, & J. A. J. Stolwijk (Eds.), *Physiological and behavioral temperature regulation*. Springfield, Ill.: C. C Thomas, 1970.

Myers, R. D. *Handbook of drug and chemical stimulation of the brain*. New York: Van Nostrand Reinhold, ©1974 by Litton Educational Publishing, Inc. (a)

Myers, R. D. Ionic concepts of the set-point for body temperature. In K. Lederis & K. E. Cooper (Eds.), *Recent studies of hypothalamic function*. Basel: Karger, 1974. (b)

Myers, R. D., & Martin, G. E. 6-OHDA lesions of the hypothalamus: Interaction of aphagia, food palatability, set-point for weight regulation and recovery of feeding. *Pharmacology Biochemistry and Behavior,* 1973, *1,* 329–345.

Myers, R. D., & Veale, W. L. Body temperature: Possible ionic mechanism in the hypothalamus controlling the set point. *Science,* 1970, *170,* 95–97.

Naftolin, F., Ryan, K. J., & Petro, Z. Aromatization of androstenedione by the anterior hypothalamus of adult male and female rats. *Endocrinology,* 1972, *90,* 295–298.

Nakayama, T., Hammel, H. T., Hardy, J. D., & Eisenman, J. S. Thermal stimulation of electrical activity of single units of the preoptic region. *American Journal of Physiology,* 1963, *204,* 1122–1126.

Nance, D. M., Shryne, J., & Gorski, R. A. Effects of septal lesions on behavioral sensitivity of female rats to gonadal hormones. *Hormones and Behavior,* 1975, *6,* 59–64.

Nauta, W. J. H. The problem of the frontal lobe: A reinterpretation. *Journal of Psychiatric Research,* 1971, *8,* 167.

Newman, H. F. Vibratory sensitivity of the penis. *Fertility and Sterility,* 1970, *21,* 791–793.

Nicolaidis, S. A possible molecular basis of regulation of energy balance. *Fifth International Conference on the Physiology of Food and Fluid Intake.* Jerusalem, October 1974.

Nicolaidis, S., & Rowland, N. Metering of intravenous versus oral nutrients and regulation of energy balance. *American Journal of Physiology,* 1976, *231,* 661–668.

Niijima, A. Afferent discharges from osmoreceptors in the liver of the guinea pig. *Science,* 1969, *166,* 1519–1520.

Norgren, R. Gustatory afferents to ventral forebrain. *Brain Research,* 1974, *81,* 285–295.

Norgren, R. Taste pathways to hypothalamus and amygdala. *Journal of Comparative Neurology,* 1976, *166,* 17–30.

Norgren, R., & Leonard, C. M. Ascending central gustatory pathways. *Journal of Comparative Neurology,* 1973, *150,* 217–237.

Novin, D. Visceral mechanisms in the control of food intake. In D. Novin, W. Wyrwicka, & G. Bray (Eds.), *Hunger: Basic mechanisms and clinical implications.* New York: Raven, 1976.

Novin, D., Vanderweele, D. A., & Rezek, M. Infusion of 2-deoxy-D-glucose into the hepatic-portal system causes eating: Evidence for peripheral glucoreceptors. *Science,* 1973, *181,* 858–860.

Novin, D., Wyrwicka, W., & Bray, G. A. *Hunger: Basic mechanisms and clinical implications.* New York: Raven, 1976.

Oatley, K. *Brain mechanisms and mind.* New York: E. P. Dutton, 1972.

Oatley, K. Simulation and theory of thirst. In A. N. Epstein, H. R. Kissileff, & E. Stellar (Eds.), *The neuropsychology of thirst: New findings and advances in concepts.* Washington, D.C.: Winston, 1973, pp. 199–223.

Oatley, K., & Dickinson, A. Air drinking and the measurement of thirst. *Animal Behavior,* 1970, *18,* 259–265.

Odum, E. P. Premigratory hyperphagia in birds. *American Journal of Clinical Nutrition,* 1960, *8,* 621–629.

Olds, J. Neurophysiology of drive. *Psychiatric Research Reports,* 1956, *6,* 15–22.

Olds, J. Effects of hunger and male sex hormone on self-stimulation of the brain. *Journal of Comparative Physiological Psychology,* 1958, *51,* 320–324. (a)

Olds, J. Satiation effects in self-stimulation of the brain. *Journal of Comparative and Physiological Psychology,* 1958, *51,* 675–678. (b)

Olds, J. Differential effects of drive and drugs on self-stimulation at different brain sites. In D. E. Sheer (Ed.), *Electrical stimulation of the brain.* Austin: University of Texas Press, 1961.

Olds, J. Hypothalamic substrates of reward. *Physiological Review,* 1962, *42,* 554–604.

Olds, J. Operant conditioning of single unit responses. Excerpta medica International Congress Series No. 87. *Proceedings of the 23rd International Congress of Physiological Sciences,* Tokyo, 1965, 372–380.

Olds, J., & Milner, P. Positive reinforcement produced by electrical stimulation of septal area and other regions of the rat brain. *Journal of Comparative and Physiological Psychology,* 1954, *47,* 419–427.

Olds, J., & Olds, M. Drives, reward and the brain. In F. Barron & C. W. Dement (Eds.), *New directions in psychology* (Vol. II). New York: Holt, Rinehart & Winston, 1965.

Olds, M. E. Comparative effects of amphetamine, scopolamine, chlondioxyepoxide, and diphenythydrantoin on operant and extinction behavior with brain stimulation and food reward. *Neuropharmacology,* 1970, *9,* 519–532.

Olds, M. E. Unit responses in the medial forebrain bundle to rewarding stimulation in the hypothalamus. *Brain Research,* 1974, *80,* 479–495.

Olds, M. E., & Olds, J. Effects of lesions in medial forebrain bundle on self-stimulation behavior. *American Journal of Physiology,* 1969, *217*, 1253–1264.

Oltmans, G. A., & Harvey, J. A. LH syndrome and brain catecholamine levels after lesions of the nigrostriatal bundle. *Physiology and Behavior,* 1972, *8,* 69–78.

Oomura, Y. Significance of glucose, insulin, and free fatty acid on the hypothalamic feeding and satiety neurons. In D. Novin, W. Wyrwicka, & G. Bray (Eds.), *Hunger: Basic mechanisms and clinical implications.* New York: Raven, 1976.

Oomura, Y., Ono, T., Ooyama, H., & Wayner, M. J. Glucose and osmosensitive neurones of the rat hypothalamus. *Nature,* 1969, *222,* 282–284.

Oomura, Y., Ooyama, H., Yamamoto, T., Nake, F., Kobayashi, N., & Ono, T. Neuronal mechanism of feeding. *Progress in Brain Research,* 1967, *27,* 1–33.

Oomura, Y., Sugimori, M., Nakamura, T., & Yamada, Y. Contribution of electrophysiological techniques to the understanding of central control systems. In G. J. Mogenson & F. R. Calaresu (Eds.), *Neural Integration of Physiological Mechanisms and behaviour.* Toronto: University of Toronto Press, 1975.

Osmond, H., & Smythies, J. R. Schizophrenia: A new approach. *Journal of Mental Science,* 1952, *98,* 309.

Panksepp, J. Central metabolic and humoral factors involved in the neural regulation of feeding. *Pharmacology, Biochemistry and Behavior,* 1975, *3,* Suppl. 1, 107–119. (a)

Panksepp, J. Metabolic hormones and regulation of feeding: A reply to Woods, Decke and Vasselli. *Psychological Review,* 1975, *82,* 158–164. (b)

Panksepp, J., & Trowill, J. A. Intraoral self-injection: I. Effects of delay of reinforcement on resistance to extinction and implication for self-stimulation. *Psychonomic Science,* 1967, *9,* 405–406. (a)

Panksepp, J., & Trowill, J. A. Intraoral self-injection. II. The simulation of self-stimulation phenomenon with a conventional reward. *Psychonomic Science,* 1967, *9,* 407–408. (b)

Parker, W., & Feldman, S. M. Effect of mesencephalic lesions on feeding behavior in rats. *Experimental Neurology,* 1967, *17,* 313–326.

Parmeggiani, P. L. Sleep behavior elicited by electrical stimulation of cortical and sub-cortical structures in the cat. *Helvetica Physiologica Pharmacologica Acta,* 1962, *20,* 347–367.

Paxinos, G. The hypothalamus: Neural systems involved in feeding, irritability, aggression, and copulation in male rats. *Journal of Comparative and Physiological Psychology,* 1974, *87,* 110–119.

Pearson, O. P. Metabolism of small mammals, with remarks on the lower limit of mammalian size. *Science,* 1948, *108,* 44.

Peck, J. W., & Novin, D. Evidence that osmoreceptors mediating drinking in rabbits are in the lateral preoptic area. *Journal of Comparative and Physiological Psychology,* 1971, *74,* 134–147.

Penfield, W., & Jasper, H. H. *Epilepsy and the Functional Anatomy of the Human Brain.* Boston: Little, Brown, 1954.

Pfaff, D. W. Autoradiographic localization of radioactivity in rat brain after injection of tritiated sex hormones. *Science,* 1968, *161,* 1355–1356.

Pfaff, D. W., & Pfaffmann, C. Olfactory and hormonal influences on the basal forebrain of the male rat. *Brain Research,* 1969, *15,* 137–156.

Pfaffmann, C. The pleasures of sensation. *Psychological Review,* 1960, *67,* 253–268.

Phillips, A. G. Enhancement and inhibition of olfactory bulb self-stimulation by odours. *Physiology and Behavior,* 1970, *5,* 1127–1131.

Phillips, A. G., & Fibiger, H. C. Substantia nigra: Self-stimulation and poststimulation feeding. *Physiological Psychology,* 1973, *1,* 233–236.

Phillips, A. G., & Mogenson, G. J. Effects of taste on self-stimulation and induced drinking. *Journal of Comparative and Physiological Psychology,* 1968, *66,* 654–660.

Phillips, A. G., & Mogenson, G. J. Self-stimulation of the olfactory bulb. *Physiology and Behavior,* 1969, *4,* 195–197.

Phillips, A. G., Morgan, C. W., & Mogenson, G. J. Changes in self-stimulation preference as a function of incentive of alternative rewards. *Canadian Journal of Psychology,* 1970, *24,* 289–297.

Phillips, M. I., & Olds, J. Unit activity: Motivation-dependent responses from midbrain neurons. *Science,* 1969, *165,* 1269–1271.

Phoenix, C. Effects of dihydrotestosterone on sexual behavior of castrated male Rhesus monkeys. *Physiology and Behavior,* 1974, *12,* 1045–1055.

Pliskoff, S. S., & Hawkins, T. D. A method for increasing the reinforcement magnitude of intracranial stimulation. *Journal for Experimental Analysis of Behaviour,* 1967, *10,* 281–289.

Poschel, B. P. H., & Ninteman, F. W. Psychotropic drug effects on self-stimulation of the brain: A control for motor output. *Psychological Reports,* 1966, *19,* 79–82.

Powley, T. L., & Keesey, R. E. Relationship of body weight to the lateral hypothalamic feeding syndrome. *Journal of Comparative and Physiological Psychology,* 1970, *70,* 25–36.

Quadagno, D. M., & Ho, G. K. W. The reversible inhibition of steroid-induced sexual behavior by intracranial cycloheximide. *Hormones and Behavior,* 1975, *6,* 19–26.

Rabedeau, R. G., & Whalen, R. E. Effects of copulatory experience on mating behavior in the male rat. *Journal of Comparative and Physiological Psychology,* 1959, *52,* 482–484.

Ranck, J. B., Jr. Behavioral correlates and firing repertoires of neurons in the dorsal hippocampal formation and septum of unrestrained rats. In R. L. Isaacson & K. H. Pribram (Eds.), *The Hippocampus* (Vol. 2). New York: Plenum Press, 1975.

Ranson, S. W., & Ingram, W. R. Hypothalamus and regulation of body temperature. *Proceedings of the Society for Experimental Biology and Medicine,* 1935,32, 1439–1441.

Reeves, A. G., & Plum, F. Hyperphagia, rage, and dementia accompanying a ventromedial hypothalamic neoplasm. *Archives of Neurology,* 1969, *20,* 616–624.

Reinberg, A., Lagoguey, M., Chauffournier, J. M., & Cesselin, F. Rythmes annuels et circadiends de la testostérone plasmatique chez cinq parisiens jeunes, adultes et sains. *Annales d'Endocrinoligie (Paris),* 1975, *36,* 44–45.

Reis, D. J. The relationship between brain norepinephrine and aggressive behavior. *Research Publications of the Association for Research in Nervous and Mental Distress,* 1972, *50,* 266–297.

Reis, D. J., & Gunne, L. M. Brain catecholamines: Relation to the defense reaction evoked by amygdaloid stimulation in the cat. *Science,* 1965, *149,* 450.

Reis, D. J., Weinbren, M., & Corvelli, A. A circadian rhythm of norepinephrine regionally in cat brain: Its relationship to environmental lighting and to regional diurnal variations in brain serotonin. *Journal of Pharmacology and Experimental Therapeutics,* 1968, *164,* 135–145.

Renaud, L. P. An electrophysiological study of amygdalo-hypothalamic projections to the ventromedial nucleus of the rat. *Brain Research,* 1976, *105,* 45–58.

Resnick, O. The role of biogenic amines in sleep. In C. D. Clemente, D. P. Purpura, & F. Mayer (Eds.), *Sleep and the maturing nervous system.* New York: Academic, 1972.

Reynolds, R. W., & Bryson, G. Effect of estradiol on the hypothalamic regulation of body weight in the rat. *Research Communications in Chemical Pathology and Pharmacology,* 1974, *7,* 715–824.

Richter, C. P. Total self-regulatory functions in animals and human beings. *Harvey Lecture Series,* 1943, *38,* 63–103.

Richter, C. P. *Biological Clocks in Medicine and Psychiatry.* Springfield, Ill.: C. C. Thomas, 1965.

Richter, C. P. Inborn nature of the rat's 24-hour clock. *Journal of Comparative and Physiological Psychology,* 1971, *75,* 1–4.

Ritter, S., & Stein, L. Self-stimulation of the locus coerulus. *Federation Proceedings,* 1972, *31,* 820.

Ritter, S., & Stein, L. Self-stimulation of noradrenergic cell group (A6) in locus coerulus of rats. *Journal of Comparative and Physiological Psychology,* 1973, *85,* 443–452.

Ritter, S., & Stein, L. Self-stimulation in the mesencephalic trajectory of the ventral noradrenergic bundle. *Brain Research,* 1974, *81,* 145–157.

Roberts, W. W., & Mooney, R. D. Brain areas controlling thermoregulatory grooming, prone extension, locomotion and tail vasodilation in rats. *Journal of Comparative and Physiological Psychology,* 1974, *86,* 470–480.

Robertson, A., Kucharczyk, J., & Mogenson, G. J. Subfornical organ, a site of brain self-stimulation. *Brain Research,* 1976, *114,* 511–516.

Robinson, T. E., Kramis, R. C., & Vanderwolf, C. H. Two types of cerebral activation during active sleep: relations to behavior. *Brain Research,* 1977, *124,* 544–549.

Robinson, B. W., & Mishkin, M. Alimentary responses to forebrain stimulation in monkeys. *Experimental Brain Research,* 1968, *4,* 330–366.

Rodgers, W. L., Epstein, A. N., & Teitelbaum, P. Lateral hypothalamic aphagia: Motor or motivational deficit? *American Journal of Physiology,* 1965, *208,* 334–342.

Rodin. J., & Slochower, J. Externality in the nonobese: Effects of environmental responsiveness on weight. *Journal of Personality and Social Psychology,* 1976, *33,* 338–344.

Roffwarg, H. P., Muzio, J. M., & Dement, W. C. Ontogenetic development of the human sleep–dream cycle. *Science,* 1966, *152,* 604–619.

Roll, S. K. Intracranial self-stimulation and wakefullness: Effect of manipulating ambient catecholamines. *Science,* 1970, *168,* 1370–1372.

Rolls, E. T. Contrasting effects of hypothalamic and nucleus accumbens septi self-stimulation on brainstem single unit activity and cortical arousal. *Brain Research,* 1971, *31,* 275–285. (a)

Rolls, E. T. Involvement of brainstem units in medial forebrain bundle self-stimulation. *Physiology and Behavior,* 1971, *7,* 297–310. (b)

Rolls, E. T. The neural basis of brain-stimulation reward. *Progress in Neurobiology,* 1974, *3,* 71–160.

Rolls, E. T. *The brain and reward.* Oxford: Pergamon Press, 1975.

Rolls, E. T. The neurophysiological basis of brain-stimulation reward. In A. Wauquier & E. T. Rolls (Eds.), *Brain stimulation reward.* Amsterdam: North-Holland, 1976.

Rolls, E. T., & Cooper, S. J. Activation of neurones in the prefrontal cortex by brain-stimulation reward in the rat. *Brain Research,* 1973, *60,* 351–368.

Rolls, E. T., & Cooper, S. J. Anesthetization and stimulation of the sulcal prefrontal cortex and brain-stimulation reward. *Physiology and Behavior,* 1974, *12,* 563–571. (a)

Rolls, E. T., & Cooper, S. J. Connection between the prefrontal cortex and pontine brain-stimulation reward sites in the rat. *Experimental Neurology,* 1974, *42,* 687–699. (b)

Rolls, E. T., & Kelly, P. H. Neural basis of stimulus-bound locomotor activity in the rat. *Journal of Comparative and Physiological Psychology,* 1972, *81,* 173–182.

Rolls, E. T., Burton, M. J., & Mora, F. Hypothalamic neuronal responses associated with the sight of food. *Brain Research,* 1976, *111,* 53–66.

Rolls, E. T., Kelly, P. H., & Shaw, S. G. Noradrenaline, dopamine, and brain-stimulation reward. *Pharmacology, Biochemistry and Behavior,* 1974, *2,* 735–740.

Rosenblatt, J. S., & Aronson, L. R. The decline of sexual behaviour in male cats after castration with special reference to the role of prior experience. *Behaviour,* 1958, *12,* 285–338.

Ross, J., Claybaugh, C., Clemens, L. G., & Gorski, R. A. Short latency induction of estrous behavior with intracerebral gonadal hormones in ovariectomized rats. *Endocrinology,* 1971, *89,* 32–38.

Rosvold, H. E., Mirsky, A. F., & Pribram, K. H. Influence of amygdalectomy on social behavior in monkeys. *Journal of Comparative and Physiological Psychology,* 1954, *47,* 173–178.

Rothballer, A. B. Studies on the adrenalin-sensitive components of the reticular activating system. *Electroencephalography and Clinical Neurophysiology,* 1956, *8,* 603–621.

Routtenberg, A. Forebrain pathways of reward in *Rattus norvegicus. Journal of Comparative and Physiological Psychology,* 1971, *75,* 269–276.

Routtenberg, A., Gardner, E. L., & Huang, Y. H. Self-stimulation pathways in the monkey, *Macaca mulatta. Experimental Neurology,* 1971, *33,* 213–224.

Routtenberg, A., & Huang, Y. H. Reticular formation and brainstem unitary activity: Effects of posterior hypothalamic and septal-limbic stimulation at reward loci. *Physiology and Behavior,* 1968, *3,* 611–617.

Routtenberg, A., & Lindy, J. Effects of the availability of rewarding septal and hypothalamic stimulation on bar pressing for food under conditions of deprivation. *Journal of Comparative and Physiological Psychology,* 1965, *60,* 158–161.

Routtenberg, A., & Malsbury, C. Brainstem pathways of reward. *Journal of Comparative and Physiological Psychology,* 1969, *68,* 22–30.

Routtenberg, A., & Sloan, M. Self-stimulation in the frontal cortex of *Rattus norvegicus. Behavioral Biology,* 1972, *7,* 567–572.

Rowell, T. Behavior and female reproductive cycles of Rhesus macaques. *Journal of Reproduction and Fertility,* 1963, *6,* 193–203.

Rowland, N. Endogenous circadian rhythms in rats recovered from lateral hypothalamic lesions. *Physiology and Behavior,* 1976, *16,* 257–266.

Rowland, N., & Nicolaidis, S. Metering of fluid intake and determinants of ad libitum drinking in rats. *American Journal of Physiology,* 1976, *231,* 1–8.

Rozin, P. Thiamine specific hunger. In C. F. Code (Ed.), *Handbook of Physiology,* Alimentary Canal Sect. 6), Food and Water Intake (Vol. 1). Washington, D.C.: American Physiological Society, 1967.

Rozin, P. Psychobiological and cultural determinants of food choice. In T. Silverstone (Ed.), *Appetite and food intake.* Dahlem Konferenzen, Berlin: Abakon Verlagsgesellschaft, 1976.

Rozin, P., & Kalat, J. Specific hungers and poison avoidance as adaptive specializations of learning. *Psychological Review,* 1971, *78,* 459–486.

Rubenstein, H. S. *The study of the brain.* New York: Grune & Stratton, 1953.

Ruch, T. C. The homotypical cortex—The "association" areas. In T. C. Ruch & H. D. Patton (Eds.), *Physiology and biophysics.* Philadelphia: Saunders, 1965.

Rusak, B., & Zucker, I. Biological rhythms and animal behavior. *Annual Review of Psychology,* 1975, *26,* 137–171.

Russek, M. Participation of hepatic glucoreceptors in the control of intake of food. *Nature (London),* 1963, *197,* 79–80.

Russell, P. J. D., Abdelaal, A. E., & Mogenson, G. J. Graded levels of hemorrhage, thirst and angiotensin II in the rat. *Physiology and Behavior,* 1975, *15,* 117–119.

Sackett, G. P. Abnormal behavior in laboratory-reared rhesus monkeys. In M. W. Fox (Ed.), *Abnormal behavior in animals.* Philadelphia: W. B. Saunders, 1968.

Sadowski, B. Physiological correlates of self-stimulation. In A. Wauquier & E. T. Rolls (Eds.), *Brain-stimulation reward.* Amsterdam: North-Holland, 1976.

Salaman, D. F. RNA synthesis in the rat anterior hypothalamus and pituitary: Relation to neonatal androgen and the oestrous cycle. *Journal of Endocrinology,* 1970, *48,* 125–137.

Salmon, U. J., & Geist, S. H. Effect of androgens upon libido in women. *Journal of Clinical Endocrinology,* 1943, *3,* 235–238.

Saller, C. F., & Stricker, E. M. Hyperphagia and increased growth in rats after intraventricular injection of 5, 7-dihydroxytryptamine. *Science,* 1976, *192,* 385–387.

Satinoff, E. Behavioral thermoregulation in response to local cooling of the rat brain. *American Journal of Physiology,* 1964, *206,* 1389–1394.

Satinoff, E. Neural integration of thermoregulatory responses. In L. V. DiCara (Ed.), *Limbic and autonomic nervous systems research.* New York: Plenum Press, 1974.

Satinoff, E., & Stanley, W. C. Effects of stomach loading on suckling behavior in neonatal puppies. *Journal of Comparative and Physiological Psychology,* 1963, *56,* 66–68.

Sawyer, C. H., & Robison, B. Separate hypothalamic areas controlling pituitary gonadotrophic function and mating behavior in female cats and rabbits. *Journal of Clinical Endocrinology and Metabolism (Philadelphia),* 1956, *16,* 914–915.

Schachter, S. Cognitive effects on bodily functioning. In D. C. Glass (Ed.), *Neurophysiology and emotion.* New York: Rockefeller University Press and Russell Sage Foundation, 1967.

Scheving, L. E., Halberg, F., & Pauly, J. E. *Chronobiology*. Tokyo: Igaku Shoin, 1974.
Schildkraut, J. J. Catecholamine metabolism and affective illnesses. In H. E. Himwich (Ed.), *Biochemistry, schizophrenias and affective illnesses*. Baltimore: Williams & Wilkins, 1971.
Schmidt, R. F. *Fundamentals of neurophysiology*. Berlin: Springer-Verlag, 1975.
Schmitt-Nielsen, K. *Animal Physiology*. Englewood Cliffs, N. J.: Prentice Hall, 1960.
Schmitt, M. Influences of hepatic protal receptors on hypothalamic feeding and satiety centers. *American Journal of Physiology,* 1973, *225,* 1089–1095.
Schon, M., & Sutherland, A. M. The role of hormones in human behavior. III. Changes in female sexuality after hypophysectomy. *Journal of Clinical Endocrinology and Metabolism,* 1960, *20,* 833–841.
Schreiner, L., & Kling, A. Behavioral changes following rhinencephalic injury in cat. *Journal of Neurophysiology,* 1953, *16,* 643–659.
Schreiner, L., & Kling, A. Rhinencephalon and behavior. *American Journal of Physiology,* 1956, *16,* 643–658.
Schwartz, M. *Physiological Psychology*. New York: Appleton-Century-Crofts, 1973.
Seeman, P., & Lee, T. Antipsychotic drugs: Direct correlation between clinical potency and presynaptic action on dopamine neurons. *Science,* 1975, *188,* 1217–1219.
Sellers, E. A., Reichman, S., & Thomas, N. Acclimatization to cold: Natural and artificial. *American Journal of Physiology,* 1957, *167,* 644–650.
Selye, H. *Stress*. Montreal: ACTA, 1950.
Seward, J. P., Myeda, A., & Olds, J. Resistance to extinction following cranial self-stimulation. *Journal of Comparative and Physiological Psychology,* 1959, *52,* 294–299.
Shealy, C. N., & Peele, T. L. Studies on amygdaloid nucleus of cat. *Journal of Neurophysiology,* 1957, *20,* 125–139.
Sherrington, C. S. *The Integrative Action of the Nervous System*. New Haven: Yale University Press, 1906.
Shettleworth, S. J. Constraints on learning. In D. S. Lehrman, R. A. Hinde, & E. Shaw (Eds.), *Advances in the study of behavior* (Vol. 4). New York: Academic, 1972.
Sidman, M., Brady, J. V., Boren, J. J., Conrad, D. J., & Schulman, A. Reward schedules and behavior maintained by intracranial self-stimulation. *Science,* 1955, *122,* 830–831.
Siegel, A., & Chabora, J. Effects of electrical stimulation of the cingulate gyrus upon attack behaviour elicited from hypothalamus in the cat. *Brain Research,* 1971, *32,* 169–177.
Siegel, A., & Skog, O. Effects of electrical stimulation of the septum upon attack behaviour elicited from the hypothalamus in the cat. *Brain Research,* 1970, *23,* 371–380.
Simpson, J. B., & Routtenberg, A. Subfornical organ: Site of drinking elicitation by angiotensin II. *Science,* 1973, *181,* 1172–1174.
Singer, J. J. Hypothalamic control of male and female sexual behavior in female rats. *Journal of Comparative and Physiological Psychology,* 1968, *66,* 738–742.
Skinner, B. F. *The Behavior of Organisms: An Experimental Approach*. New York: Appleton-Century-Crofts, 1938.
Snyder, S. H. Catecholamines as mediators of drug effects in schizophrenia. In F. O. Schmitt & F. G. Worden (Eds.), *The neurosciences: Third study program*. Cambridge, Mass: MIT Press, 1974.
Snyder, S. H., Banerjee, S. P., Yamamura, H. J., & Greenberg, D. Drugs, neurotransmitters and schizophrenia. *Science,* 1974, *184,* 1243–1249.
Södersten, P. Estrogen-activated sexual behavior in male rats. *Hormones and Behavior,* 1973, *4,* 247–256.
Sorenson, J. P., Jr., & Harvey, J. A. Decreased brain acetylcholine after septal lesions in rats: Correlation with thirst. *Physiology and Behavior,* 1971, *6,* 723–725.
Soulairac, J. L., & Soulairac, A. Monoaminergic and cholinergic control of sexual behavior in the male rat. In M. Sandler & G. L. Gessa (Eds.), *Sexual behaviour: Pharmacology and biochemistry*. New York: Raven, 1975.

Spies, G. Food versus intracranial self-stimulation reinforcement in food deprived rats. *Journal of Comparative Physiological Psychology*, 1965, *60*, 153–157.

Steffens, A. B. Plasma insulin content in relation to blood glucose level and meal pattern in the normal and hypothalamic hyperphagic rat. *Physiology and Behavior*, 1970, *5*, 147–151.

Steffens, A. B. Influence of reversible obesity on eating behavior, blood glucose, and insulin in the rat. *American Journal of Physiology*, 1975, *228*, 1738–1744.

Steffens, A. B., Mogenson, G. J., & Stevenson, J. A. F. Blood glucose, insulin, and free fatty acids after stimulation and lesions of the hypothalamus. *American Journal of Physiology*, 1972, *222*, 1446–1452.

Stein, L. Effects and interactions of imipramine, chlorpromazine, reserpine, and amphetamine on self-stimulation: Possible neurophysiological basis of depression. In J. Wortis (Ed.), *Recent advances in biological psychiatry*. New York: Plenum Press, 1962.

Stein, L. Chemistry of purposive behavior. In T. J. Tapp (Ed.), *Reinforcement and behavior*. New York: Academic, 1969.

Stein, L. Neurochemistry of reward and punishment: Some implications for the etiology of schizophrenia. *Journal of Psychiatric Research*, 1971, *8*, 345–361.

Stein, L., & Wise, C. D. Release of norepinephrine from hypothalamus and amygdala by rewarding medial forebrain bundle stimulation and amphetamine. *Journal of Comparative and Physiological Psychology*, 1969, *67*, 189–198.

Stein, L., & Wise, C. D. Possible etiology of schizophrenia: Progressive damage to the noradrenergic reward system by 6-hydroxydopamine. *Science*, 1971, *171*, 1032–1036.

Stein, L., Belluzzi, J. D., & Wise, C. D. Norepinephrine self-stimulation pathways: Implications for long-term memory and schizophrenia. In A. Wauquier & E. T. Rolls (Eds.), *Brain-stimulation reward*, Amsterdam: North-Holland, 1976.

Stellar, E. The physiology of motivation. *Psychological Review*, 1954, *61*, 5–22.

Stellar, E. Drive and motivation. In J. Field (Ed.), *Handbook of physiology* (Vol. 3). Baltimore: Williams & Wilkins, 1960.

Stellar, E., & Hill, J. H. The rat's rate of drinking as a function of water deprivation. *Journal of Comparative and Physiological Psychology*, 1952, *45*, 96–102.

Stephan, F. K., & Zucker, I. Circadian rhythms in drinking behavior and locomotor activity are eliminated by hypothalamic lesions. *Proceedings of the National Academy of Science (U.S.A)*, 1972, *69*, 1583–1586.

Sterman, M. B., & Clemente, C. D. Forebrain inhibitory mechanisms: Sleep patterns induced by basal forebrain stimulation in the behaving cat. *Experimental Neurology*, 1962, *6*, 103–117.

Stern, W. C., & Morgane, P. J. Theoretical view of REM sleep function: Maintenance of catecholamine systems in the central nervous system. *Behavioral Biology* 1974, *11*, 1–32.

Stetson, M. H., & Watson-Whitmyre, M. Nucleus suprachiamaticus: The biological clock in the hamster. *Science*, 1976, *191*, 197–199.

Stevens, C. F. *Neurophysiology: A primer*. New York: Wiley, 1966.

Stevens, J. R., Mark, V. H., Ervin, F. R., Pacheco, P., & Suematsu, K. Long latency, long lasting psychological changes induced by deep temporal lobe stimulation in man. *A.M.A. Archives of Neurology and Psychiatry*, 1969, *21*, 157–169.

Stevenson, J. A. F. The hypothalamus in the regulation of energy and water balance. *The Physiologist*, 1964, *7*, 305–318.

Stevenson, J. A. F. Current concepts of the regulation of water and salt exchange. *Applied Therapeutics*, 1965, *7*, 713–718.

Stevenson, J. A. F. Central mechanisms controlling water intake. In C. F. Code (Ed.) *Handbook of Physiology*, Alimentary Canal (Sect. 6), Food and Water Intake (Vol. 1). Washington, D.C.: American Physiological Society, 1967.

Stevenson, J. A. F. Neural control of food and water intake. In W. Haymaker, E. Anderson, & W. J. H. Nauta (Eds.), *The hypothalamus*. Springfield, Ill.: C. C. Thomas, 1969.

Stone, C. P. The effects of cerebral destruction on the sexual behavior of rabbits: II. The frontal and parietal regions. *American Journal of Physiology,* 1925, *72,* 372–385.

Stricker, E. M. Extracellular fluid volume and thirst. *American Journal of Physiology,* 1966, *211,* 232–238.

Stricker, E. M. Drinking by rats after lateral hypothalamic lesions: A new look at the lateral hypothalamic syndrome. *Journal of Comparative and Physiological Psychology,* 1976, *90,* 127–143.

Stricker, E. M., & Zigmond, M. J. Brain catecholamines and the lateral hypothalamic syndrome. In D. Novin, W. Wyrwicka, & G. A. Gray (Eds.), *Hunger: Basic mechanisms and clinical implications.* New York: Raven, 1976.

Strümpell, A. Ein beitrag zur theorie des Schlafs. *Archiv für die Gesamte Physiologie des Menschen und der Tiere,* 1877, *15,* 573–574.

Sundsten, J. W. Alterations in water intake and core temperature in baboons during hypothalamic thermal stimulation. *Annals of the New York Academy of Sciences,* 1969, *157,* Art. 2, 1018–1029.

Swanson, L. W., & Sharpe, L. G. Centrally induced drinking: comparison of angiotensin-II and carbachol sensitive sites in rats. *American Journal of Physiology,* 1973, *225* 556–573.

Tagliamonte, A., Fratta, W., Del Fiacco, M., & Gessa, G. L. Possible stimulatory role of brain dopamine in the copulatory behavior of male rats. *Pharmacology Biochemistry and Behavior,* 1974, *2,* 257–260.

Takigawa, M., & Mogenson, G. J. A study of inputs to antidromically identified neurons of the locus coeruleus. *Brain Research,* 1977 (in press).

Teitelbaum, P. The use of operant methods in the assessment and control of motivational states. In W. K. Honig (Ed.), *Operant behavior: Areas of research and application.* Englewood Cliffs, N.J.: Prentice-Hall, 1966.

Teitelbaum, P. Levels of integration of the operant. In W. K. Honig, & J. E. R. Staddon (Eds.), *Handbook of operant behavior.* Englewood Cliffs, N.J.: Prentice-Hall, 1976.

Teitelbaum, P., & Epstein, A. N. The lateral hypothalamic syndrome: Recovery of feeding and drinking after lateral hypothalamic lesions. *Psychological Review,* 1962, *69,* 74–90.

Ter Haar, M. B. Circadian and estrual rhythms in food intake in the rat. *Hormones and Behavior,* 1972, *3,* 213–219.

Terman, M., & Terman, J. S. Circadian rhythm of brain self-stimulation behavior. *Science,* 1970, *168,* 1242–1244.

Terman, M., & Terman, J. S. Control of the rat's circadian self-stimulation rhythm by light–dark cycles. *Physiology and Behavior,* 1975, *14,* 781–789.

Terzian, H., & Ore, G. D. Syndrome of Kluver and Bucy reproduced in man by bilateral removal of the temporal lobes. *Neurology,* 1955, *5,* 373–380.

Teyler, T. T. *A primer of psychobiology: Brain and behavior.* San Francisco: W. H. Freeman & Co., 1975.

Thompson, M. L., & Edwards, D. A. Experiential and strain determinants of the estrogen–progesterone induction of sexual receptivity in spayed female mice. *Hormones and Behavior,* 1971, *2,* 299–305.

Thompson, R. F. *Foundations of physiological psychology.* New York: Harper & Row, 1967.

Thompson, R. F. *Introduction to physiological psychology.* New York: Harper & Row, 1975.

Thompson, W. R. Early experience—Its importance for later behavior. In P. H. Hock & J. Zulin (Eds.), *Psychopathology of children.* New York: Grune & Stratton, 1955.

Tinbergen, N. *The study of instinct.* London: Oxford University Press, 1951.

Tolman, E. C. A behavioristic account of the emotions. *Psychological Review,* 1923, *30,* 217–227.

Towbin, E. J. Gastric distention as a factor in the satiation of thirst in esophagostomized dogs. *American Journal of Physiology,* 1949, *159,* 533–541.

Trimble, M. R., & Herbert, J. The effect of testosterone or oestradiol upon the sexual and associated behaviour of the adult female Rhesus monkey. *Journal of Endocrinology,* 1968, *42,* 171–185.

Trowill, J. A., Panksepp, J., & Gandelman, R. An incentive model of rewarding brain stimulation. *Psychological Review,* 1969, *76,* 264–281.

Turner, B. H. Sensorimotor syndrome produced by lesions of the amygdala and lateral hypothalamus. *Journal of Comparative and Physiological Psychology,* 1973, *82,* 37–47.

Udry, J. R., & Morris, N. M. Distribution of coitus in the menstrual cycle. *Nature,* 1968, *220,* 593–596.

Udry, J. R., Morris, N. M., & Waller, L. Effect of contraceptive pills on sexual activity in the luteal phase of the human menstrual cycle. *Archives of Sexual Behavior,* 1973, *2,* 205–214.

Ungerstedt, U. Adipsia and aphagia after 6-hydroxydopamine induced degeneration of the nigrostriatal dopamine system. *Acta Physiologica Scandinavica Supplement,* 1971, *367,* 95–122.

Ungerstedt, U. Functional dynamics of central monoamine pathways. In F. O. Schmitt & F. G. Worden (Eds.), *The neurosciences: Third study program.* Cambridge, Mass.: M.I.T. Press, 1974.

Valenstein, E. S. *Brain stimulation and motivation.* Glenview, Ill.: Scott, 1973.

Valenstein, E. S., & Beer, B. Continuous opportunity for reinforcing brain stimulation. *Journal for Experimental Analysis of Behavior,* 1964, *7,* 183–184.

Valenstein, E. S., Cox, V. C., & Kakolewski, J. W. Polydipsia elicited by the synergistic action of a saccharine and glucose solution. *Science,* 1967, *157,* 552–554.

Valenstein, E. S., Cox, V. C., & Kakolewski, J. W. Modification of motivated behavior elicited by electrical stimulation of the hypothalamus, *Science,* 1968, *159,* 1119–1121.

Vander, A. J., Sherman, J. H., & Luciano, D. S. *Human physiology: The mechanisms of body function.* New York: McGraw-Hill, 1970.

Vanderwolf, C. H., Kramis, R., Gillespie, L. A., & Bland, B. H. Hippocampal rhythmic slow activity and neocortical low-voltage fast activity: Relation to behavior. In R. L. Isaacson & K. H. Pribram (Eds.), *The hippocampus* (Vol. 2). New York: Plenum Press, 1975.

van Dis, H., and Larsson, K. Induction of sexual arousal in the castrated male rat by intracranial stimulation. *Physiology and Behavior,* 1971, *6,* 85–86.

Van Sommers, P. *The biology of behavior.* New York: Wiley, 1972.

Vaughan, E., & Fisher, A. E. Male sexual behavior induced by intracranial electrical stimulation. *Science,* 1962, *137,* 758–759.

Veale, W. L., & Cooper, K. E. Evidence for the involvement of prostaglandins in fever. In K. Lederis & K. E. Cooper (Eds.), *Recent studies of hypothalamic function.* Basel: Karger, 1974.

Velasco-Suárez, M. M. Electrical and chemical stimulation of limbic structures within the temporal lobe. *Bibliotheca Psychiatrica (Basel),* 1970, *143,* 187–196.

Verney, E. B. The antidiuretic hormone and factors which determine its release. *Proceedings of the Royal Society of London,* Series B: Biological Sciences, 1947, *135,* 25–106.

Villablanca, J., & Myers, R. D. Fever produced by microinjection of typhoid vaccine into hypothalamus of cats. *American Journal of Physiology,* 1965, *208,* 703–707.

Vincent, J. D., Arnauld, E., & Bioulac, B. Activity of osmosensitive single cells in the hypothalamus of the behaving monkey during drinking. *Brain Research,* 1972, *44,* 371–384.

Von Euler, U. S. A specific sympathomimetic ergone in adrenergic nerve fibers (sympathin) and its relation to adrenaline and noradrenaline. *Acta Physiologica Scandinavica,* 1946, *12,* 73–97.

Wade, G. N., & Zucker, I. Development of hormonal control over food intake and body weight in female rats. *Journal of Comparative and Physiological Psychology,* 1970, *70,* 213–220. (a)

Wade, G. N., & Zucker, I. Modulation of food intake and locomotor activity in female rats by diencephalic hormone implants. *Journal of Comparative and Physiological Psychology,* 1970, *72,* 328–336. (b)

Wagener, J. W. Self-stimulation of preoptic and lateral hypothalamus during behavioral thermoregulation in the albino rat. *Journal of Comparative and Physiological Psychology,* 1973, *84,* 652–660.

Wallen, K., Goldfoot, D. A., Joslyn, W. D., & Paris, C. A. Modification of behavioral estrus in the guinea pig following intracranial cycloheximide. *Physiology and Behavior,* 1972, *8,* 221–224.

Warden, G. J. *Animal motivation studies. The albino rat.* New York: Columbia University Press, 1931.

Wasman, M., & Flynn, J. P. Directed attack elicited from hypothalamus. *Archives of Neurology (Chicago),* 1962, *6,* 220–227.

Watson, J. B. Psychology as the behaviorist views it. *Psychological Review,* 1913, *20,* 158–177.

Watson, J. B. A schematic outline of the emotions. *Psychological Review,* 1919, *26,* 165–196.

Wauquier, A., & Niemegeers, C. J. Intracranial self-stimulation in rats as a function of various stimulus parameters. II. Influence of haloperidol, pimozide and pipamperone on medial forebrain bundle stimulation with monopolar electrodes. *Psychopharmacologia (Berlin),* 1972, *27,* 191–202.

Wauquier, A., & Rolls, E. T. *Brain-stimulation reward.* Amsterdam: North-Holland, 1976.

Waxenberg, S. E., Drellich, M. G., & Sutherland, A. M. The role of hormones in human behavior. I. Changes in female sexuality after adrenalectomy. *Journal of Clinical Endocrinology,* 1959, *19,* 193–202.

Weiner, I. H., & Stellar, E. Salt preference of the rat determined by a single-stimulus method. *Journal of Comparative and Physiological Psychology,* 1951, *44,* 394–401.

Weiss, A. D. Sensory functions. In J. E. Birren (Ed.), *Handbook of aging and the individual.* Chicago: University of Chicago Press, 1959.

Weiss, B., & Laties, V. G. Behavioral thermoregulation. *Science,* 1961, *133,* 1338–1344.

Weitzman, E. D., & Graziani, L. Sleep and the sudden infant death syndrome: A new hypothesis. In E. D. Weitzman (Ed.), *Advances in sleep research* (Vol. 1). New York: Spectrum Publications, 1974.

Welch, B. L., & Welch, A. S. Fighting: Preferential lowering of norepinephrine and dopamine in the brainstem, concomitant with a depletion of epinephrine from the adrenal medulla. *Communications in Behavioral Biology,* 1969, *3,* 125.

Whalen, R. E. The initiation of mating in naive female cats. *Behaviour,* 1963, *11,* 461–463.

Whalen, R. E. The ontogeny of sexuality. In H. Moltz (Ed.), *The Ontogeny of Vertebrate Behavior.* New York: Academic Press, 1971, pp. 229–261.

Whalen, R. E., Battie, C., & Luttge, W. G. Anti-estrogen inhibition of androgen-induced sexual receptivity in rats. *Behavioral Biology,* 1972, *7,* 311–320.

Whalen, R. E., Beach, F. A., & Kuehn, R. E. Effects of exogenous androgen on sexuality responsive and unresponsive male rats. *Endocrinology,* 1961, *69,* 373–380.

Whalen, R. E., & Gorzalka, B. B. The effects of progesterone and its metabolites on the induction of sexual receptivity in rats. *Hormones and Behavior,* 1972, *3,* 221–226.

Whalen, R. E., & Gorzalka, B. B. Effects of an estrogen antagonist on behavior and on estrogen retention in neural and peripheral target tissues. *Physiology and Behavior,* 1973, *10,* 35–40.

Whalen, R. E., & Gorzalka, B. B. Estrogen–progesterone interactions in uterus and brain of intact and adrenalectomized immature and adult rats. *Endocrinology,* 1974, *94,* 214–223.

Whalen, R. E., Gorzalka, B. B., & DeBold, J. F. Methodologic considerations in the study of animal sexual behavior. In M. Sandler & G. L. Gessa (Eds.), *Sexual behavior: Pharmacology and biochemistry.* New York: Raven, 1975. (a)

Whalen, R. E., Gorzalka, B. B., DeBold, J. F., Quadagno, D. M., Ho, G. K. W., & Hough, J. C. Studies on the effects of intracerebral actinomycin-D implants on estrogen-induced receptivity in rats. *Hormones and Behavior,* 1974, *5,* 337–343.

Whalen, R. E., & Hardy, D. F. Induction of receptivity in female rats and cats with estrogen and testosterone. *Physiology and Behavior,* 1970, *5,* 529–533.

Whalen, R. E., Luttge, W. G., & Gorzalka, B. B. Neonatal androgenization and the development of estrogen responsivity in male and female rats. *Hormones and Behavior,* 1971, *2,* 83–90.

Whalen, R. E., Neubauer, B. L., & Gorzalka, B. B. Potassium chloride induced lordosis behavior in rats is mediated by the adrenal glands. *Science,* 1975, *187,* 857–858. (b)

Whishaw, I. Q., Bland, B. H., Robinson, T. E., & Vanderwolf, C. H. Neuromuscular blockade: The effects on two hippocampal RSA (theta) systems and neocortical desynchronization. *Brain Research Bulletin,* 1977, *1,* 573–581.

White, N. M., & Fisher, A. E. Relationship between amygdala and hypothalamus in the control of eating behavior. *Physiology and Behavior,* 1969, *4,* 199–205.

Wiepkema, P. R. Positive feedbacks at work during feeding. *Behaviour,* 1971, *39,* 266–273.

Wise, C. D., & Stein, L. Facilitation of brain self-stimulation by central administration of norepinephrine. *Science,* 1969, *163,* 299–301.

Wolf, A. V. Osmometric analysis of thirst in man and dog. *American Journal of Physiology,* 1950, *161,* 75–86.

Wong, R. *Motivation: A biobehavioral analysis of consummatory activities.* New York: Macmillan, 1976.

Woods, S. C., Decke, E., & Vasselli, J. R. Metabolic hormones and regulation of body weight. *Psychological Review,* 1974, *81,* 26–43.

Woodworth, R. S. *Dynamic psychology.* New York: Columbia University Press, 1918.

Wooldridge, P. E. *The machinery of the brain.* Toronto: McGraw-Hill, 1963.

Wyrwicka, W., & Doty, R. W. Feeding induced in cats by electrical stimulation of the brainstem. *Experimental Brain Research,* 1966, *1,* 152–160.

Yokel, R. A., & Wise, R. A. Increased lever pressing for amphetamine after pimozide in rats: Implications for a dopamine theory of reward. *Science,* 1975, *187,* 547–549.

Young, P. T. The role of affective processes in learning and motivation. *Psychological Review,* 1959, *66,* 104–125.

Young, P. T. Palatability: The hedonic response to foodstuffs. In C. F. Code (Ed.) *Handbook of Physiology,* Alimentary Canal (Sect. 6), Food and Water Intake (Vol. 1). Washington, D.C.: American Physiological Society, 1967.

Young, P. T. *Emotion in man and animal.* New York: Robert E. Krieger, 1973.

Young, S. P., & Goldman, E. A. *The wolves of North America.* Washington: The American Wildlife Institute, 1944.

Young, W. C. The hormones and mating behavior. In W. C. Young (Ed.), *Sex and internal secretions.* Baltimore: Williams & Wilkins, 1961.

Zeigler, H. P. Feeding behavior in the pigeon. In A. Rosenblatt, R. Hinde, A. Shaw, & B. Beer (Eds.), *Advances in the study of behavior.* New York: Academic, 1976.

Zeigler, H. P., & Karten, H. J. Central trigeminal structures and the lateral hypothalamic syndrome in the rat. *Science,* 1974, *186,* 636–638.

Zeigler, H. P., Miller, M., & Levine, R. R. Trigeminal nerve and eating in the pigeon (*Columba livia*): Neurosensory control of the consummatory response. *Journal of Comparative and Physiological Psychology,* 1975, *89,* 845–858.

Zeman, W., & King, F. A. Tumors of the septum pellucidum and adjacent structures with abnormal affective behavior: An anterior midline structure syndrome. *Journal of Nervous and Mental Diseases,* 1958, *127,* 490–502.

Zigmond, M. J., & Stricker, E. M. Deficits in feeding behavior after intraventricular injection of 6-hydroxydopamine in rats. *Science,* 1972, *177,* 1211–1214.

Zigmond, M. J., & Stricker, E. M. Ingestive behavior following damage to central dopamine neurons: Implications for homeostasis and recovery of function. In E. Usdin (Ed.), *Neuropsychopharmacology of monoamines and their regulatory enzymes.* (Advances in Biochemistry and Psychopharmacology, Vol. 12). New York: Raven, 1974.

Glossary

ablation—Surgical removal of neural tissue.
acetylcholine—A chemical transmitter substance released at synapses in the CNS and at the neuromuscular junction.
ACTH—Adrenocorticotrophic hormone released from the anterior pituitary gland whose target is the adrenal cortex.
action potential—The nerve impulse or transient depolarization that travels along the axon of a neuron.
adipsia—Absence of drinking behavior, which may result for example from damage to the lateral hypothalamus.
adrenal cortex—The outer shell of the adrenal gland, which produces a number of steroid hormones. They are released continuously but especially during stress.
adrenal medulla—The inner portion of the adrenal glands, which produces adrenaline and noradrenaline, designated catecholamines.
adrenaline—Hormone produced by the adrenal medulla.
adrenergic neurons—Neuron that releases the neurotransmitter, noradrenaline. Adrenergic neurons have been visualized and the fiber projections mapped using the technique of histofluorescence.
affective—Pertaining to the emotions.
agonist—A drug that mimics the effect of a neurotransmitter—e.g., apomorphine is a dopamine agonist.
amino acid—Type of organic acid that is the major ingredient of protein molecules.
amphetamine—A stimulant drug that is addicting and in large doses can mimic paranoia.
amygdala—Structures of the brain within the temporal lobe, which is part of the limbic system and which is associated with aggressive and other emotional behaviors.
angiotensin II—A hormone, released by the kidneys, that stimulates thirst when injected intracranially and intravenously.
anorexia—A condition in which animals will eat, but only reluctantly, and will not eat enough to sustain themselves. It occurs when animals are recovering from a lesion of the lateral hypothalamus.
antagonist—A drug that blocks the effects of a neurotransmitter—e.g., haloperidol is a dopamine antagonist.
aphagia—Absence of eating behavior sufficient to meet body needs, for example resulting from damage to the lateral hypothalamus.

autonomic nervous system—The portion of the nervous system that controls the activity of the viscera, consisting of the sympathetic nervous system and the parasympathetic nervous system.

axon—The generally long and single extension of the neuron that conducts impulses away from the cell body to other neurons; it can release chemical transmitters from its end to affect an adjoining neuron or muscle.

baroreceptors—Receptors that transduce blood pressure into neural signals.

basal ganglia—The caudate nucleus, putamen, and globus pallidus—neural structures involved in motor control.

basal metabolic rate (BMR)—A measure of the metabolic rate taken when subject is fasting and at rest. The oxygen consumed under these conditions indicates the rate of chemical reactions in the body.

biogenic amines—Neurotransmitters formed from amino acids (e.g., dopamine, noradrenaline, serotonin).

blood brain barrier—A mechanism that filters the extracellular fluid of the brain so that its composition is different from that of the rest of the body.

cannula—Fine tube usually implanted with the stereotaxic procedure for infusing drugs or hormones into the cerebral ventricles or brain tissue.

capillaries—Thin-walled blood vessels through whose walls oxygen, carbon dioxide, nutrients, etc., are exchanged with the extracellular fluid.

castration—Operative removal of the male gonads.

catecholamines—Neurotransmitters, including dopamine, adrenaline, and noradrenaline, that are formed by amino acids.

caudate nucleus—An area in the forebrain that is part of the basal ganglia and is involved in motor control. The nigrostriatal dopamine pathway projects from the substantia nigra to the caudate.

cell—Protoplasmic material that constitutes the basic unit of life in all living organisms.

central nervous system (CNS)—Consists of the brain and spinal cord.

cerebellum—A large, convoluted structure dorsal to the pons that contributes to the coordination of movement.

cerebral cortex—Neurons on the outer surface of the brain of man and higher organisms.

chemoreceptors—Receptors that specialize in transducing chemical stimuli (e.g., CO_2, taste, smell) into nerve impulses.

cholinergic—Pertaining to nerve fibers that release acetylcholine as a transmitter at their synapses with other nerve cells.

circadian rhythms—From the Latin *circa* (about) and *dies* (day). Refers to the 24-hour cycle of various biological rhythms.

cognitive—Pertaining to the use of perceptual, conceptual, and language processes.

dendrites—Processes of the cell body of neurons that receive synaptic inputs from axons of other neurons.

diabetes insipidus—A disorder characterized by excessive water loss through urination, accompanied by an increase in drinking behavior (polydipsia).

diencephalon—Portion of the forebrain, including thalamus and hypothalamus.

dopamine—A CNS neurotransmitter. Parkinson's disease is associated with a deficiency of dopamine in the substantia nigra.

EEG—The abbreviation for an electroencephalogram for rhythmical waves recorded from the brain.

efferent—Used to describe neurons that carry impulses away from the CNS.

electrolyte—A chemical compound that dissociates into ions when dissolved in water (e.g., sodium chloride).

endocrine gland—An internal organ such as the pituitary gland or thyroid that releases a hormone into the blood to act on a target organ.

estrogen—A female gonadal hormone.

estrous cycle—Cyclic changes in gonadal hormone levels and sexual receptivity.

ethology—The study of animal behavior.

extracellular fluid—The medium that surrounds all cells; about one third of the body water is contained in the extracellular fluid compartment.

fistula—A valvelike apparatus that can be implanted surgically to make an opening between an internal hollow organ and the exterior.

forebrain—The largest subdivision of the brain of man and primates, consisting of the cortex, thalamus, hypothalamus, pituitary, and limbic systems.

frontal lobe—One of the four main divisions of the cortex involved in motor control and emotional behavior.

functional coupling—The concept that autonomic, endocrine, and skeletomotor responses are coordinated by the hypothalamus and limbic system.

GABA—A neurotransmitter that exerts inhibitory effects by hyperpolarizing the postsynaptic membrane.

ganglia—Groups of neural cell bodies and dendrites outside the brain and the spinal cord. Ganglia in the autonomic nervous system include sympathetic ganglia, which are located immediately outside the spinal cord and form the sympathetic cord, and parasympathetic ganglia, found near their target organs.

gland—An organ that produces and secretes hormones that affect the activity of other organs.

globus pallidus—A part of the basal ganglia involved in spontaneous movement.

glucoreceptors—Neurons that monitor blood glucose concentration.

histofluorescence—An anatomical technique used by histologists to trace nerve pathways differentially according to their neurotransmitter. A fluroescent compound is produced in the nerve axon by a chemical reaction with its specific neurotransmitter.

homeostasis—The relatively stable state of the internal environment.

hormone—The chemical product of an endocrine gland that is carried by the bloodstream to various target organs.

hyperphagia—A condition characterized by excess food intake.

hypothalamus—A small structure at the base of the brain dorsal to the pituitary gland concerned with physiological regulations and behavior.

imprinting—Refers to the observation that newly hatched birds and newly born animals tend to fixate themselves behaviorally to a person, other animal, or object that is the first stimulus they see.

intracellular fluid—Fluid inside the cells. Two-thirds of body water is in the intracellular fluid compartment.

isotonic—Of the same tonicity or strength; of a solution having the same concentration as a solution of reference. A .9% solution of sodium chloride is isotonic as compared with body fluid.

lateral hypothalamus—An area in the hypothalamus believed to play a role in the control of feeding, drinking, and brain-stimulation reward.

L-dopa—A drug that promotes the synthesis of dopamine in the substantia nigra, used in treatment of Parkinson's disease.

lesion—Destruction of a region of neural tissue, for example by passing a DC current (electrolytic lesion).

limbic system—Functionally related structures in the forebrain—including amygdala, hippocampus, septal area, cingulate gyrus—that contribute to motivated and emotional behaviors.

mania—An affective disorder characterized by hyperactivity, euphoria, and increased vocal output.

medial forebrain bundle—A neural pathway running through the lateral hypothalamus and connecting limbic forebrain structures and the midbrain.

medulla—Lower brainstem region associated with most of the cranial nerves and with autonomic functions such as heart rate and respiration.

microelectrode—An electrode with a tip diameter of 1–3 microns used for recording action potentials from single neurons.

micron—One-millionth of a meter.

midbrain—One of the three main subdivisions of the brain, between the hindbrain and the forebrain, also called the *mescencephalon*.

motor neurons—A neuron that connects the CNS with effectors (muscles and glands).

neural impulse—See *action potential*.

neuron—A nerve cell, specialized for the transmission of information as nerve impulses and for integration.

neurotransmitter—A chemical substance released from the axon terminal involved in synaptic transmission.

noradrenaline—A neurotransmitter released by postganglionic fibers of the sympathetic nervous system and by noradrenergic fibers in the brain.

olfactory tract—A bundle of axons that travels from the olfactory bulb to various areas of the brain (rhinencephalon or limbic system).

oscilloscope—An electronic instrument that uses a cathode ray tube for recording electrical activity of the nervous system.

osmolality—The concentration of the solute in a solution per unit of solvent.

osmolarity—The concentration of the solute in a solution per unit of total volume of solution.

osmoreceptors—Neurons that specialize in transforming osmolality of the intracellular fluid into neural impulses.

paradoxical sleep—Sleep characterized by low-voltage, fast-frequency EEG similar to that recorded during waking. Rapid eye movements occur, and therefore it is sometimes designated as REM sleep. Dreaming is associated with paradoxical sleep.

parasympathetic division—Portion of the autonomic nervous system that operates to conserve and maintain bodily resources.

Parkinson's disease—Disorder characterized by muscle tremors and rigidity, related to a deficiency of dopamine in the substantia nigra.

pituitary—A small endocrine gland at the base of the brain that functions closely with the hypothalamus to regulate such functions as reproductive cycles, growth, temperature, energy, and water balance.

polydipsia—An increase in thirst and drinking behavior characteristic of diabetes insipidus, a disorder caused by a deficiency of the antidiuretic hormone, vasopressin.

portal system—Blood vessels through which releasing hormones from the hypothalamus travel to the pituitary gland.

preoptic region—A group of cell bodies in the rostral hypothalamus involved in temperature regulation and other functions.

presynaptic membrane—Neural membrane of axon terminal from which neurotransmitters are released in the synapse.

procaine—A local anesthetic agent.

progesterone—A female gonadal hormone.

psychosis—Behavioral disorder characterized by delusions, hallucinations, affective breakdown, and cognitive breakdown.

psychosomatic—Pertaining to any bodily illness associated with stress and psychological factors.

psychosurgery—Brain surgery to treat behavioral disorders, such as prefrontal lobotomy and lesions in the amygdala.

raphé nucleus—An area in the rostral pons containing high levels of the neurotransmitter, serotonin, and involved in sleep.

receptor—A structure that specializes in the detection of stimuli, transducing light, sound, etc., into nerve impulses.

reinforcement—The process of increasing the strength of behavioral responses through the use of a reinforcer (e.g., food).

REM sleep—Rapid eye movement sleep, associated with periods of desynchronized EEG and dreaming (see *paradoxical sleep*).

reticular formation—A diffuse area in the core of the brainstem concerned with arousal and alertness.

retrograde degeneration—An anatomical method of tracing neural pathways by destroying the axon terminals and then locating the cell bodies of origin that show degenerative changes.

schizophrenia—A psychosis characterized by cognitive abnormalities (hallucinations, incoherent patterns of thought and speech).

sensory neurons—A neuron that carries nerve impulses from receptors toward the CNS.

septal area—Part of limbic system. A lesion in the septal area produces ferocious behavior (septal rage syndrome).

serotonin—A neurotransmitter composed of amino acids, produced by the raphé nucleus implicated in sleeping and temperature regulation.

sham rage—Violent behavior observed in decorticate animals. Differs from normal rage in that it occurs with the slightest provocation, lasts only as long as the provoking stimulus, and is not directed at the provoking stimulus.

slow-wave sleep—See *paradoxical sleep* or *REM sleep*.

stereotaxic instrument—An instrument that holds an animal's head in a fixed position, used for implanting electrodes in the brain.

supraoptic nucleus—Nucleus in the anterior hypothalamus that synthesizes vasopressin or antidiuretic hormone (ADH).

sympathetic nervous system—The subdivision of the autonomic nervous system associated with stress, emotions, and expenditure of energy.

synapse—The microscopic gap between the axon of one neuron and the dendrite or cell body of another. The neurotransmitter from the axon is released into the synapse.

target organs—Organs such as the adrenal cortex whose functioning is regulated by hormones.

thermode—A probe-type device for heating or cooling the brain or other parts of the body.

thyroxine—A hormone that regulates metabolism, produced by the thyroid gland. A deficiency causes lethargy in adults and cretinism in infants.

vasoconstriction—Decrease in the caliber of blood vessels.

ventricles—Cavities in the brain, called the lateral, third, and fourth ventricles, that are filled with cerebral spinal fluid (CSF).

ventromedial nuclei—Paired nuclei in the hypothalamus that are involved in the regulation of feeding behavior.

volemic thirst—Drinking in response to a deficit in extracellular fluid. Angiotensin II is believed to be a hormonal mediator of volemic thirst. Used to denote drinking in response to extracellular hypovolemia or dehydration.

Author Index

Numbers in *italics* refer to the pages on which complete references are listed.

A

Abdelaal, A. E., 129, 273, *282*, *305*
Abrahams, V. C., 240, *282*
Adair, E. R., 76, 83, 106, *282*
Adams, D. W., 217, *288*
Adolph, E. F., 66, 88, 125, 126, 133, *282*
Aghajanian, G., 230, *286*
Agnati, L., 263, 264, *290*
Ahlskog, J. E., 55, 110, 117, *282*
Akert, E., 250, *293*
Albert, D. J., 46, *282*
Anand, B. K., 17, 103, 115, 117, 137, 148, *282*
Anderson, K. M., 181, *282*
Andersson, B., 48, 75, 76, 77, 138, 143, 148, *283*
Antelman, S. M., 173, *286*
Appley, M. H., 1, *287*
Arbuthnott, G. W., 231, *287*
Arnauld, E., 138, 139, 140, 148, *309*
Aronson, L. R., 166, *304*
Aserinsky, E., 190, *283*
Assaf, S., 140, 143, 144, *283*, *295*, *299*
Astic, L., 196, *294*

B

Baile, C. A., 105, *283*
Baker, R., 168, *298*
Balagura, S., 109, 219, *283*
Baldwin, B. A., 76, *283*
Ball, J., 168, *283*
Banerjee, S. P., 263, *306*
Barbeau, A., 144, 180, *283*, *291*
Barchas, J. D., 208, 258, *283*, *293*
Bard, P., 250, *283*
Barfield, R. J., 173, *283*
Barnes, R. H., 97, *296*
Barnwell, G. M., 118, *283*
Batini, C., 205, 206, 276, *283*
Battic, C., 168, 183, *310*
Baum, M. J., 181, *283*
Baxter, L., 233, *283*
Beach, F. A., 158, 166, 167, 168, 169, 174, 178, 179, *283*, *284*, *310*
Beagley, G., 218, 223, *290*
Bean, S. M., 105, *283*
Beaton, J. R., 45, *297*
Beer, B., 216, *309*
Behn, C., 124, 143, *291*
Belluzzi, J. D., 232, *307*
Benevenga, N. J., 93, 106, *292*
Benkert, O., 181, *284*
Benzinger, T. H., 58, 72, *284*
Berger, B. D., 114, *284*
Berger, H., 190, 203, 207, 210, *284*
Berkley, M. A., 216, *287*
Berlyne, D. E., 11, *284*
Bermant, G., 178, *284*

Bernard, C., 16, 127, *284*
Bernardis, L. L., 109, *284*
Berthoud, H., 104, *284*
Beyer, C., 168, 182, *284, 298*
Bindra, D., 11, *284*
Bioulac, B., 138, 139, 140, 148, *309*
Bixler, E. O., 202, *295*
Bland, B. H., 193, 274, *309, 310*
Blandau, R. J., 152, 167, *285*
Blass, E. M., 126, 140, *285*
Bligh, J., 63, 65, 71, 81, 82, 83, *285*
Blundell, J. E., 55, 110, *285*
Boling, J. L., 152, 167, *285*
Bolles, R. C., 1, 11, 217, 218, *285*
Bolme, P., 144, *290*
Booth, D. A., 90, 106, *282, 285*
Borbely, A. A., 225, 230, *293*
Boren, J. J., 217, 218, *285, 306*
Boucher, R., 144, *291*
Bowden, D. M., 222, 232, *291*
Boyd, E. S., 224, *285*
Boyle, D., 93, *298*
Boyle, P. C., 99, *292*
Brady, J. V., 217, 218, 249, 252, 257, *285, 306*
Bray, G. A., 97, 105, 118, *285, 301*
Brecht, H. M., 144, *291*
Breese, G. R., 114, *285*
Bremer, F., 166, 203, 204, 205, 207, 210, *285*
Brien, B., 168, *298*
Brimley, C. C., 148, *285*
Brobeck, J. R., 17, 115, 137, 148, *282, 286*
Brodie, D. A., 216, *286*
Brodie, H. K. H., 258, *283*
Brodie, J. J., 216, *286*
Brook, A. H., 77, *283*
Brookhart, J. M., 17, 174, *286*
Brooks, C. McC., 179, *286*
Brooks, V. B., 46, *286*
Brown, J. L., 17, *286*
Brown, S., 217, *286*
Bryson, G., 108, *303*
Bucy, P. C., 44, 114, 177, 251, *295*
Buggy, J., 140, 142, *294*
Bunney, B. S., 230, *286*
Burton, M. J., 183, 227, 274, *286, 297, 304*
Butter, C. M., 234, *286*

C

Cabanac, M., 77, 94, 111, *286*
Caggiula, A. R., 173, 176, 222, *286*
Calaresu, F. R., 89, 90, *299*
Calloway, J. W., 97, *296*
Campbell, B. A., 100, *296*
Campbell, H. J., 220, *286*
Campbell, J. C., 181, *296*
Campfield, L. A., 97, *285*
Cannon, W. B., 3, 16, 89, 127, 240, 242, 249, 257, *286*
Cantor, M. B., 217, *286*
Carlton, P. L., 231, *293*
Carr, W. J., 158, *286*
Carrington, P., 198, *288*
Casby, J. U., 76, *282*
Case, B., 98, *294*
Casey, K. L., 224, 230, *295*
Cesselin, F., 155, 156, *303*
Chabora, J., 252, *306*
Chase, M. H., 204, *286*
Chauffournier, J. M., 155, 156, *303*
Chhina, G. S., 103, 117, *282*
Chi, C. C., 251, *286*
Chouvet, G., 200, 209, *287, 294*
Chow, M. I., 231, *293*
Christensen, J., 98, 109, *294*
Ciaranello, R., 258, *283*
Cioe, J., 212, 220, 233, 235, *299*
Clark, T. K., 173, *286*
Clavier, R. M., 222, *286*
Claybaugh, C., 173, *304*
Clemens, L. G., 173, 179, *286, 304*
Clemente, C. D., 102, 177, 206, 207, 281, *287, 291, 307*
Cloutier, R. J., 107, *287*
Cofer, C. N., 1, *287*
Coindet, J., 209, *287*
Collier, G., 7, 89, 95, *287*
Comhaire, F., 162, *287*
Conrad, D. J., 217, *218, 285, 306*
Coons, E. F., 219, *287*
Cooper, B. R., 114, *285*
Cooper, K. E., 63, 77, 78, *287, 309*
Cooper, R. M., 224, *287*
Cooper, S. J., 222, 230, 231, *287, 304*
Corbit, J. D., 76, *287*
Corvelli, A., 109, *303*
Coury, J. N., 148, *289*
Cowgill, U., 155, *287*
Cox, V. C., 102, 132, *309*
Craig, C., 94, *298*
Craig, W., 6, *287*
Cranston, W. I., 77, *287*
Crow, T. J., 222, 231, 232, *287*
Cruce, J. A. F., 219, *287*

Culberton, J. L., 216, *287*
Curtis, B. A., *57*

D

Dahlström, A., 264, *287*
Darwin, C., 2, 3, 4, 15, 237, 238, 239, 258, 264, *287*
Davidson, J. M., 166, 175, 178, *284, 287, 294*
Davis, J. M., 155, *294*
Deaux, E., 98, 109, *294*
DeBold, J. F., 179, 180, 182, *310*
Debons, A. F., 107, *287*
Decke, E., 109, *311*
De Groot, J., 102, 177, *291*
Del Fiacco, M., 180, *308*
Delgado, J. M. R., 102, 111, 252, *297, 290*
Dement, W. C., 190, 195, 202, *287, 292, 304*
DeRuiter, L., 89, 91, *288*
Deutsch, J. A., 216, 217, 235, *288*
Devenport, L. D., 109, *283*
Dey, F. L., 17, 174, *286*
DiCara, L. V., 216, *288*
Dickinson, A., 133, *301*
Didiergeorges, F., 257, *295*
Dissinger, M. L., 158, *286*
Doty, R. W., 42, 47, 102, 212, *288, 311*
Douglas, R. J., *294*
Dowling, M. T., 94, *289*
Dragovich, J. A., 111, 221, *297*
Dreese, A., 232, *288*
Dreifuss, J. J., 111, *288, 291*
Drellich, M. G., 165, *310*
Drezner, N., 164, *289*

E

Eccles, J. C., 22, *288*
Edholm, O. G., 89, *288*
Edwards, D. A., 183, *308*
Egger, M. D., 252, *288*
Ehrhardt, A. A., 159, 160, *299*
Eibl-Eibesfeldt, I., 5, *288*
Eichelman, B., 257, 258, *288*
Eisenman, J. S., 51, *301*
Ekman, L., 77, *283*
Ellingson, R. J., 195, *288*
Elliott, G. R., 208, *293*
Ephron, H. S., 198, *288*
Epstein, A. N., 46, 48, 61, 91, 92, 99, 100, 102, 104, 105, 115, 116, 121, 127, 133, 136, 140, 143, 144, 148, 272, *285, 288, 289, 292, 293, 294, 298, 304, 308*
Epstein, A. W., 179, *289*
Ervin, F. R., 252, *268, 307*
Evarts, E. V., 52, 192, 194, 274, *289*
Evered, M. D., 134, 147, *289*
Everitt, B. J., 165, 167, 175, *289*

F

Falk, J. L., 216, 217, *289*
Fann, W. E., 155, *294*
Feinier, L., 166, *289*
Feldberg, W., 78, *289*
Feldman, S. M., 102, *302*
Fenton, P. P., 94, *289*
Fibiger, H. C., 114, 232, *289, 302*
Filler, W., 164, *289*
Fisher, A. E., 48, 111, 148, 175, *289, 309, 311*
Fisher, C., 174, *289*
Fishman, J., 97, 108, *289*
Fitzsimons, J. T., 48, 128, 129, 130, 131, 133, 138, 143, 144, 146, 279, *289, 290*
Fletcher, J. G., 89, *288*
Flourens, P., 44, *290*
Flynn, J. P., 17, 251, 252, *286, 288, 290, 310*
Folk, G. E., Jr., 60, 64, 87, *290*
Folkow, B., 278, *290*
Folkow, F., 240, *290*
Fonberg, E., 102, 111, *290*
Fowler, H., 167, *284*
Fowler, S. J., 73, *290*
Frankenhaeuser, M., 241, *290*
Frankland, M., 93, *298*
Fratta, W., 180, *308*
Fredriksen, P. K., 264, *290*
Freund, K., 222, *297*
Fritsh, G., 46, *290*
From, A., 107, *287*
Fuller, J. L., 102, *290*
Fuxe, K., 144, 263, 264, *287, 290, 293*

G

Gagge, A. P., 273, *292*
Gale, C. C., 75, 76, 77, 78, *283, 290, 300*
Gallistel, C. R., 218, 223, *290*
Gandelman, R., 216, *309*
Ganten, D., 144, *290, 291*
Garattini, S., 110, *291*
Gardner, E., 33, *291*
Gardner, E. L., 225, *304*
Gardner, L. C., 224, *285*
Gauer, O. H., 124, 143, *291*

Gebhard, P. H., 163, *295*
Geist, S. H., 167, *305*
Genest, J., 144, *291*
German, D. C., 222, 232, *291*
Gessa, G. L., 180, *308*
Gibson, W. E., 216, *291*
Gillespie, L. A., 274, *309*
Girvin, J. P., 258, 259, *291*
Glickman, S. E., 178, 218, 235, *284, 291*
Gloor, P., 111, 146, *288, 291*
Glowinski, J., 208, *293*
Gluckman, M. I., 233, *283*
Gold, R. M., 46, *291*
Goldblum, C., 127, *289*
Goldfoot, D. A., 183, *309*
Goldman, E. A., 2, *311*
Goldstein, M., 258, *291*
Gomez, C. J., 184, *299*
Gorski, R. A., 173, 177, 179, *286, 301, 304*
Gorzalka, B. B., 152, 161, 168, 173, 174, 176, 179, 180, 181, 182, 183, 184, *291, 310*
Goy, R. W., 174, *291*
Grace, J. E., 70, *299*
Granger, P., 144, *291*
Grant, L. D., 114, *285*
Graybiel, A. M., 280, *291*
Graziani, L., 196, *310*
Green, J. D., 102, 177, *291*
Green, J. H., *291*
Greenberg, D., 263, *306*
Greenhill, R., 168, *298*
Greer, M. A., 137, 148, *291*
Gregersen, M. A., 128, *293*
Greipel, M., 93, *298*
Grodins, F. S., 269, *291*
Grofova, I., 222, *297*
Grossman, R., 111, *292*
Grossman, S. P., 48, 55, 102, 110, 111, 114, 140, 148, 197, *291, 292*
Grunt, J. A., 167, *292*
Guilleminault, C., 202, *292*
Gunne, L. M., 257, *303*
Guyton, A. C., *57,* 59, 60, *292*

H

Hagamen, W. D., 178, *292*
Hainsworth, F. R., 61, *292*
Halberg, F., 110, *305*
Hall, N. R., 181, *296*
Halpern, N. M., 102, *296*
Hamburg, D., 258, *283*
Hamer, J. D., 165, 167, *289*
Hamilton, C. L., 94, *292*
Hammel, H. T., 51, *301*
Hardy, D. F., 168, *310*
Hardy, J. D., 51, 76, 83, 273, *282, 292, 300, 301*
Hare, W. K., 75, *295*
Harlow, H. F., 2, 14, 18, 157, 158, 246, *292*
Harper, A. E., 93, 99, 106, *292*
Harris, G. W., 175, *292*
Harris, V. S., 178, *292*
Hart, B. L., 172, *292*
Harvey, J. A., 110, 114, 257, 258, *296, 302, 306*
Haugen, C. M., 172, *292*
Hawkins, T. D., 217, *303*
Hayduk, K., 144, *291*
Heath, R. G., 177, 252, *292*
Hebb, D. O., 4, 12, 14, 18, 22, 200, 204, 213, 238, 239, 242, 243, 245, 247, 249, 257, *292*
Heimer, L., 173, 174, *292, 293*
Heller, A., 114, *293*
Henneman, E., 38, *293*
Henry, J. P., 124, 143, *291*
Herbert, J., 164, 165, 167, 175, *289, 298, 308*
Hervey, G. R., 89, 105, *293*
Hery, F., 208, *293*
Hess, W. R., 46, 206, 240, 250, *293*
Hill, J. H., 121, *307*
Hilton, S. M., 240, *282*
Himwich, H. E., 262, *293*
Hinde, R. A., 5, 242, 275, *293*
Hirsch, E., 7, 89, 95, *287*
Hitt, J. C., 177, *298*
Hitzig, E., 46, *290*
Ho, G. K. W., 182, 183, *303, 310*
Hoebel, B. G., 55, 97, 103, 110, 111, 112, 117, 146, 218, 220, 221, 222, *282, 283, 293, 300*
Hökfelt, B., 77, *283*
Hökfelt, T., 144, 263, 264, *290, 293*
Hollander, F., 93, *294*
Holman, B. R., 208, *293*
Holmes, J. H., 128, *293*
Honour, A. J., 77, *287*
Hornykiewicz, O., 232, *293*
Horowitz, Z. P., 231, *293*
Hough, J. C., 182, *310*
Houpt, K. A., 99, 100, *293*
House, E. L., 31, 101, *293*
Howarth, C. I., 216, 235, *288*
Huang, Y. H., 19, 142, 146, 225, 230, 254, *299, 304*

Hudson, J. W., 124, *293*
Hull, C. L., 3, 8, 218, *293*
Hull, D., 73, *293*
Hunsperger, R. W., 17, *286*
Huston, J. P., 225, 230, *293*
Hutchinson, R. R., 251, *293*

I

Ibuka, N., 209, *294*
Ingram, D. L., 76, *283*
Ingram, W. R., 75, *294, 303*
Isaacson, R. L., 22, 252, *294*
Ito, M., 227, *294*
Iversen, L. L., 52, 55, *294*
Iversen, S. D., 52, 55, *294*
Iwasa, K., 172, *298*

J

Jacobs, H. L., 105, *283*
Jacobsen, C. F., 44, *294*
Jacobson, S., *57*
James, W., 242, *294*
Janowitz, H. D., 93, *294*
Janowsky, D. S., 155, *294*
Jasper, H. H., 191, *302*
Jaynes, J., 179, *284*
Johansson, O., 263, 264, *290, 293*
Johnson, A. K., 140, 142, *294*
Johnson, R. N., 2, *294*
Johnson, V. E., 162, *297*
Johnston, P., 175, *294*
Jones, B., 113, 234, *294*
Jones, H., 198, *294*
Joslyn, W. D., 183, *309*
Jouvet, M., 192, 200, 208, *294*
Jouvet-Mounier, D., 196, *293*
Jovanović, U. J., 260, *294*

K

Kaada, B. R., 254, *294*
Kakolewski, J. W., 98, 102, 109, 132, *294, 309*
Kalat, J., 95, *305*
Kalbermann, L. E., 184, *299*
Kales, A., 202, *295*
Kales, J. D., 202, *295*
Kanarek, R., 7, 89, 95, *287*
Karli, P., 257, *295*
Karten, H. J., 116, 148, 270, *311*
Kawamura, H., 209, *294*
Kayser, C., *295*
Keene, J. J., 224, 230, *295*
Keesey, R. E., 93, *303*
Keller, A. D., 75, *295*
Kellogg, C., 73, *290*
Kelly, P. H., 230, 232, *304*
Kennedy, G. C., 103, *295*
Kenney, N. J., 97, *299*
Kety, S. S., 261, 262, 263, *295*
Keys, N. W., 216, *295*
King, F. A., 252, 253, 259, *295, 311*
Kinsey, A. C., 159, 163, *295*
Kissileff, H. R., 131, 132, 133, 145, 146, *295*
Kitching, P., 168, *298*
Kleitman, N., 188, 189, 190, 197, 200, 203, 204, *283, 287, 295*
Kline, F., 96, *295*
Kling, A., 177, 252, *295, 306*
Kling, J. W., 216, *287*
Klüver, H., 44, 114, 177, 251, *295*
Kobayashi, N., 50, *302*
Kolb, B., 114, *295*
Konorski, J., 21, *295*
Kramis, R. C., 192, 274, *304*
Krimsky, I., 107, *287*
Kucharczyk, J., 140, 141, 142, 143, 144, 224, 233, *295, 299, 304*
Kuehn, R. E., 166, 167, 174, *310*

L

Lacote, D., 196, *294*
Lagoguey, M., 155, 156, *303*
Larsen, T., *295*
Larson, K., 173, 174, 175, 178, *292, 293, 296, 309*
Lashley, K. S., 16, 18, *296*
Latham, C. J., 110, *285*
Laties, V. G., 68, 69, *296, 310*
Law, T., 174, *296*
Lee, T., 263, *306*
Leeper, R. W., 238, *296*
LeMagnen, J., 89, 91, 94, 95, 131, *290, 296*
Leonard, C. M., 222, 230, *296, 301*
Leshem, M. B., 110, *285*
Levi, L., 240, *296*
Levin, R., 98, 100, *296*
Levine, R. R., 116, *311*
Levine, S., 242, *296*
Lewin, R., 30, 34, 35, 37, *296*
Lewis, M. J., 111, 221, *297*
Liao, S. 181, *282*

Libertun, C., 184, *299*
Liebeskind, J. C., 224, *297*
Likuski, H. J., 107, *287*
Lindsley, D. B., 4, 256, 281, *296*
Lindy, J., 214, 216, *305*
Lints, C. E., 257, *296*
Lipton, J. M., 77, *296*
Lisk, R. D., 174, 175, *296*
Ljungdahl, A., 263, 264, *290, 293*
Loeb, L. S., 158, *286*
Long, C. N. H., 17, *286*
Lopez, M., 208, *293*
Lorenz, K., *296*
Louis-Sylvestre, J., 118, *296*
Lubar, J. F., *294*
Luciano, D. S., 82, *309*
Luttge, W. G., 152, 161, 168, 181, 183, *296, 310*
Lyon, M., 102, *296*
Lytle, L. D., 100, *296*

M

MacKay, E. M., 97, *296*
MacKinnon, P. C. B., 183, *297*
MacLean, P. D., 111, 177, 251, 252, *297*
Macon, J., 208, *293*
Madlafousek, J., 222, *297*
Magalhoes, H., 153, *297*
Magnes, J., 205, *297*
Magoun, H. W., 174, 203, 204, 210, *289, 297, 300*
Malamud, N., 259, *297*
Malis, J. L., 216, *286*
Malmo, R. B., 140, 142, *297*
Malsbury, C. W., 175, 224, 232, *297, 305*
Manshardt, J., 109, *297*
Marcus, E. M., *57*
Margules, A. S., 111, 221, *297*
Margules, D. L., 111, 218, 221, *297*
Mark, V. H., 252, *268, 307*
Marshall, J. F., 115, 270, *297*
Marshall, N. B., 103, *298*
Martin, C. E., 159, 163, *295*
Martin, G. E., 56, 115, *297, 300*
Maslow, A. H., 22, *297*
Masters, W. H., 162, *297*
Matsumoto, K., 172, *298*
Matthews, M., 76, *290*
May, K. K., 45, *297*
Mayer, D. J., 224, *297*
Mayer, F. E., 281, *287*
Mayer, J., 103, *297, 298*
McCance, R. A., 89, *288*
McCleary, R. A., 254, *298*
McConnell, R. A., 173, *286*
McDonald, P. G., 168, 173, *283, 298*
McFarland, D. J., 4, 66, 90, 91, *298*
McGeer, E. G., 114, *289*
McGeer, P. L., 280, *298*
McGinty, D. J., 197, *298*
Meagher, W., 174, *296*
Mellinkoff, S. M., 93, *298*
Melzack, R., 241, 245, *298*
Mendels, J., 262, *298*
Mercer, P. F., 273, *282*
Mettler, F. A., 266, *298*
Metzner, R. J., 217, *288*
Meyer, P. M., 253, *295*
Michael, R. P., 164, 167, 175, *292, 298*
Michal, E. K., 178, *298*
Mickelsen, O., 94, *298*
Mickle, W. A., 252, *292*
Miller, M., 116, *311*
Miller, N. E., 48, 102, 106, 144, 218, *282, 298*
Milner, P. M., *57*, 68, 203, 205, 212, 214, 216, 234, *298, 301*
Minnich, J. L., 144, *291*
Mintz, E. Y., 102, *296*
Miselis, R. R., 104, 272, *289, 298*
Mishkin, M., 111, 113, 234, *294, 304*
Mizutani, S., 172, *298*
Modianos, D. T., 177, *298*
Mogenson, G. J., 17, 19, 36, 39, 47, 53, 70, 89, 90, 103, 104, 105, 129, 134, 137, 140, 141, 142, 143, 144, 146, 147, 148, 177, 212, 214, 215, 216, 217, 218, 219, 220, 220, 222, 223, 224, 233, 235, 236, 254, 272, 273, 280, *282, 283, 284, 285, 289, 295, 299, 300, 302, 303, 304, 305, 307, 308*
Moguilevsky, J. A., 184, *299*
Mok, A. C. S., 70, 177, *299*
Money, J., 159, 160, 172, *299*
Monroe, R. R., 252, *292*
Montemurro, D. G., 137, *299*
Mook, D. G., 97, *299*
Mooney, R. D., 77, *304*
Moorcroft, W. H., 100, *296*
Moore, R. Y., 109, 114, 209, 254, *293, 298, 300*
Mora, F., 227, 274, 232, 233, *286, 300, 304*
Morali, G., 182, *284*
Moreno, O. M., 216, *286*
Morgan, C. T., 16, 18, *300*

Morgan, C. W., 214, 216, 217, *303*
Morgane, P. J., 197, 198, 208, *300, 307*
Morishima, M. S., 78, *300*
Morris, N. M., 155, 167, *309*
Morrison, S. D., 131, 145, *300*
Moruzzi, G., 187, 198, 203, 204, 205, 206, 210, 239, 276, *283, 297, 300*
Mount, G., 218, *300*
Mount, L. E., 63, *300*
Mountcastle, V. B., 52, 188, 199, 204, 206, 207, *300*
Mouret, J., 200, 209, *287, 294*
Moyer, K. E., 241, *300*
Mrosovsky, N., 95, 98, *300*
Mundl, W. J., 140, *297*
Murgatroyd, D., 76, *300*
Murphy, J. T., 111, *288, 291*
Muzio, J. M., 195, *304*
Myeda, A., 216, *306*
Myers, R. D., 48, 56, *57,* 77, 78, 79, 115, *289, 297, 300, 309*

N

Naftolin, F., 168, *301*
Nakayama, T., 51, *301*
Nake, F., 50, *302*
Nance, D. M., 177, *301*
Nauta, W. J. H., 252, 254, 255, 256, 257, *285, 301*
Neil, E., 240, *290*
Neubauer, B. L., 179, 180, *310*
Newman, H. F., 162, *301*
Newton, F., 168, *298*
Nicolaidis, S., 92, 104, 105, 134, 135, 144, 272, *289, 290, 301, 305*
Niemegeers, C. J., 232, *310*
Niijima, A., 104, 143, *301*
Ninneman, A. J., 114, *295*
Ninteman, F. W., 231, *303*
Norgren, R., 146, 222, *301*
Novin, D., 104, 105, 140, *301, 302*
Nussbaum, A. I., 97, *299*

O

Oatley, K., 10, 72, 95, 121, 126, 133, 146, *301*
Odum, E. P., 95, *301*
Olds, J., 212, 214, 216, 218, 221, 224, 227, 230, 274, *294, 297, 301, 302, 303, 306*
Olds, M. E., 52, 224, 227, 231, *301, 302*
Oltmans, G. A., 110, 114, *302*
Ono, T., 50, 51, 102, 117, 139, 148, *302*
Oomura, Y., 50, 51, 102, 104, 106, 107, 117, 139, 148, *302*
Ooyama, H., 50, 51, 102, 117, 139, 148, *302*
Ore, G. D., 177, 259, *308*
Osmond, H., 262, *302*
Oswald, I., 198, *294*

P

Pacheco, P., 252, *307*
Palestini, M., 205, 206, 276, *283*
Panksepp, J., 104, 105, 106, 109, 118, 216, *302, 309*
Pansky, B., 31, 101, *293*
Paris, C. A., 183, *309*
Parker, W., 102, *302*
Parmeggiani, P. L., 206, *302*
Pauly, J. E., 110, *305*
Paxinos, G., 176, *302*
Pearson, O. P., 73, *302*
Peck, J. W., 140, *302*
Peele, T. L., 177, *306*
Penfield, W., 191, *302*
Perez de la Mora, M., 263, 264, *290*
Petro, Z., 168, *301*
Pfaff, D. W., 176, *302*
Pfaffmann, C., 11, 14, 111, 133, 146, 176, 218, 219, 220, 235, *302*
Phillips, A. G., 39, 53, 134, 217, 219, 222, 224, 232, 233, 236, 280, *299, 302, 303*
Phillips, M. I., 274, *303*
Phoenix, C. H., 166, 174, *291, 303*
Platt, S. V., 90, *285*
Pliskoff, S. S., 217, *303*
Ploog, D. W., 177, *297*
Plum, F., 101, *303*
Pomeroy, W. B., 159, 163, *295*
Pompeiano, O., 205, *297*
Popolow, H. B., 177, *298*
Porter, P. B., 216, *291*
Poschel, B. P. H., 231, *303*
Powley, T. L., 93, *303*
Pribram, K. H., 102, *290*
Pritchard, D., 168, *298*
Pujol, J. F., 208, *293*
Purpura, D. P., 281, *287*

Q

Quadagno, D. M., 182, 183, *303, 310*

R

Rabedeau, R. G., 157, *303*
Ranck, J. B., Jr., 52, 273, 274, *303*
Ranson, S. W., 17, 75, 174, *286, 289, 303*
Reeves, A. G., 101, 178, *292, 303*
Reichman, S., 69, *306*
Reid, L. D., 216, *291*
Reinberg, A., 155, 156, *303*
Reis, D. J., 109, 257, *303*
Renaud, L. P., 111, *303*
Renfrew, J. W., 251, *293*
Resnick, O., 197, *303*
Reynolds, R. W., 108, *303*
Rezek, D. L., 168, 181, *291*
Rezek, M., 104, *301*
Richardson, J. S., 115, 270, *297*
Richter, C. P., 3, 8, 9, 16, 59, 65, 66, 67, 69, 89, 97, 109, 125, 131, 200, 210, 275, *303*
Ritter, S., 222, 231, *303*
Roberts, S., 97, *299*
Roberts, W. W., 77, *304*
Robertson, A., 224, 233, *304*
Robinson, B. W., 111, *304*
Robinson, T. E., 192, 193, *304, 310*
Robison, B., 17, *305*
Rodgers, W. L., 116, *304*
Rodier, W. I., 97, *299*
Rodin, J., 95, *304*
Roffwarg, H. P., 195, *304*
Roll, S. K., 231, *304*
Rolls, B. J., 48, 143, *289*
Rolls, E. T., 52, 112, 113, 114, 118, 146, 220, 222, 223, 224, 225, 226, 227, 229, 230, 231, 232, 234, 236, 274, *286, 287, 304, 310*
Rolls, G. T., 232, 233, *300*
Rosenblatt, J. S., 166, *304*
Ross, J., 173, *304*
Rossi, G. F., 205, 206, 276, *283*
Rosvold, H. E., 17, 102, *286, 290*
Rothballer, A. B., 199, 242, *304*
Rothman, T., 166, *289*
Routtenberg, A., 143, 214, 216, 222, 224, 225, 230, 232, *286, 304, 305, 306*
Rowell, T., 155, *305*
Rowland, N., 92, 109, 134, 135, *301, 305*
Rozin, P., 93, 95, 96, 148, *305*
Rubenstein, H. S., 40, *305*
Rubinstein, E. H., 278, *290*
Ruch, T. C., 255, *305*
Rusak, B., 10, 81, 98, 201, 209, *305*
Russek, M., 104, 109, *305*
Russell, P. J. D., 129, *305*
Ryan, K. J., 168, *301*

S

Saayman, G. M., 167, *298*
Sachs, B. D., 178, *292*
Sackett, G. P., 246, *305*
Sadowski, B., 220, *305*
Sakai, M., 216, *291*
Salaman, D. F., 183, *305*
Saller, C. F., 117, *305*
Salmon, U. J., 167, *305*
Samanin, R., 110, *291*
Samman, A., 127, *289*
Sampson, C., 168, *298*
Sanguinetti, A. M., 232, 233, *300*
Satinoff, E., 76, 81, 83, 99, *305*
Sawyer, C. H., 17, *305*
Scerni, A., 233, *283*
Schachter, S., 94, 248, 254, *305*
Scheving, L. E., 110, *305*
Schiff, B. B., 218, 235, *291*
Schildkraut, J. J., 262, *306*
Schmaltz, L. W., *294*
Schmidt, R. F., 278, *306*
Schmitt, M., 104, 108, 143, *306*
Schneider, A. M., *306*
Schon, M., 165, *306*
Schreiner, L., 177, 252, *306*
Schulman, A., 217, *306*
Schwartz, M., 198, *306*
Seeman, P., 263, *306*
Sellers, E. A., 69, *306*
Selye, H., 241, *306*
Seward, J. P., 216, *306*
Sharpe, L. G., 143, *308*
Shaw, S. G., 232, 233, *300, 304*
Shealy, C. N., 177, *306*
Sherman, J. H., 82, *309*
Sherrington, C. S., 18, *306*
Shettleworth, S. J., 217, *306*
Shryne, J., 177, *301*
Sidman, M., 217, 218, *285, 306*
Siegel, A., 252, 253, *306*
Siffre, M., 200, *294*
Simons, B. J., 130, *290*
Simpson, C. W., 105, *283*
Simpson, J. B., 143, *306*
Singer, J. J., 174, *306*
Singh, B., 103, 117, *282*
Skinner, B. F., 6, *306*

Skinner, J. E., *57*
Skog, O., 253, *306*
Sloan, M., 222, *305*
Slochower, J., 95, *304*
Smith, R. D., 114, *285*
Smythies, J. R., 262, *302*
Snyder, S. H., 258, 263, *306*
Södersten, P., 168, *306*
Sonoda, T., 172, *298*
Sorenson, J. P., Jr., 258, *306*
Soulairac, A., 180, 181, *306*
Soulairac, J. L., 180, 181, *306*
Spear, P. J., 231, *287*
Spector, N. D., 127, *289*
Spies, G., 216, *307*
Stanley, W. C., 99, *305*
Steffens, A. B., 46, 47, 97, 108, 273, *307*
Stein, K., 233, *283*
Stein, L., 56, 114, 222, 224, 231, 232, 262, *284, 303, 307, 311*
Stellar, E., 19, 111, 121, 132, *307, 310*
Stephan, F. K., 81, 109, 145, *307*
Sterman, M. B., 206, 207, *307*
Stern, J. M., 98, 100, *296*
Stern, W. C., 197, 198, 208, *307*
Stetson, M. H., 81, *307*
Stevens, C. F., 49, *307*
Stevens, J. R., 252, *307*
Stevenson, J. A. F., 17, 47, 70, 88, 90, 111, 122, 137, 148, 215, 218, 219, 223, *299, 307*
Stolk, J., 258, *283*
Stolwijk, J. A. J., 76, 273, *282, 292*
Stone, C. P., 179, *308*
Storlien, L. H., 46, *282*
Stricker, E. M., 105, 114, 115, 116, 117, 130, 141, 270, 272, *305, 308, 311*
Strümpell, A., 203, *308*
Suematsu, K., 252, *307*
Sundsten, J. W., 71, 75, 76, *283, 308*
Sutherland, A. M., 165, *306, 310*
Swanson, L. W., 143, *308*

T

Tagliamonte, A., 180, *308*
Takahashi, S., 94, *298*
Takigawa, M., 223, *308*
Tan, H. S., 168, *298*
Taylor, L. H., 224, *287*
Teitelbaum, P., 6, 17, 91, 92, 97, 103, 115, 116, 218, 270, 271, *289, 293, 297, 304, 308*
Tepperman, J., 17, *286*
ter Haar, M. B., 98, 183, *297, 308*
Terman, J. S., 214, 221, *308*
Terman, M., 214, 221, *308*
Terzian, H., 177, 259, *308*
Teyler, T. T., 190, 244, *308*
Thomas, N., 69, *306*
Thompson, M. L., 183, *308*
Thompson, R. F., 204, 261, 265, *308*
Thompson, W. R., 245, *308*
Tinbergen, N., 5, *308*
Toates, F. M., 90, *285*
Tolman, E. C., 238, *308*
Towbin, E. J., 126, *308*
Trimble, M. R., 167, *308*
Trowill, J. A., 216, 217, *286, 302, 309*
Turner, B. H., 270, *298, 309*

U

Udry, J. R., 155, 167, *309*
Ungerstedt, U., 53, 54, 55, 102, 110, 114, 115, 270, *309*

V

Valenstein, E. S., 102, 132, 216, 258, 259, *309*
Vander, A. J., 82, *309*
Vanderweele, D. A., 104, *301*
Vanderwolf, C. H., 192, 193, 274, *304, 309, 310*
van Dis, H., 175, *309*
Van Sommers, P., 37, 62, 91, 128, 193, *309*
Vargas, R., 182, *284*
Vasselli, J. R., 109, *311*
Vaughan, E., 175, *309*
Veale, W. L., 63, 78, *300, 309*
Veening, J. G., 91, *288*
Velasco-Suarez, M. M., 260, *309*
Vergnes, M., 257, *295*
Vermeulen, A., 162, *287*
Verney, E. B., 129, 138, *309*
Villablanca, J., 77, *309*
Vincent, J. D., 138, 139, 140, 148, *309*
von Euler, U. S., 257, *309*
Vreeburg, J. T. M., 181, *283*

W

Wade, G. N., 97, 108, *309*
Wagener, J. W., 222, *309*
Wallen, K., 179, 183, *286, 309*
Waller, L., 167, *309*

Wallis, C. J., 181, *296*
Warden, G. J., 152, *310*
Wasman, M., 251, *310*
Watson, J. B., 16, 242, *310*
Watson-Whitmyre, M., 81, *307*
Wauquier, A., 232, 236, *310*
Waxenberg, S. E., 165, *310*
Wayner, M. J., 51, 102, 117, 139, 148, *302*
Weinbren, M., 109, *303*
Weiner, I. H., 132, *310*
Weiss, A. D., 162, *310*
Weiss, B., 68, 69, *296, 310*
Weitzman, E. D., 196, *310*
Welch, A. S., 257, *310*
Welch, B. L., 257, *310*
Welegalla, J., 164, *298*
Whalen, R. E., 152, 157, 161, 166, 167, 168, 174, 176, 179, 180, 181, 182, 183, 184, *291, 303, 310*
Whishaw, I. Q., 193, *310*
White, N. M., 111, *311*
Widdowson, E. M., 89, *288*
Wiepkema, P. R., 91, *288, 311*
Wilson, C., 173, *283*
Wise, C. D., 56, 114, 231, 232, 262, *284, 307, 311*
Wise, R. A., 233, *311*
Wohlhueter, R. M., 93, 106, *292*
Wolf, A. V., 128, *311*
Wong, R., 198, *311*
Woods, S. C., 109, *311*
Woodworth, R. S., 16, *311*
Wooldridge, P. E., 255, *311*
Wurtman, R. J., 109, *297*
Wyrwicka, W., 102, 105, *301, 311*

Y

Yamamoto, T., 50, *302*
Yamamura, H. J., 263, *306*
Yokel, R. A., 233, *311*
Young, J., 76, *290*
Young, P. T., 11, 94, 133, 219, 220, 243, 248, *311*
Young, S. P., 2, *311*
Young, W. C., 152, 167, *285, 292, 311*

Z

Zanchetti, A., 205, 206, 276, *283*
Zbrozyna, A., 240, *282*
Zeigler, H. P., 116, 148, 270, *311*
Zeman, W., 252, 259, *311*
Zigmond, M. J., 114, 115, 116, 270, 272, *308, 311*
Zis, A. P., 114, *289*
Zitrin, A., 179, *284*
Zitzmann, E. K., 178, *292*
Zucker, I., 10, 81, 97, 98, 108, 109, 145, 201, 209, *305, 307, 309*
Zumpe, D., 167, *298*

Subject Index

A

Ablation, 44
Acclimation, 69
Acetylcholine, 148
Actinomycin D, 182
Action potentials, 50, 112
  during sleep, 192
Adipsia, 114–115, 137
Adrenal cortex, 164
Adrenalectomy, 131, 165, 167, 168
Adrenal gland
  emotion and, 241–242
Aggression, 240, 241, 250, 258–261
Aging
  sexual behavior and, 162
Aldosterone, 131, 143
Amino acids, 93, 106
Aminostatic signals, 90, 106–107
Amphetamine, 110–111, 233
  and nigrostriatal lesions, 53, 55
Amygdala, 111, 113, 118, 146, 220
  electrical stimulation, 102
  emotional behavior, 252
  feeding, 102
  lesions, 102
  self-stimulation, 229–230, 234
  sexual behavior, 177–178
Androgens, 160–161, 164, 167–168, 176
Angiotensin, 130, 141, 143–144
Anorexia nervosa, 108
Anticipatory drinking, 146
Antidiuretic hormone, 123 124, 129–130, 138, 143
Aphagia, 114–115
Aphrodisiac effect, 180
Apomorphine, 53, 180, 233
Arcuate nucleus, 173
Aromatization, 168, 181
Arousal, 115, 118
Atropine, 180
Autonomic nervous system, 31

B

Baboon
  temperature regulation, 76
Baroreceptors, 28
Baroreceptors, *see* Volume receptors
Biogenic amines, *see also* Monoamine systems
Biogenic amines, 109–110
Biological clock, 81, 98, 109, 145
  sleep and, 199–201, 209–211
Birth rates, 153
Blinding
  effect on sleep-wake cycle, 209
Block
  reversible, 46, 109
Blocking
  of neural pathways, 46
Blood brain barrier, 144
Blood glucose, 93, 97, 100, 103–104, 107–108, 118
Blood sugar, *see* Blood glucose
Body composition, 122
Body water, 122–123
Body weight, 122

Body weight regulation, 94, 97
Brainstem
  self-stimulation, 230–231
  sleep, 205
Brainstem, *see also* Reticular formation

C

Carbachol, 148
Cariboo, 72
Carotid artery, 138, 142
Castration, 165–168
Catelepsy, 232
Catecholamine pathways
  self-stimulation, 222
Catecholamines, *see also* Dopamine, Noradrenalin, Serotonin
Catecholamines
  feeding, 110, 114–115
  temperature regulation, 76–78
Catheters, 134
Caval ligation, 130, 143
Cerebral cortex, 113
  emotional behavior, 254–255
  sexual behavior, 178–179
Cerebral ventricles, 138
Cerveau isolé, 203
Cholecystokinin, 90
Circadian rhythm
  of body temperature, 70, 81
  of drinking, 131
  of food intake, 97–98, 109–111
  of self-stimulation, 214
  of sleep, 200–201
  of sleep and waking, 209–210
  of water intake, 131, 145
  role in motivation, 9–10
  self-stimulation, 221
Cognition, *see also* Experiential Factors
  role in motivation, 12
Coitus, 166
Colchicine, 46, 109
Cold exposure, 115
Collision, 226
Consciousness, 199
Coolidge effect, 156–157
Copulation, 152, 156–157, 169, 171, 174–176
Cortisol, 36
Cycloheximide, 183

D

Dehydration, 125, 130–131
  cellular, 128, 136
Dehydration (*contd.*)
  extracellular, 143–145
  intracellular, 138–143
2-Deoxy-D-glucose, 100, 104–105
Deoxyribonucleic acid (DNA), 182
Dexamethasone, 179
Diabetes mellitus, 107
Disinhibition, 235
Diuresis, 130–138
Diruesis, *see also* Antidiuretic hormone
Dopamine
  brain levels, 115
  fibers, 115, 137
  sexual behavior, 180
Dopaminergic systems, *see also* Neurons, Nigrostriatal pathway
Dopaminergic systems, 230
  nigrostriatal pathway, 53–55, 148
  self-stimulation, 232
Dreaming, 190
Drinking behavior
  anticipatory, 146
  associated with feeding, *see* Prandial drinking
  pattern, 120–122
  rate, 121
  volume, 125
  water temperature, 127, 133
Drinkometer, 121
Dual-mechanism model, 102, 108, 119, 275, 277

E

Eating, *see* Feeding
EEG
  during sleep, 190–191
Electrical recording, 48–51
Electrical stimulation, 46
  brain effect on feeding, 111
  brain
    sexual behavior, 171
Electroencephalogram (EEG), 50
Emergency theory, 240
Emotion
  autonomic responses, 240
  behavior and, 238–239
  biological rhythms of, 243–244
  characteristics of, 237–240
  cognitive factors, 247–249
  conditioned response, 245–246
  experiential factors, 244–247
  external stimuli, 242–243
  homeostatic signals, 241

Emotion (*contd.*)
skeletomotor responses, 240
theories, 249–250
Encéphale isolé, 203–204
Endocrine glands, *see* Endocrine system
Endocrine system, 34
Energy
basal requirements, 88
Energy
diet, 88
expenditure, 88–90
homeostasis, 87
Epilepsy, 179
Estradiol, *see* Estrogen
Estrogen, 154–155, 160, 167, 168, 181
role in feeding, 97, 108
Estrus, 152, 154
Estrus cycle
effect on feeding, 97
Estrus cycle, *see* Estrus
Experiential factors
role in drinking, 135
role in feeding, 95–96, 113
role in motivation, 12
role in thermoregulation, 71–72
External stimuli
role in brain self-stimulation, 219–220
role in drinking, 131–134
role in emotional behavior, 242–243
role in feeding, 93
role in motivation, 10–11
role in sexual behavior, 156–157
role in sleep and waking, 199–200
role in thermoregulation, 71–72
Extinction, 216
Extracellular fluid, 124–126, 129, 143–144

F

Fat, 94–95, 105
Fear, 242, 245
Feedback mechanisms
in feeding, 91
in thermoregulation, 72
Feeding
acquired aversions, 95
anticipation of deficits, 95
circadian rhythms, 97–98
in infant rats, 98
cultural factors, 96
2-Deoxy-D-glucose
into hepatic portal system, 104–105
intracranial, 104
intraventricular, 104
Feeding (*contd.*)
estrus cycles, 97–98
experiential factors, 95–96
external stimuli, 93–95
glucose
intracranial injection, 104
long-term regulation, 105
olfactory stimuli, 94
oropharyngeal factors, 91, 94
patterns, 86–87
role of hormones, 97, 107–109
signals, 90
social factors, 96
systems analysis, 90–93
Feeding center, 100–102
Fenfluramine, 55, 110–111
Fever, 77, 80
Fiber pathways
self-stimulation, 225
"Fight or flight", 240
Fistula
cheek, 146
esophageal, 91, 126–127
gastric, 91, 93
Food deprivation, 94
Food intake, *see* Feeding
Forebrain, 113–114, 118, 146, 206
Free fatty acids (FFA), 106
Frontal lobe
lesions, 114
Functional coupling, 149, 278

G

Gastric distension, 126–127
Genital stimulation, 159, 171–172
Genome, 159, 182
Glucoprivation, 100, 115
Glucoreceptor, *see also* Neuron
Glucoreceptors
central, 103–105, 107–108
hepatic, 105
peripheral, 97, 108
Glucose, 93, 103, 111
intracranial injection, 104
plasma, 108
Glucostatic signals, 90
Goat
drinking, 138
temperature regulation, 75–76
Gold thioglucose, 103, 107
Gonadal secretions, 160
Gonadectomy, 160

Gonadotropins, 155, 173
Gonads, 159–160
Growth hormone, 97, 108–109
Gustation, *see* Taste

H

Habenular nucleus
  sexual receptivity, 177
Haloperidol, 53, 180
Heat gain, 59
Heat loss, 60–61
Hedonic values
  in drinking, 133
Hemorrhage, 129–130, 143
Hibernation, 70, 98
Hippocampus
  sexual behavior, 177
Histofluorescence, 53, 207
Homeostasis, 3, 8, 16, 29
  in temperature regulation, 68, 70, 82
Homeostatic signals
  role in motivation, 8
Hormone implantation,
  intracerebral, 171
Hormones
  emotion and, 241–242
  role in feeding, 97, 107–109
  role in motivation, 8–9
  role in sexual behavior, 163–169
  role in temperature regulation, 59–61, 69–70, 74
  self-stimulation and, 220–221
Huddling, 63, 73
Hunger, 113, 220
6-Hydroxydopamine, 55, 114–115
Hyperphagia, 103, 107, 109
Hypertonic saline, 128, 136, 138, 140–142, 147
Hypophysectomy, 66–67, 97, 165
Hypothalamus, *see also* Sexual behavior
Hypothalamus, 15–20, 30–37, 276–278
  drinking behavior, 137–149
  emotional behavior, 250–251
  feeding behavior, 100–107
  self-stimulation, 226–229
  sexual behavior, 173–176
  thermoregulatory behavior, 74–82
Hypothalamus, *see also* Ventromedial hypothalamus, Lateral hypothalamus
Hypothermia, 75
Hypovolemia, 115, 128, 130, 136, 141, 143–144

I

Immunohistochemistry, 44
Impotency, 181
Imprinting, 246–247
Incentive stimuli, 219
Insomnia, 202
Insulin, 97, 100, 103, 107–109, 118, 219
Insulin sensitivity, 104, 107
Integration
  of body systems, 27–41
Intercourse, 155–156
Intracellular fluid, 123, 125, 138–143
Intravenous infusion
  of food, 92
Ions
  temperature regulation, 79
Iontophoresis, 51, 103, 106, 117
Isolation, 246
  social
    effect on mating, 157, 159
Isotonic saline, 129
Isotonicity, 128

K

Kidney, 128, 131, 144
Klüver-Bucy syndrome, 177, 251
Knife cuts, 44, 75

L

Lactation, 98
Lap volume, 147–148
Lateral hypothalamic syndrome
  recovery, 145
Lateral hypothalamus, 100, 103, 106, 108, 114–118
  chemical stimulation, 114
  electrical stimulation, 102, 137, 217
    drinking, 148
  electrophysiological recordings, 220
  electrophysiology, 102
  lesions, 105, 109, 137, 143, 148
    adipsia, 148
  osmoreceptors, 138–143
  self-stimulation, 218–219, 234
Learning
  self-stimulation, 234
Lesions, *see* Ablation
Limbic system, 41
  emotional behavior, 251–253

Limbic system (*contd.*)
  self-stimulation, 223–224
  sexual behavior, 176–178
Lipostatic signals, 90, 105
Liquid diet, 133, 147
Liver
  hepatic glucoreceptors, 104
  hepatic osmoreceptors, 154
Locus coeruleus, 208, 230–231
Lordosis, 152, 160
Lordosis quotient, 152
LSD, 261

M

Mammillary bodies, 174
Masturbation, 156
Medial forebrain bundle, 114
  self-stimulation, 222–226
  sexual behavior, 176
Medulla
  temperature regulation, 77
Menopause, 241
Menstrual cycle, 154–155, 241
Menstruation, *see* Menstrual cycle
Metabolic rate, *see also* Metabolism
Metabolic rate, 70
Metabolism
  role in temperature regulation, 59–62
Microelectrode
  recording techniques, 50–52
Microiontophoresis, 51, 139
Midbrain
  drinking, 142, 144
  feeding, 102
Milk, 99–100, 135–136
Monoamine systems, 110, 117, 207
  noradrenergic pathway, 55, 110
Motivated behaviors, *see also* Motivational deficits
  characteristic of, 3–5
  definition of, 4
  factors influencing, 7–12
  history of, 1–3
  measurement of, 5–7
Motivational deficits, 116
Motor deficits, 116
Motor system, 37–39
  during sleep, 193

N

Narcolepsy, 202
Neocortex, 157
Neostriatum, 53
Nephrectomy, 128
Nest-building, 63, 66, 69
Neurons
  activity during self-stimulation, 225–229
  catecholaminergic, 108, 114
  discharge rate, 112
  dopaminergic, 102, 110, 114
  firing rate, 227
  food reward, 147
  glucoreceptive, 102, 103, 117
  glucosensitive, 102, 103, 108, 117
  noradrenergic, 110, 114
  thermosensitive, 79
  water reward, 146
Neurotransmitters
  emotional behavior, 257–258, 261–264
  measurement of, 55
  sexual behavior, 180
  sleep, 207–209
  temperature regulation, 78–79
Nigrostriatal pathway, 102, 114–115
Nigrostriatal pathway, *see also* Dopaminergic systems
Noradrenalin, 69, 110
  assay, 55–57
  brain levels, 114
  feeding, 144
  intracisternal, 114
  intracranial, 114
  temperature regulation, 78
Noradrenergic pathway, *see* Monoamine systems
Norepinephrine, *see* Noradrenalin
Nucleus accumbens, 220

O

Obesity, 94–95, 103, 107
  produced by insulin, 97
Odors, *see* Olfaction
Olfaction, 41, 94, 111, 219
Olfactory bulb, 219
Ontogeny
  of drinking, 135–137
  of feeding, 99–100
  of sexual behavior, 158–163
  of sleep, 194–197
  of thermoregulation, 72–74
  role in motivation, 14–15
Oral metering, 126

Oropharyngeal factors
in drinking, 126–127, 131–134, 145–146
Orbitofrontal cortex, 114
Orgasm, 155
Osmolality, 129, 138
Osmoreceptors, 36, 124, 138–143
Osmosensitive neurons, 138–140, 148
Osmotic gradient, 128
Osmotic thirst
experimental production, 128
Ovariectomy, 97, 155, 164–165, 167
Ovary, 159
Ovulation, 153
Oxygen consumption, *see* Metabolic rate

P

Palatability, 94
drinking solutions, 127, 132–133
Parachlorophenylalanine (PCPA), 180, 208
Paradoxical sleep, 190
Paraplegics, 172
Parasympathetic system, 34
Parkinson's disease, 180, 232
Pathways
dopaminergic, 102
dopaminergic, *see also* Dopaminergic systems
Pelvic thrust, 159
Penile erection, 159
Penile sensitivity, 162
Perseverative tendency, 114
Pilocarpine, 180
Piloerection, 61
Pleasure, 220
Polyethylene glycol, 125, 130, 136, 143, 147
Polysynaptic activation, 226
Pons
temperature regulation, 77, 78
Posterior hypothalamus
temperature regulation, 75, 79
Posterior pituitary, 124, 130
Prandial drinking, 127, 131, 145
Prefrontal cortex
lesions and feeding, 114
self-stimulation, 230, 234
Preoptic area
drinking, 140–141, 143–144
sexual behavior, 173, 175–176
sleep, 206
temperature regulation, 63, 68, 75–81, 84
Primary drinking, 133, 138, 146–147
Procaine, 46, 102
Progesterone, 154–155, 160, 167–168, 183
Prostaglandins, 105
fever, 78
Protein synthesis, 182–183
Psychiatric disorders, 258–264
Psychosomatic disorders, 249
Psychosurgery, 261, 265–268
Puberty, 97, 160
Push-pull cannula, 55–57
Pyriform cortex
damage to, 177
Pyrogens, 63, 77, 80

Q

Quinine
diet adulteration, 94
Quinine solutions
drinking by rats, 132

R

Rabbit
sexual behavior, 179
sleep, 195
Radioactive tracers
sex hormones and, 172, 176
Raphé nuclei, 208–209
Reinforcement, 146, 216–217
Releasing hormones, 36–37
REM sleep, 190–197, 207–209, 211
Renin-angiotensin system, 130, 136, 143–144
Reserpine, 261
Respiration, 27–28
Reticular activating system
emotional behavior, 256–257
Reticular formation
sleep, 203, 210
Reversible block, *see* Block, reversible
Reward, 215–220
neural mechanisms, 227–229
noradrenergic hypothesis, 231–232
Ribonucleic acid (RNA), 182–183
Rotational behavior, 53–55

S

Saccharin solutions
drinking by rats, 132, 146, 219
Salicylates, 80
Salivary glands, 127, 145

Salivation, 145–146
  role in temperature regulation, 61
Salty solutions
  drinking of, 219
Satiety, 90, 94, 119
Satiety center, 102–103, 119
Schizophrenia, 234, 262–264
Secondary drinking, 133, 147, 150
Self-stimulation
  characteristics, 214–215
  drug studies, 230–233
  effect of deprivation, 218
  of the LH, 111
Self-stimulation sites, 222–224
Sensory-motor deficits, 147
Sensory neglect, 115
Septal rage, 257–258
Septum
  emotional behavior, 252–253
  sexual behavior, 177
Serotonin, 109–110, 180
  pathways, 55
  sleep, 197, 208
  temperature regulation, 78
Set-point, 62, 70, 94, 97
  for body weight, 109
  temperature regulation, 79, 81
Sex center, 169
Sex chromosomes, 159–160
Sex drive, 152
Sexual arousal, 156
Sexual behavior
  brain lesions and, 170
  brainstem lesions and, 173
  cerebral cortex and, 178–180
  hypothalamus and, 173–176
  limbic system and, 176–178
  midbrain lesions and, 173
  neurophysiological activity, 171–172
  olfactory stimulation and, 158
  role of experience, 157
  role of hormones, 153–154, 163–169
  spinal reflex responses, 172
Sexual receptivity, 152, 179
Shivering, 61–62, 72
Sleep
  behavior, 188
  electrical activity of brain, 190–194, 203–207
  emotional factors, 201–202
  homeostatic factors, 197–198
  hormones, 199
  periodicity, 188
Sleep (*contd.*)
  physiological responses, 189
  physiology, 188–190
  slow wave, 193, 208
  species difference, 195–197
  tonic component, 192
Sleep clinics, 201–202
Sodium chloride solutions
  drinking by rats, 131–132
Sodium ions, 128
Sodium receptors, 143
Somnolence, 203
Specific dynamic action (SDA), 59
Spinal cord
  temperature regulation, 77
Splanchnic nerve, 104
Spreading depression, 179
Stereotaxic research, 43, 44, 100
Steroid hormones, 105
Steroid hormones, *see also* Androgens, Estrogens
Stimulation
  chemical, 48
  electrical, *see* Electrical stimulation
Stretch receptors, *see* Volume receptors
Stria terminalis, 111
Subfornical organ, 143–144
  self-stimulation, 233
Substantia nigra, 53, 102, 117
Suckling, 99, 136
Sucrose, 128, 140
Supersensitivity, 53
Suprachiasmatic nucleus
  lesions, 109, 145
  sleep and waking, 209, 211
Supraoptic nucleus, 124, 130, 138
Sweating, 123
Sympathetic nervous system, 32–33

T

Tandem copulation, 178
Taste, 111–113, 118–119
  relation to drinking, 127, 131–134, 145–146
  relation to feeding, 91
Taste, *see also* Palatability
Techniques, 41–57
Tegmentum
  feeding, 102
Temperature receptors, 75, 80
Temperature regulation
  lesion studies, 75–83

Temporal lobe
  ablation, 114
  sexual behavior, 177
Testis, 159
Testosterone, 156, 160, 162, 164, 167, 181
Thalamus, 148
  sleep, 206-207
Thermode, 68-69, 75-77, 80
Thirst
  produced experimentally, 125-127
Thirst center, 137
Thyroidectomy, 69
Thyroid gland
  humans, 241
Thyroid hormone, *see* Thyroxine
Thyroxine, 36, 69
Tongue, 148
Trigeminal nucleus, 116-117, 148
Tryp taminergic pathways, *see* Serotonin

U

Urine concentration, 124-125

V

Vagus nerve, 104
Vasoconstriction, 29, 63
Vasodilatation, 61, 63
Ventral tegmental area, 115
Ventricles
  osmoreceptor site, 142, 144
Ventromedial hypothalamus, 100, 103-109, 117, 119
  electrophysiology, 102, 111
  feeding rhythms, 98
Vision
  neural projections, 111
Visual cues, 112-113, 118
Volume receptors, 124, 143-144

W

Wakefulness
  sleep and, 204-207
Water balance, 122-125
Water deprivation, 125, 128
Weaning, 99-100, 135-137

X

Xylocaine, 224

Z

Zona incerta, 134, 147-148